AGRICULTURALLY IMPORTANT MICROORGANISMS

Mechanisms and Applications for Sustainable Agriculture

About the Editors

Dr. Bibhuti Bhusan Mishra is presently working as the ICAR-Emeritus Professor at the P.G. Department of Microbiology, College of Basic Science & Humanities, Odisha University of Agriculture and Technology, Bhubaneswar, Odisha, India after his superannuation from the University in May, 2018. He obtained M. Phil. and Ph.D. Degree in 1983 and 1987 from Berhampur University, Odisha respectively. He has more than 37 years of teaching and research experience. A total of 12 students have been awarded doctoral degree under his supervision from various Universities across India and currently four more are actively working in the field of environmental and soil microbiology. In addition, he has guided more than 25 P.G. students. He has 61 research publications and more than 20 book chapters & research manuscript in various journals of national and international repute. He is credited with 25 accession numbers submitted to NCBI, USA. He has contributed research article as book chapter to the encyclopedia 'Environmental Engineering' published by Gulf Publishing, USA and book chapters in many more edited books pertaining to soil and environmental microbiology. He has edited 9 books in microbiology & biotechnology published by national and international publishers and 1 book on Botany practical. Currently 3 more edited books are in press with Wiley and Springer publishers. He has successfully completed 1 UGC Major project from Govt. of India and was the Chief Nodal Officer of project 'Biofertilizer Production Unit' under RKVY (Rastriya Krishi Vikash Yojana), Govt. of Odisha. He is awarded as 'Best Teacher award' from the University in 2012 and from the college in 2015. For significant contributions in microbiology, he was conferred with Prof. Harihar Pattnaik Memorial award by the Orissa Botanical Society in 2016. In addition he was awarded as 'Best Teacher' from Orissa Botanical Society in 2018.

Dr. Suraja Kumar Nayak obtained Ph.D. from Odisha University of Agriculture and Technology in 2013 and is presently working as Asst. Prof., Department of Biotechnology, College of Engineering and Technology, Biju Patnaik University of Technology, Bhubaneswar, Odisha, India. Dr. Nayak has 10 years of teaching and research experience in field of Microbiology and Biotechnology. His areas of teaching and research include general and environmental microbiology, soil microbiology, industrial & food biotechnology, microbial biotechnology. Dr. Nayak has published 14 scientific papers including book chapters in various journals and national & international books. He has edited one book from CRC Press of Taylor & Francis group and 1 more edited book is in press from Wiley publishers. He has also submitted 05 accession no. to NCBI, USA and has presented papers in various national and international seminars. Four numbers of M.Tech and more than 10 B.Tech students have successfully completed their dissertation under his supervision.

Dr. Avishek Pahari is presently working as Research Fellow, Centre for wildlife health, College of Veterinary science & Animal husbandry, Odisha University of Agriculture and Technology, Bhubaneswar, Odisha. He obtained his M.Sc. and Ph.D. degree in 2013 and 2017 respectively from Department of Microbiology, College of Basic Science & Humanities, Odisha University of Agriculture and Technology, Bhubaneswar, Odisha, India. Dr. Pahari has more than 07 years of research experience in the field of Microbiology and his area of specialization is agricultural microbiology. During his doctoral research, he has isolated 04 siderophore producing Plant Growth Promoting Rhizobacteria which can increase the growth and productivity of different vegetable significantly reducing application of chemical fertilizer. He has published 08 research article and 05 book chapters in various national & international journals and books. Dr. Pahari has also submitted 35 accession no. to NCBI, USA and has presented papers & posters in various national and international symposiums.

AGRICULTURALLY IMPORTANT MICROORGANISMS

Mechanisms and Applications for Sustainable Agriculture

Bibhuti Bhusan Mishra
ICAR-Emeritus Professor
P.G. Department of Microbiology
College of Basic Science & Humanities
Odisha University of Agriculture and Technology
Bhubaneswar, Odisha

Suraja Kumar Nayak
Asst. Professor
Department of Biotechnology
College of Engineering and Technology
Biju Patnaik University of Technology
Bhubaneswar, Odisha

Avishek Pahari
Centre for Wildlife Health
College of Veterinary Science & Animal husbandry
Odisha University of Agriculture and Technology
Bhubaneswar, Odisha

nipa®

NEW INDIA PUBLISHING AGENCY
New Delhi – 110 034

NEW INDIA PUBLISHING AGENCY

101, Vikas Surya Plaza, CU Block, LSC Market
Pitam Pura, New Delhi – 110 034, India
Email: info@nipabooks.com
Web: www.nipabooks.com

For customer assistance, please contact
Phone: + 91-11-27 34 17 17 Fax: + 91-11- 27 34 16 16
E-Mail: feedbacks@nipabooks.com

ISBN : 978-93-90175-03-1

Composed and Designed by NIPA.

Preface

India has been predominantly a country of agriculture and more than 70% of people depend on agriculture and agricultural produce. Exponential rise in population coupled with industrialization and urbanization has demanded increase in agricultural produce with depleting land coverage. Under the Green revolution-I, enhanced application of agrochemicals significantly increased crop productivity. However, excessive use of chemical fertilizers not only adversely affects the soil health & the quality of produce but also contaminate the soil. It is also not cost effective. But over the past couples of decades, both developed and developing countries are harvesting the impact of Green Revolution I. The residuals of agrochemicals applied in the crop fields drastically affected the soil health; in terms of nutrient availability, soil microbial diversity and increased pathogen attack on the crops. In India states like Punjab and Haryana, which are otherwise called as the kitchen house of the country, fail to develop crop without fertilizer application. In Green Revolution-II, emphasis is being given on organic farming. In this perspective, soil microorganisms with advantageous effect on soil & plant health vis-a-vis growth & yield, correspond to be an effective alternative to conventional agricultural practices. Biofertilizers are live formulation of microbial inoculants that enhances soil fertility and crop productivity, thus, can be an ideal supplement / substitute to chemical fertilizers. The commercial exploration of such biological organisms is of recent interest.

In the era of sustainable agriculture, organic farming has globally emerged as a priority area in view of the growing demand for safe & healthy food and long term sustainability. It not only ensures food safety, but also adds to the biodiversity of soil. The additional advantages of biofertilizers include longer shelf life causing no adverse effects to ecosystem. Organic farming includes application of compost, biofertilizers, biopesticides etc. which may not serve as a complete substitute but can be effective supplement to decrease application of agrochemicals. Organic farming is mostly dependent on the natural microflora of the soil which constitutes all kinds of useful bacteria, cyanobacteria, fungi including the arbuscular mycorrhiza fungi (AMF) and plant growth promoting rhizobacteria (PGPR). These bacteria also plays an important role in agricultural crops and termed as Agriculturally Important Microorganisms (AIMs). AIMs have diverse applications in agriculture, horticulture, and forestry. These microbes also play a key role in crop protection through enhancing the disease resistance capacity of plants against pathogens, exhibiting antagonistic activities or acting as biotic elicitors.

The book, *Agriculturally Important Microorganisms: Mechanisms and Applications* enlighten the inherent potential of AIMs in inter and intra community interactions, metabolite production and plant protection with latest information available in the relevant field. The field is so large and the interest in PGPR microbiota is so varied that the topics covered will make the book more informative & meaningful to be accepted by scientists, academia and researchers. This book can be largely informative on the biofertilizers in sustainable agriculture by enhancing soil fertility, plant tolerance and crop productivity, plant-microbe interactions and their inherent principles for plant protection, mitigation of salt stress, biological N_2 fixation, plant growth promotion, Potassium Solubilizing Microorganisms for advances in nutrition for rice and other crops. In addition, nanofertilizers for sustainable agriculture will provide new frontiers for future research. It will also provide key knowledge to cutting edge biotechnological methods applied in soil and agricultural microbiology.

The editors express sincerely gratitude to all contributors for their excellent cooperation, critical thoughts and contribution to complete this timely edited volume. We also sincerely thank New India Publishing Agency, New Delhi, India and their team for providing us with an opportunity in publishing the book. Last but not the least, we wish the ongoing and upcoming scientific generations to use this text knowledge for social benefit and development. We will definitely appreciate any comments on the book for future prospective.

Bibhuti Bhusan Mishra
Suraja Kumar Nayak
Avishek Pahari

Contents

Preface .... *vii*

1. Biofertilizers: A Representative Illustration of Plant-Microbe Interaction **1**

Soumya Lipsa Rath, Swaraj Mohanty and Debakanta Tripathy

1.1. Introduction 2
1.2. Biofertilizers 3
1.3. Mechanism of Biofertilizer 15
1.4. Advantages of using biofertilizers over chemical fertilizers 18
1.5. Conclusion 20
References *20*

2. Plant Growth Promoting Microbes of Rice and Their Application for Sustainable Agriculture **27**

Avishek Banik

2.1. Introduction 28
2.2. Rice microbiome 28
2.3. Different plant growth promoting (PGP) traits 30
2.4. Microbial distribution at different rice niches 35
2.5. Futuristic approaches for application of PGP 37
References *38*

3. Role of Biofertilizer in Sustainable Agriculture Enhancing Soil Fertility, Plant Tolerance and Crop Productivity **41**

Priyanka Verma, Dheer Singh, Ishwar Prasad Pathania and Komal Aggarwal

3.1. Introduction 42
3.2. Importance of sustainable agriculture 43
3.3. Efficient soil microbes for sustainable agriculture and environment 44

3.4. Impact of biofertilizers on crop productivity 45
3.5. Biofertilizers 46
3.6. Inoculation of Biofertilizers 56
3.7. Mechanisms of biological fertilizers to increase crop yield 56
3.8. Improve Soil Fertility utilized Biofertilizers 57
3.9. Economic Development in Agriculture 58
3.10. Conclusion 58
References *59*

4. Role of Plant Growth Promoting Rhizobacteria in Mitigating Salt Stress 65

Priyanka Chandra, Awtar Singh, Madhu Choudhary and R. K. Yadav

4.1. Introduction 66
4.2. Salt affected soils 67
4.3. Adverse effect of salt stress on crops 68
4.4. Salt stress mitigation by plant growth-promoting rhizobacteria (PGPR) 70
4.5. Benefits of plant growth-promoting rhizobacteria (PGPR) 70
4.6. Mechanism of mitigating salt stress 75
4.7. Conclusion 82
References *83*

5. Multiple potential traits of PGP *Pseudomonas* sp. 91

Padmavathi Tallapragada and Kavyashree B M

5.1. Introduction 92
5.2. Plant Growth Promoting Rhizobacteria (PGPR) 92
5.3. Biofertilizer 94
5.4. PGP *Pseudomonas* 95
5.5. Multiple Potential Traits of PGP *Pseudomonas* 99
5.6. Conclusion 110
References *110*

6. Use of Biofertilizers for Sustainable Agriculture 119

Arup Sen, Kaushik Batabyal, Sanchita Mondal Arindam Sarkar and Dibyendu Sarkar

6.1. Introduction 120
6.2. Efficient soil microbes used as biofertilizers 120
6.3. Conclusion 132
References *133*

7. Biological Nitrogen Fixation: Mechanism and Applications ... 137

Avishek Pahari, Suraja Kumar Nayak, Avishek Banik Priti Binita Lakra and Bibhuti Bhusan Mishra

7.1. Introduction ... 138
7.2. Biological nitrogen fixation (BNF): Key to sustainable agriculture ... 139
7.3. Types of Nitrogen fixation ... 140
7.4. Nitrogenase enzyme complex ... 144
7.5. Assimilation and transportation of biologically fix ammonia ... 145
7.6. Nitrogen fixing bacteria: Biofertilizer formulation and technology ... 146
7.7. Conclusion ... 148
References ... *148*

8. Potassium Solubilizing Microorganisms (KSM) A Very Promising Biofertilizers ... 153

Padmavathi Tallapragada and Titus Matthew

8.1. Introduction ... 153
8.2. Biofertilizer ... 154
8.3. Potassium in the soil ... 155
8.4. Potassium solubilizing microorganisms (KSM) ... 156
8.5. Potassium solubilizing fungi (KSF) ... 160
8.6. Mechanism of potassium solubilization by microorganisms ... 161
8.7. Mass production and quality control ... 161
8.8. Examples of researched application of KSM ... 167
8.9. Production of biofertilizer in India ... 170
References ... *172*

9. Advances in Potassium Nutrition in Crop Plants by Potassium Solubilizing Bacteria ... 175

Nur Uddin Mahmud, Musrat Zahan Surovy, Dipali Rani Gupta and Md. Tofazzal Islam

9.1. Introduction ... 176
9.2. Discovery of the elite strains of KSB ... 177
9.3. Mechanisms of plant K nutrition by bacteria ... 179
9.4. Formulation and practical delivery of KSB to the crop field ... 184
9.5. Performances of KSB in promoting growth and yield of crop plants ... 186
9.6. Molecular genetics of KSB and mechanism of their action ... 189
9.7. Challenges and opportunities ... 190
9.8. Conclusion and perspective ... 192
References ... *192*

10. Nano Biofertilizer for Sustainable Agriculture 201

S. Pati, A. Dash, S. Maity, S. Pattnaik, S. Mohapatra and D. P. Samantaray

10.1. Introduction 201
10.2. Biofertilizer in agriculture 202
10.3. Nano-biofertilizer 203
10.4. Advantages of Nano-biofertilizer 204
10.5. Future prospects of nano-biofertilizer 206
References *206*

11. Role of Phosphorus Solubilizing Microorganisms for Sustainable Crop Production 209

Sushanta Saha, Bholanath Saha, Ayon Alipatra, Biplab Pal, Partha Deb Roy and Abhas Kumar Sinha

11.1. Introduction 210
11.2. Functions of phosphorus in plants 210
11.3. Deficiency of phosphorus in plants 211
11.4. Sources of phosphorus in soils 212
11.5. Soil solution phosphorus 212
11.6. Inorganic soil phosphorus 213
11.7. Organic soil phosphorus 213
11.8. Phosphorus cycle 213
11.9. Phosphate Solubilizing Microorganisms (PSM) 214
11.10. Conclusion 218
References *219*

12. Plant Growth Promoting Traits Exhibited by *Rhizobium* Isolated from Different Regions of Odisha 221

Aiswarya Panda and Bibhuti Bhusan Mishra

12.1. Introduction 221
12.2. Materials and Methods 223
12.3. Results And Discussion 224
12.4. Conclusion 234
References *235*

Index 237

1

Biofertilizers: A Representative Illustration of Plant-Microbe Interaction

Soumya Lipsa Rath[1], Swaraj Mohanty[2] and Debakanta Tripathy[3]

[1]*Department of Biotechnology, National Institute of Technology, Warangal Telangana, India*
[2]*College of Engineering and Technology (CET), Bhubaneswar, Odisha, India*
[3]*Indira Gandhi Institute of Technology Sarang (IGIT), Sarang, Dhenkanal Odisha, India*

Abstract

Microbes are ubiquitously found in the environment. Their interaction with other living organisms could either be pathogenic or beneficial in nature. Soil as a rich source of microbes, researchers have extensively studied the plant-microbe interactions. The soil microbes supply the nutrients deficit plants with essential growth elements such as nitrogen, phosphorous, potassium, etc. and improve the texture and quality of the soil. Due to the destructive nature of chemical fertilizers to soil and environment, it was required to generate a more ecofriendly means of increasing the soil fertility. This led to the development of microbes based biofertilizers. The primary benefit of biofertilizers is to promote a healthy soil environment along with supplying essential nutrients to the plant. Again, biofertilizers can be tailor-made according to the specific plant or the climatic conditions. Fungal and bacteria based biofertilizers, either alone or in combinations are formulated for the plants. Lately, liquid fertilizers are increasingly finding their use in the biofertilizer industry due to their ease in manufacture and prolonged effect on plant growth. The scope of this chapter lays on the currently popular biofertilizers and recent trends in the biofertilizer industry.

Keywords: Nitrogen fixing microbes (NFM), Mycorrhiza, PGPR, Cyanobacteria, Liquid Biofertilizers

1.1 Introduction

Agriculture is one of the primary sources of livelihood of a country where nearly 50% of the country's population depends on it as it's the main source of income. "Nowadays", agriculture not only includes the crop production, but also forestry, dairy, fruit cultivation, bee keeping, mushroom cultivation and poultry etc. Conventionally, agriculture has been referred to as the one which meets the growing food demand of the human population (Madhusudhan, 2015). Over time we have been trying to improvise the quality and productivity of crops by various mechanisms such as genetic engineering, use of fertilizers and pesticides etc. Biotechnology has helped in the generation of a new variety of crops with enhanced resistance to disease, drought and salt tolerance. The end-products are of better nutritional value (Conijn *et al.*, 2018). While genetic manipulation might not hamper the soil environment of the crop, the unregulated use of pesticides and chemical fertilizers adversely affects the soil health and the biogeochemical cycle. Similarly, the chemical fertilizers cause air and water pollution harming the environment. Leaching due to rain and irrigation causes loss of runoff nutrients, especially the phosphorous and nitrogen from the soil, degrading the environment (Clark *et al.*, 2004). The prolonged use of fertilizers is one of the main reasons for this kind of environmental pollution (Xuan, 2018). The countries with the highest fertilizer use according to the International Fertilizer Industry Association are China and India (Connor, 2018). Measures are being taken by these developing nations to promote sustainable agriculture and reduce the environmental effect of chemical fertilizers.

The newest trend in the formulation of fertilizers includes incorporating environment-friendly biofertilizers. Organic farming or biofertilizers utilize the soil microflora. The plant-microbe interactions are required for the smooth functioning of the plant-soil ecosystem. Presence of microbes enhance the stress tolerance in plants, they provide disease resistance, aid in the availability of essential nutrients such as phosphate, potassium solubilization and nitrogen fixation encourage healthy biodiversity (Paswan *et al.*, 2017; Singh, 2018). Thus, they do not affect the ecosystem and have a relatively long shelf life compared to that of chemical fertilizers. The most significant advantage of using biofertilzers is that of promoting healthy plant growth. Again, not all microbial species enhance the growth of all plants, i.e., specific microbial communities have been found to have colonized specific type of plants. Bacterial and fungal species interact with the host plants, promoting their growth and might also provide necessary defence against the pathogens. Thanks to the extensive research on the identification of useful microbial communities, people now have access to readily available microbial inoculants for their crops. Microbial inoculants are safe, environment and human health friendly, effective in small quantities, decompose quicker than chemical pesticides, reduce resistance development and control pests (Calvo *et al.*, 2017; Zaidi *et al.*, 2017).

1.2 Biofertilizers

Nature creates harmony between plant and soil environment by furnishing certain useful microorganisms which foster the plant grown and in turn benefit themselves. This unique relationship between plants and microbes is explored by "Biofertilizers" which supply beneficiary living cells (Paswan *et al.*, 2017; Singh, 2018). Under favourable conditions, these microbes secrete metabolites and enzymes enabling the lacking nutrient accessible to the plant. At times, these beneficial microorganisms are packed in special carrier packages carrying microbial cultures (inoculants) (Calvo *et al.*, 2017). The microbial inoculants which act as biofertilizers may carry living or latent strains of microbes capable of fixing nitrogen, solubilizing phosphates and degrading cellulose etc. However, unlike conventional chemical fertilizers or pesticides, they are environment-friendly.

Biofertilizers also include organic fertilizers such as manure, compost or vermicompost because they include the plant to microbe interaction and vice versa. However, broadly one can classify the biofertilizers into bacterial and fungal biofertilizers. In the following parts, we would discuss in length more about the bacterial and fungal biofertilizers and a brief account of the latest liquid fertilizers. This would be followed by an overview of the mechanism of action of biofertilizers and their advantages over chemical fertilizers.

1.2.1 Types of Biofertilizers

The main nutrients essential for the growth of plants include nitrogen fixation, phosphate solubilization and mobilization along with supply of additional micronutrients. Plants require nitrogen for their growth and metabolism. Most of this nitrogen is obtained from the molecular nitrogen of atmosphere, which is limited. Moreover, this molecular nitrogen cannot be directly taken by plants. They are fixed by certain prokaryotic organisms who perform the "nitrogen fixation". Nitrogen fixation refers to generating ammonia from the molecular nitrogen. This reaction is catalyzed by the enzyme nitrogenase, found in free-living and symbiotic diazotrophs. The synthesized ammonia is utilized further to synthesize certain metabolic products in microbial cells, or converted to an amino acid by reacting with an organic acid (reaction with α-keto glutamic acid gives rise to glutamic acid) or other organic molecules (giving rise to alanine or glutamine etc.). This ammonia is generally not accumulated in the microbial cell (Sahu *et al.*, 2012; Mishra and Pabbi, 2004). The organisms that perform nitrogen fixation in plants include *Azotobacter*, *Azospirillum*, *Anabaena*,

Nostoc, *Rhizobium*, *Frankia*, *Anabaena azollae*, *Beijerinkia*, *Clostridium*, *Klebsiella* etc.

Soils are deficient in soluble forms of phosphate, which is required for the plant growth and metabolic processes such as cell division, photosynthesis, etc. 0.2% of the total dry weight of the plant is due to phosphate. Phosphate solubilizing and mobilizing bacteria help in solubilizing the diverse forms of phosphates available in the soil into a form that can be easily assimilated by the plants. They produce phosphatase enzyme which hydrolyses organic phosphates. Metal ions act as a chelating agent facilitating the release of adsorbed phosphate through ligand exchange reactions (Nagesh, 2012; Satyaprakash *et al.*, 2017). Microorganisms include; *Pseudomonas*, *Flavobacterium*, *Bacillus*, *Rhizobium*, *Micrococcus*, *Aerobacter*, *Agrobacterium* and *Enterobacter* etc. Although phosphate mobilization in soil is poorly understood, it is suggested that the formation of mycorrhiza like structure helps in modifying the plant architecture for ease in the uptake of phosphates. Other forms of classification of microbes according to their function are given in **Fig 1.1**.

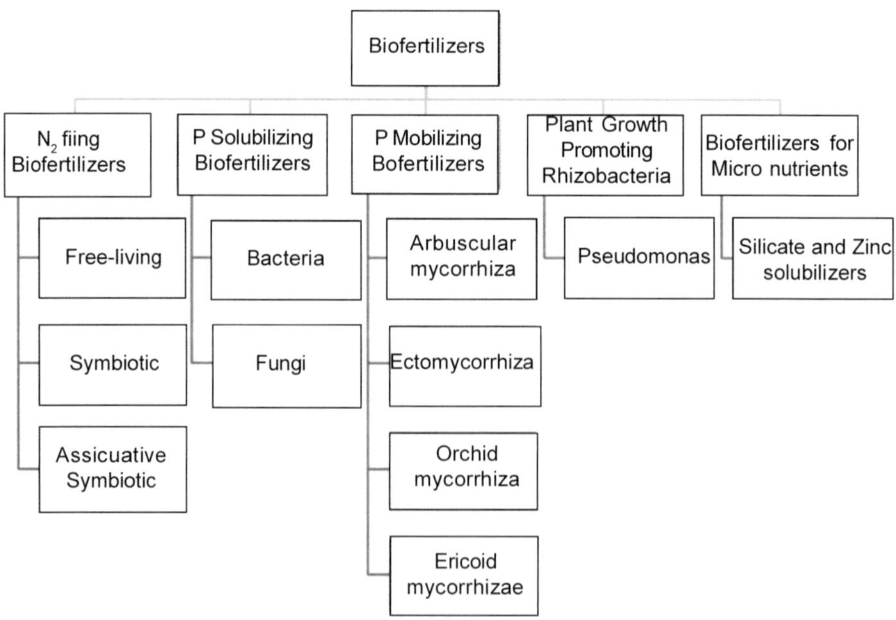

Fig. 1.1 Classification of Biofertilizers according to their specific function

1.2.1.1 Fungal Biofertilizers

Fungal biofertilizers comprise fungal inocula either alone or in combination. They benefit the plant growth and crop yield either directly or indirectly through different mechanisms (**Table 1.1**).

Table 1.1 List of Fungal biofertilizers

Sl.No.		Sub species
1.	Mycorrhiza i. Endomycorrhiza ii. Ectomycorrhiza	*Glomus* sp., *Pisolithus* sp., *Acaulospora* sp., *Scutellospora* sp., *Gigaspora* sp., *Sclerocystis* sp., *Laccaria* sp., *Boletus* sp., *Amanita* sp., *Pezizella ericae*, *Rhizoctonia solani*
2.	Trichoderma	*T. harzianum*, *T. virens*, *T. viride*, *T. lignorum*, *T. asperellum*
3.	Ampelomyces	*Ampelomyces quisqualis*
4.	Chaetomium	*Ch. globosum*, *Ch. cochliodes*, *Ch. cupreum*
5.	Giocladium	*Gliocladium catenulatum*, *Gliocladium virens*

1.2.1.1.1 *Mycorrhiza*

Mycorrhiza is a distinct morphological structure where the fungus associated with the plant's roots. Mycorrhiza (Greek: *mycos*-fungus; *rhiza*-roots) increase the root surface and mineral uptake efficiency and helps the plants to tolerate stress conditions like increased temperature, changing climatic conditions. In more than 80% of plant roots including many angiosperms and gymnosperms, mycorrhiza has been found to establish a symbiotic relationship (Trabelsi and Mhamdi, 2013; Barman *et al.*, 2016). Mycorrhizae forms a filamentous network that associate with plants at their roots. Here they absorb the soil nutrients which was not accessible by the plant root system earlier. Such associations also promote the growth of plants especially at the roots. The total surface area of tiny hyphae extends to wellover 1km in length. The fine nature of the filaments allows to access water and minerals from the smallest soil pores. The plants also become tolerant to different forms of stresses such as drought and saline conditions or attack from the soil borne pathogens. The fungi, absorbs the carbohydrates and other essential nutrients from the plants for growth and synthesizes necessary molecules such as glomalin (glycoprotein) which improves the soil structure and organic matter content. Primarily plants which suffer from nutrient scarcity, especially P, N, Zn, Cu, Fe, S and B develop mycorrhiza. Although mycorrhizae can be classified into as many as seven varieties, i.e., endo, ecto, ectendo, arbutoid, monotropoid, ericoid, and orchidaceous while endomycorrhizae and ectomycorrhizae are most abundant and widespread (Trabelsi and Mhamdi, 2013).

Endomycorrhizae

The endomycorrhizal fungi, penetrate the root cortical cells. They form clusters of finely divided hyphae that form small ellipsoidal structures known as arbuscules. Such endomycorrhizae forming vesicular-arbuscular mycorrhiza is also known as arbuscular mycorrhizal (AM) fungi. The AM fungi are widely present in agricultural and crop ecosystems. They grow from the plant root cortex and develop highly branched structures known as hyphae that are thinner than the roots and able to penetrate deeper. The hyphae are the functional sites for exchange of carbohydrates and other mineral nutrients (Hadi and Kalantar, 2015). They are very efficient in the uptake of specific nutrients like P, Ca, S, N, B, Zn, etc. AM are the prominent phosphate mobilizing biofertilizers. The mycorrhiza facilitates phosphate mobilization from remote places, which are not accessible to the plant roots cannot reach and increase its availability.

Apart from the primary function of nutrient uptake, AM fungi can provide resistance against the pathogenic fungi of the soil, reduce metal toxicity due to Fe, Cu and Zn, reduce the soluble N in the soil and limit denitrification and also regulate emissions of greenhouse gas. Since they also improve the structure of the soil and help in building plant communities and increase productivity, interest in using these fungi as biofertilizers is increasing day by day (Soytong and Ratanacherdchai, 2005). *Ambispora, Archaeospora*, *Intraspora*, *Diversispora*, *Entrophospora*, *Gigaspora*, *Scutellospora*, *Glomus* (*Sclerocystis*), *Pacispora*, *Paraglomus*, *Acaulospora*, *Kuklospora* constitute the AM fungi.

Ectomycorrhizae

The main difference between endo and ectomycorrhizal fungi is the mechanism of symbiosis. While, endomycorrhizal fungi form a symbiosis with a particular plant root, ectomycorrhizal (ECM) fungi form mutualistic symbioses with several species of trees. They colonize the tree roots and help in the absorption and accumulation of essential elements required for the growth of plants such as nitrogen, phosphorus, potassium and calcium, etc. They break down complex minerals and substrates and transfer them to the plants. They increase the stress tolerance of plants in drought conditions, high temperatures and in soils with high toxins or high pH (Soytong and Ratanacherdchai, 2005). The ECM fungi partially invade the host root, form a thick mantle around it, however, they do not penetrate the cortical cells. ECM fungi include species from Basidiomycetes, Ascomycetes and few Zygomycetes.

1.2.1.1.2 Trichoderma

Trichoderma are commonly available saprophytic organisms feeding on dead and decaying wood and organic materials from where they can be easily isolated. They defend plants against pathogenic fungal species such as *Rhizoctonia*, *Pythium*, *Fusarium*, *Botrytis cinerea*, *Phytophthora palmivora*, *P. parasitic* etc. *Trichoderma* species which are used as biofertilizer includes *T. harzianum*, *T. viride* and *T. virens*. *Trichoderma harzianum* is the most potent and effective biological control agent among them. It is effective against a wide range of fungal plant pathogens namely *Botrytis cinerea*, *Fusarium*, *Pythium*,

Phytophthora erythroseptica, *Rhizoctonia* that cause plant diseases like pink rot, stem rot or blight disease (**Table 1.2**). *Trichoderma harzianum* affects crops like corn, soybean, potato, tomato, beans, cotton, peanut, and various trees (Keswani *et al.*, 2013).

The mechanism of action of the *Trichoderma* species comprises of a combination of mechanisms. The main mechanism is mycoparasitism and antibiosis. Mycoparasitism acts by recognition of the host, binding and then enzymatic disruption of the host fungus cell wall. Lytic enzymes are produced by the *Trichoderma* species, which helps in the coiling of *Trichoderma* mycelium around the mycelium of the pathogen, triggering cell wall lysis and death of the host cell. The major classes of enzymes released include proteolytic enzymes, β-1, 3-glucanase and chitinases etc. *Trichoderma* species have faster growth and high capacity to reproduce. Thus, they have been successfully used as mycofungicides. They are able to constrain the growth of a large variety of fungal diseases and especially have strong aggressiveness against phytopathogenic fungi. They act as competitors in the rhizosphere and can in turn modify the rhizosphere, and colonize and grow in association with plant roots which is known as rhizosphere competence. They are resistant to fungicides and can survive in unfavorable conditions. They utilize the soil nutrients and promote plant growth (Bisen *et al.*, 2016). Antibiosis is a process of secretions from antagonist fungi for suppress and kill pathogenic fungi around its growth area. Such antibiotics disrupt the cell membranes, inhibit the metabolic activity and induce plant defence system (Singh *et al.*, 2016).

Table 1.2 Biological control of some major plant pathogens by *Trichoderma* Species

Trichoderma sp.	Plant Pathogen	Disease Caused
T. harzianum	*Crinipellis perniciosa*	Witches'-broom disease of cocoa
T. lignorum	*Rhizoctonia solani*	Damping-off of bean
T. harzianum	*Sclerotium rolfsii*	Rotting of common vegetables
T. virens	*Serpula lacrymans*	Wood decay
T. viride	*Rhizopus oryzae*	Cotton seedling disease
T. harzianum	*Pythium ultimum*	Damping-off of cucumber
T. virens	*Colletotrichum truncatum*	Brown blotch disease of cowpea
T. asperellum	*Fusarium udum*	Pigeon pea wilt
T. harzianum	*Sclerotinia sclerotiorum*	Sunflower head rot

1.2.1.1.3 Ampelomyces

Ampelomyces quisqualis, a mycoparasitic anamorphic ascomycete that kills powdery mildews hence reduces their growth. The pathogen can be affected through antibiosis and parasitism. Hyphae of *Ampelomyces* penetrate the hyphae, conidiophores and immature cleistothecia of powdery mildews and grow internally then kill all the parasitized cells (Kiss *et al.*, 2004). It takes approximately 7 – 10 days for the microparasites to spread within the hyphae of the mildew colony without killing it. The process of pycnidial formation begins subsequently and gets completed within 2 – 4 days. The infected cells generally die as soon as the pycnidia are formed. For better result, it is necessary to develop and spread mildews colonies which are facilitated by repeated applications of the micro parasite, high humidity and rain fall. The pycinidia which are formed from the infected mildew under adverse climatic condition i.e. low humidity & low temperature are more resistant and survive in the environment for a relatively longer time at least into next season (Aly *et al.*, 2008). These, in turn, may give rise to viable spore when the condition becomes favourable again. *Ampelomyces* species can cure the Mildew affected crops like cucurbits, grapes, apple, peas, beans, tomato, pulses, cumin, chilies, coriander, mango, ber, peas, strawberry, medicinal and aromatic crops and roses (Liang *et al.*, 2007). Other parasitic infections of *Botrytis cinerea*, *Alternaria solani*, *Colletotrichum coccodes* and *Cladosporium cucumerinum* can also be treated by *Ampelomyces* species.

1.2.1.1.4 *Chaetomium*

Chaetomium species are normally found in soil and organic compost. They act as potential antagonists for soil-borne and seed borne plant pathogens. They suppress bacterial and fungal growth by competing for nutrients and substrates or by mycoparasitism, antibiosis, or a combined mechanism. Its mycofungicide effect not only protects the plant but also has a curative effect on the plant (Soytong, 2014). *Chaetomium* can produce ergosterol promoting increased high organic matter in the soil and improving its fertility. Primarily *Ch. globosum* and *Ch. cupreum* have been studied extensively. They help the plants stay off from a large number of diseases such as root rot disease in citrus, black pepper, strawberry or raspberry spur blight disease by *Didymella applanata* or disease of potato plants by rigid porous *solani*. *Ch. cupreum* increases the growth of tomatoes. It can also cause inhibition of wilt causing *Fusarium oxysporum*. and *Phytophthora parasitica* that causes rot of the citrus root. The white root disease of rubber trees caused by *Rigidoporus microporous* can be cured by *Ch. cupreum*. Localized and sub-systemic oxidative burst can be induced by the *Ch. cuperum* in plants like carrot, potato, tomato and tobacco. *Ch. globosum* acts against *Diaporthe phaseolorum* that causes soybean stem canker disease. Bis-spiro-Azaphilones and azaphilones are produced as an active metabolite by *Ch. cochliodes* species that promotes its defence mechanism. The *Chaetomium* species are generally formulated in the form powder, pellets or spores to act as broad-spectrum mycofungicide (Tann and Soytong, 2017). At times, they are administered as a combination with other fungal species, such as *Trichoderma harzianum* and *Chaetomium globosum* for increased antifungal effect.

1.2.1.1.5 *Gliocladium*

Gliocladium is a counterpart of *Penicillium,* albeit comprising of slimy conidia. Their colonies grow fast, in texture they appear suede-like to downy. Their color ranges from white to pink to salmon and later with sporulation becomes pale to dark green. The genus is recognized by its distinctive erect and dense penicillate conidiophores. These conidiophores with phialides bear slimy, one-celled hyaline to green, smooth-walled conidia in heads or columns. At times they might also produce verticillate branching conidiophores similar to those present in *Verticillium* or *Trichoderma*. These species are common soil saprobes and several species have been reported to be parasites of many plant pathogens, for example, *Gliocladium catenulatum* parasites *Sporidesmium sclerotiorum* and *Fusarium* sp., reduce the incidence of damping-off disease caused by *Pythium ultimum* and *Rhizoctonia solani* in the greenhouse. *Gliocladium virens* has been used as a biological control agent against a wide range of soil-borne pathogens such as *Pythium* and *Rhizoctonia* under greenhouse and field conditions (Santi *et al*., 2013). *Gliocladium virens* produces antibiotic metabolites such as gliotoxin, a sulphur containing epipolythiodioxopiperazine metabolite which have antibacterial, antifungal, antiviral and antitumor activities.

1.2.1.1.6 Other fungi as biofertilizers

While the above-mentioned organisms constitute the primary fungi based biofertilizers, many other Fungi impart their mycofungicide effect and can be used as biofertilizers. *Coniothyrium minitans,* a coelomycete, is a mycoparasite of *Sclerotinia* species. It controls the *Sclerotinia* based disease in crops like lettuce, oilseed rape, peanut and alfalfa. It uses sclerotia of *S. sclerotiorum* as the source of food for survival. Commercially, the products of *C. minitans* can be directly applied to the soil or sprayed. These organisms can survive for several years. Its efficiency can be improved in combination with *Trichoderma* species (Brahmaprakash and Sahu, 2012; Spadaro and Gullino, 2005). *Fusarium* includes both plants pathogenic and non-pathogenic races. The non-pathogenic species have effective biocontrol activity such as control of Fusarium wilt. They act either by competition or by induction of host defence. Due to the lack of understanding of their biology, commercial production has not materialized. Similarly, the genus *Rhizoctonia* contains both plants pathogenic and non-pathogenic species and the latter can act as biocontrol agents. *Clonostachys rosea* is applied as a single strain or as a mixture of strains against *Moniliophthora roreri*. By using the fungal sprays on cocoa crops it was found that the sporulation was reduced (Heydari and Pessarakli, 2010; Benhamou *et al*., 2012). *Pythium oligandrum* can control the soil-borne pathogens. Its oospores reduce the damping-off disease caused by *P. ultimum* in sugarbeet (Druzhinina and Kubicek, 2007). The hypha of the fungi coils around the host. It then forms appressoria-like structures and penetrates the hyphae of the host. It can control the pathogens in the rhizosphere and induce resistance as well as phosphorous uptake in plants. *P. nunn* is a mycoparasite of *Rhizoctonia solani*, *Phytophthora cinnamomi*, *P. parasitica*, *P. aphanidermatum*, *P. ultimum* and *P. vexans* (Deshmukh *et al.*, 2016).

Other fungi that can be used as mycofungicides are *Aspergillus* and *Penicillium* species. *Aspergillus* species are effective against the white-rot basidiomycetes (Pindi and Satyanarayana, 2012). Reports suggest that *Penicillium adametzioides* and *P. expansum* help in the reduction of Ochratoxin A (OTA), a secondary metabolite production and mycelial growth in grapes. The fungal antagonistic *Aureobasidium pullulans*, and *Ulocladium atrum* have also been tested for the control of *Botrytis aclada* which causes onion neck rot (Pal *et al.*, 2015).

1.2.2 Plant Growth Promoting Rhizobacteria (PGPR)

The use of agrochemicals in farming is very common to get high yields as these are used as fungicides, pesticides, insecticides and weedicides. But it has an indirect impact on the ecosystem and our health so scientists are trying to overcome this situation by using some alternative sources to chemical fertilizers. The research and development in the biological science helped a lot to understand the interaction between the plant and rhizospheric microorganism which opened a new door for the development of biofertilizers by the help of plant growth promoting rhizobacteria (PGPR) (Bhattacharyya and Jha, 2012). This PGPR is found to efficient enough as it enhanced the soil fertility (Prasad *et al.*, 2015), root growth, seed germination rate (Kaymak *et al.*, 2009), nutrient uptake capacity, protein content along with act as biocontrol of plant diseases (Guo *et al.*, 2004) and helps in delaying of senescence of leaf (Puppo *et al.*, 2005). The PGPR is one of the most important constituents for the formulation of biofertilizers (Ibiene *et al.*, 2012).

Based on the habitat of the PGPR has been categorised into intracellular plant growth promoting rhizobacteria (iPGPR) and extracellular plant growth promoting rhizobacteria (ePGPR). The iPGPR includes the *Mesorhizobium*, *Allorhizobium*, *Bradyrhizobium* and *Rhizobium* which are mainly found inside the nodular region of root cells and have the ability to fix the atmospheric nitrogen with the higher plants symbiotically. The ePGPR includes the species of *Bacillus*, *Micrococcus*, *Azospirillum*, *Agrobacterium*, *Serratia*, *Flavobacterium*, *Azotobacter* and *Pseudomonas* which are found in the rhizospheric region (Jha and Saraf, 2015; Saraf *et al.*, 2011). This is because the rhizosphere zone is enriched with nutrient sources required for microbes compared to the region of bulk soil (**Table 1.3**).

Table 1.3 List of plant growth promotion Rhizobacteria (PGPR)

Si.No.	PGFR Microorganism	Applicable in crops	Mechanism of Action	Results	References
1.	*Azotobacter*, *Azorhizobium*, *Azoarcus*, *Burkholderia*, *Herbaspirillum*, *Rhizobia* and *Rhizobium*	Wheat, Rice, Beetroot, Coffee, Coconuts, Tea, Maize, Tobacco and Legumes	Nitrogen Fixation in plant roots	Short lateral roots development occur which facilitates the growth of plants	Tejera et al., 2005; Wani, et al., 2013; Vejan et al., 2016; Yanni Youssef et al., 200)
2.	*Bacillus*, *Mycobacterium*, *Pseudomonas* and *Rhizobia*	Peanuts, Maize and Cotton	Stress resistance in plants	The level of Potassium, Phosphorous and Nitrogen increases in the calcareous soil	Beckers and Conrath, 2007; Yang et al., 2009
3.	*Bacillus, Chryseobacterium*, *Phyllobacterium*, *Rhizobium* and *Streptomyces*	Pepper, Maize, Tomato, Lettuce, Strawberries, Potato, Carrot and Neem	Production of siderophore	Increases the transportation of iron into the plant by the high-affinity iron-chelating compound	Sayyed et al., 2005; Husen, 2003
4.	*Pseudomonas* and *Sinorhizobium*	Pigeon pea and several crops	Production of β-glucanases and chitinase	B-1,3-glucanase activity helps in plant growth	Prathuangwong and Buensanteai, 2007; Li et al., 2003
5.	*Bacillus*, *Penibacillus* and *Phyllobacterium*	Black pepper, Cucumber and Strawberries	Potassium and Phosphate solubilization	Increases the level of Potassium and Phosphorous in the soil which helps in plant growth	Sindhu et al., 2016; Yazdani et al., 2009
6.	*Bacillus*, *Penibacillus*, *Rhizobium* and *Streptomyces*	Pepper, Carrot, Lettuce, Tomato, Potato and Lodgepole pine	By synthesising Auxin	Increases the concentration of Nitrogen, Calcium and Phosphorous in plants which ensure the higher potential to uptake the nutrient in plants.	Merzaeva and Shirokikh, 2010; Mohite, 2013
7.	*Azobacter* and *Bacillus*	Cucumber	By synthesising Cytokinin	The formation of lateral roots in plants	Gupta et al., 2015; de Garcia Salamone et al., 2005

It has been observed that an association of PGPR with the host plants is very much essential to show a lucrative effect. The association between the host plants and PGPR has been classified into two levels such as rhizospheric and endophytic (Bhattacharyya and Jha, 2012). In rhizospheric association, the PGPR colonize near the rhizosphere region of the soil, root surfaces and the intracellular spaces of the plants (Shrivastava and Kumar, 2013). Due to the change in the physical and chemical composition of the soil, no other microbial colonization appears in these areas. In the endophytic association, the PGPR will start colonizing in the apoplastic region in the host plants which is seen by nodules formation in the areas of roots which helps in the nitrogen fixation process in plants (Piccoli and Bottini, 2013).

1.2.3 Other Biofertilizer

This includes the fertilizers which are prepared by the help of living form of organisms and enhance the health of soil, increase the yield and creates a nutrient-rich environment for the growing crops. Based on the involvement of the organism this can be broadly classified into the followings.

1.2.3.1 *Azolla*

Azolla is a type of fern which is very commonly found in the aquatic environment like ditches, ponds and rice fields and consists up short, branched, floating stem, hanging roots into the water. This plant has a diverse use in the agricultural sector in ancient aged but the use of this plant as biofertilizer came into light by the people of Northern Vietnam and Southern China. Due to its nitrogen fixation ability, they used it in the rice crop as the eco-friendly green fertilizer and it shows a better yield (Giller, 2001). *Azolla* has the ability to double its biomass within 3-10 days in the ambient environmental condition as it can grow in the temperate as well as tropical countries. *Azolla* can be grown in the field as a monocrop and intercrop i.e. incorporation of *Azolla* before the plantation of rice and along with the plantation of the rice crop in the soil of agricultural fields respectively (Sarangi *et al.*, 2014).

It has been reported earlier that *Azolla* has the high protein content, herbicidal property, producible in the right environment, ability to carry out bioremediation which can decrease the heavy metals, pesticides, radionuclides present in the soil, fix nitrogen present in the atmosphere and decreases the volatilization of nitrogen fertilizer. *Azolla* covers the soil from sunlight hence prevent the germination of weeds (Temmink *et al.*, 2018).

1.2.3.2 Blue-Green Algae (BGA) or Cyanobacteria

Cyanobacteria are commonly known as blue-green algae (BGA), commonly found in the stagnant and slightly warm water areas like ponds, lakes, brackish and slow-moving streams which are rich in nitrogen and phosphorous. This organism helps in soil fertility by the process of phosphorous solubilization and biological nitrogen fixation (BNF) which improves the yield (Sahu *et al.*, 2012). Due to its photoautotrophic property, they can synthesize their own food and can survive without depending upon the availability of nutrients in the agricultural field (Mishra and Pabbi, 2004). Cyanobacteria has two different types of cells i.e. vegetative cells, which is responsible for activities like reproductive growth and photosynthesis and heterocysts, responsible for the synthesis of ammonia (Saadatnia and Riahi, 2009).

The filamentous structure of *Cyanobacteria* creates pores in the soils which increases water holding capacity of the soil, excreted growth promoting matters which contains auxin, gibberellin, amino acids and vitamins which are essential for plant growth. It also reduces the salinity level of the soil, weed growth in the agricultural fields and secretes the organic acids increases phosphorous content in the soil. It has been found to be an efficient inoculant which hikes the yield for the crops like sugarcane, maize, radish, barley, tomato, chilli, cotton and lettuce. The genera like *Aulosira*, *Anabaena*, *Nostoc*, *Scytonema* sp., *Calothrix* sp., *Tolypothrix* sp. have the ability to fix the atmospheric nitrogen and hence use during the process of farming of paddy & rice (Dey *et al.*, 2010).

1.2.3.3 Others

This includes the biofertilizers like compost which are prepared by the help of *Azotobacter* culture, Phosphotika and fungal cultures showing cellulolytic activity. Vermicompost consists up carbon, nitrogen, potassium, phosphorus, sulphur, vitamins, hormones, antibiotics and enzymes which are necessary for the growth of crops and hence increases the yield. The use of vermicompost by natural farming eradicate the loss of soil fertility and saline soil condition which occurs due to the excessive use of chemical fertilizers in farming (Bharti *et al.*, 2016).

Biocompost is another form of biofertilizers prepared by harmless forms of fungi and bacteria which helps in the decomposition process of the wastes from sugar industry and various plant parts. The use of biocompost works by increasing soil fertility and prevents the soil-borne diseases in the crops (Kannahi and Ramya, 2015). The common microorganisms use in this process is bacterial groups with nitrogen fixing and phosphate solubilizing ability and some decomposing fungus.

1.2.4 Liquid Biofertilizers

Liquid biofertilizers are a formulation of microorganisms which are dormant. The formulation also contains nutrients and other necessary elements needed to form resting spores or cysts in order to increase their shelf life and stress tolerance. Once they reach the soil, such organisms grow, multiply and colonize around the plant using elements from plant roots and soil and produce new active microorganisms. These fertilizers have more strength than conventional biofertilizers because of the longer shelf life of microorganism and sustained efficiency (Pindi and Satyanarayana, 2012). The plants that have been inoculated with liquid biofertilizers exhibited an improved biomass production. The basic characteristics of the formulation include stabilizing the organism during production distribution and storage, easy delivery, protection from adverse climatic conditions, increasing activity by increasing viability, interaction with the crop and reproducing potential (García-Fraile *et al.*, 2015). Broadly, liquid fertilizers can be categorized into: Nitrogen Fixing Microbes (NFM), Phosphorus Solubilizing Microbes (PSM) and Phosphate Mobilizing Microbes (VAM) and Potash Mobilizing Microbes (*Frateuria aurentia*) (Pal *et al.*, 2015).

Compared to conventional biofertilizers, liquid biofertilizers have the following advantages:

i. Longer shelf life (12- 24 months) compared to conventional biofertilizers (which lasts upto 3 months). The formulation might be sufficient for the entire crop cycle.

ii. There is no effect of high temperature in liquid fertilizers while the carrier based biofertilizers are not so temperature tolerant.

iii. The contamination of liquid fertilizers can be controlled constructively by means of proper sterilization and maintaining hygienic conditions. Normal biofertilizers have a high rate of contamination.

iv. No loss of properties due to storage at high temp. up to 45°C.

v. High moisture retaining capacity of liquid biofertilizers increasing cell viability.

vi. High populations can be maintained more than 10^9 cells/ ml up to 12 to 24 months.

vii. Easy to be administered and used by the farmers.

viii. Easy to export.

ix. Lower dosages than the carrier-based.

x. Easier methods to perform quality control.

Liquid biofertilizers are again classified into a) Dry inoculum (including dust, granules and briquettes) and, b) Liquid suspensions (oil or water-based and emulsions).

1.2.4.1 Dry inoculum products

They include inoculums in semi-dry condition such as dust, granules and briquettes, classified according to the particle size. They also take account of wettable dry powders which require liquid/water before application. The organism is fed into an air stream for mixing with a blender during its preparation. In such inocula, size, its bulk density and flow are important factors (Mazid and Khan, 2014).

These products may also cover inert carriers like clay minerals, starch polymers and ground plant residues, charcoal, lignite, vermiculite, holding the organisms. Various materials can be used in order to encapsulate the product and promote controlled release rate, varying with unit size. Generally, the concentration of organisms in the granules ranges from 20- 30% (Chesti *et al.*, 2013).

1.2.4.2 Liquid suspensions

They include aqueous or oil-based suspensions which are dormant in nature. Growth and contaminants suppressant like sodium azide, sodium benzoate, butanol, acetone, fungicides, and insecticides etc. are used for the long-term viability. However, when aqueous suspensions are used, they take a very long time for reactivation. Therefore, one can restrict the use of preservatives by using oil. The medium is well suited for supplying the inoculants in viable condition. Microorganisms in oil remain viable at high concentration in various degrees of dehydration. Initially the organisms are delivered at a dormant state such that there is no growth of contaminants during storage. By continuous aeration as a suspension in oil, the bacterial or fungal is dried, such that their shelf life increases for several years. The solution viscosity is almost similar to the setting rate of the particles achieved by using colloidal clays, polysaccharide gums, starch, cellulose or synthetic polymers (Kannaiyan and William, 2001).

1.3 Mechanism of Biofertilizer

There are various mechanisms by which plant growth and yielding of products can be enhanced by using biofertilizers. Generally, the growth-promoting biofertilizers and the PGPR supplies nutrients, prevent the crops from various pathogen infection, modulates the level of plant hormones, act as biocontrol agents and gives an eco-friendly environment for the agriculture (Bhattacharyya and Jha, 2012).

The growth of agricultural crops can be achieved by two different processes like direct method and indirect method. Our article mainly focuses on the direct method which includes growth by; atmospheric nitrogen fixation, solubilization of phosphate and potassium, production of siderophore and production of phytohormones.

1.3.1 Atmospheric Nitrogen Fixation

Nitrogen is the second essential elements after water which is highly necessary for plant growth and productivity. Although nitrogen is readily available in the atmosphere no plant species are capable to utilize the nitrogen directly for their growth so conversion of this atmospheric nitrogen to utilizable forms by the help of microorganisms is known as biological nitrogen fixation (BNF). This biochemical process converts the nitrogen to a simpler form i.e. ammonia by the help of nitrogen-fixing bacteria with the help of nitrogenase enzyme (Tejera *et al.*, 2005; Vejan *et al.*, 2016). This can be possible by two different methods based on the microorganism involved i.e. Symbiotic nitrogen fixation and Asymbiotic nitrogen fixation.

In the symbiotic mode of nitrogen fixation there exist a mutual interaction of microbes with the plant as it enters into the plant through the roots and forms the nodules where nitrogen fixation occurs (Wani *et al.*, 2013). This process can be possible with the help of symbionts like *Rhizobium* sp. for leguminous plants and *Frankia* for non-leguminous shrubs and tree (Mus *et al.*, 2016). On the other hand, the asymbiotic nitrogen fixation involves the genus of bacteria like *Azospirillum* sp., *Azoarcus* sp., *Azatobacter* sp., *Enterobacter* sp., *Pseudomonas* sp. and *Cyanobacteria* for non-leguminous plants such as Ficus, Margo, Rice, Mango and Radish (Santi *et al.*, 2013).

Both the symbiotic and asymbiotic nitrogen fixation occurs due to the presence of nitrogenase (nif) gene which activates the enzyme by biosynthesizing the iron-molybdenum co-factor, iron protein and electron donors.

1.3.2 Solubilization of Phosphate and Potassium

Phosphorus is the most vital element among the nutrients required for plant growth after nitrogen as it is highly necessary to carry out the metabolism as respiration, photosynthesis, generation of energy and transduction of signals in plants. The precipitated, insoluble and immobilized form of phosphorous is abundantly found in soil but plant needs the soluble form of phosphorous only (Walia *et al.*, 2017). So, the PGPR converts the insoluble phosphorous to soluble forms i.e. monobasic (H_2PO_4) and diabasic (HPO_4^{2-}) which can be absorbed through the plant roots. This phosphorous solubilization process mainly occurs by the production of extracellular enzymes, phosphate liberation during the substrate degradation, release of acidic proton and anions by the PGPR. The genera such as *Rhizobium*, *Azotobacter*, *Enterobacter*, *Bacillus*, *Pseudomonas*, *Serratia*, *Flavobacterium* and *Rhodococcus* has been reported earlier which enhances the plant growth (Satyaprakash *et al.*, 2017; Manzoor *et al.*, 2017).

Potassium is also an essential element next to phosphorous for plant growth and its concentration is also very low in the soil environment. The invasive use of chemical fertilizers in farming creates the deficiency in potassium level in the agricultural soil and affect the production of crops (Gontia-Mishra *et al.*, 2017). The deficiency of potassium can be observed with slow growth, dormancy in the seed size, undeveloped roots and hence low yields. The genera of PGPR such as *Acidothiobacillus*, *Burkholderia*, *Bacillus*, *Pseudomonas* and *Paenibacillus* has been reported to release the soluble form of potassium which can be uptake by the plants and used directly for plant growth (Gupta *et al.*, 2015).

1.3.3 Production of Siderophore

The ferric (Fe^{+3}) ion are readily available in the soil which is sparingly soluble by the plants as well as bacteria hence the assimilation is very low. In order to conquer this problem, the PGPR produces a low molecular weight compound known as siderophores which can chelate the iron (Tariq *et al.*, 2017). More than 500 types of siderophores have been reported out of which structural characterization of 270 siderophores has been completed successfully (Lambrese *et al.*, 2018). The PGPR genera such as *Azotobacter*, *Aeromonas*, *Bacillus*, *Rhizobium*, *Streptomyces*, *Pseudomonas* and *Serratia* secretes the siderophores and helps in the process of plant growth by uptake of the ferric-siderophores which has been observed by radiolabelling techniques (Adnan *et al.*, 2016).

1.3.4 Production of Phytohormones

Phytohormones are the chemical substances synthesize by the plant with a very minute quantity and travel to different parts in a plant for its applications which are directly and indirectly helpful for plant growth (Goswami *et al.*, 2016; Maheshwari *et al.*, 2015). The phytohormones have been commonly classified into four different classes like Auxin, Gibberellins, Cytokinins and Ethylene based on its functions.

Auxin is the naturally available form of indole acetic acid (IAA), which increases the uptake of nutrients and minerals from the soil, helps in the proliferation of the cells, well differentiation of cell occurs along with the development in xylem and root, promotes tuber and seed germination, vegetative growth of plant, formation of lateral and adventitious roots in plant and provide resistance against stress condition (Shah *et al.*, 2017). Plant growth promoting bacteria are found to secrete tryptophan which has been identified as the precursor molecule for indole-3-pyruvic acid biosynthesis (Ahmad *et al.*, 2005). PGPR also produces the minute amount of gibberellins and cytokinins which stimulates the growth of plants (Bhattacharyya and Jha, 2012; Joo *et al.*, 2005). Ethylene is a class of plant hormone which promotes root initiation, seed germination and fruit ripening in plants (Bal *et al.*, 2013). The above processes are mainly carried out by the following genera such as *Bradyrhizobium*, *Pseudomonas*, *Klebsiella*, *Agrobacterium*, *Rhizobium* and *Enterobacter* (Hayat *et al.*, 2010) (**Fig 1.2**).

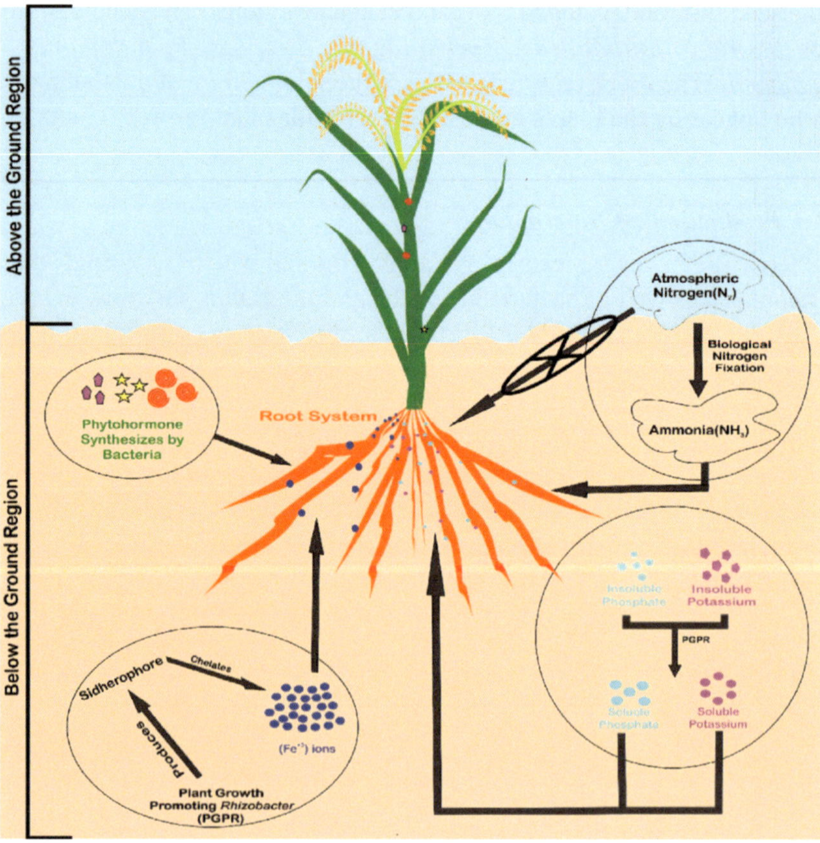

Fig. 1.2: Shows the overall mechanism of biofertilizer to increase productivity

1.4 Advantages of using biofertilizers over chemical fertilizers

Most of the plants and crops are adversely affected due to nitrogen and phosphorous deficiency, particularly in degraded soils. Fertilizers can overcome these deficiencies are started being used. However, the continuous use of chemical fertilizers without taking into account the climatic and environmental conditions of the soil began to affect the soil quality and ecosystem. The beneficial microflora and microfauna and their activities were affected too (Dasgan *et al.*, 2016). Continuous erosion washes away essential elements from the soil. Biofertilizers were a boon the agricultural sector. They aid in deriving benefits of the fertilizers and being environmental friendly at the same time. They are not only improve the soil fertility, crop productivity and production in agriculture, but also cost-effective (Kawalekar, 2013; Barman *et al.*, 2017). Some of the important advantages of biofertilizers in the agriculture industry are as follows:

1.4.1 Advantages to the soil

- Biofertilizers are eco-friendly, non-pollutants and cost-effective.
- Biofertilizers increase the availability of nutrients and improve the yield without affecting the soil.
- They produce growth promoting substances and vitamins and help to maintain soil fertility.
- They help in improving the texture of the soil, its structure and the water holding capacity.
- They help in the decomposition of organic matter and mineralization of soil.
- Biofertilizers play an important role in the recycling of plant nutrients and thus, act as a renewable source of nutrients.
- AMF can enhance the uptake of P, Zn, S and water, leading to uniform crop growth and increased yield.
- They help in solubilization and help in mobilizing the nutrients.
- They solubilize the insoluble forms of phosphates from the soil layer, like tricalcium, iron and aluminium phosphates into available forms by secreting acids.

1.4.2 Advantages in production

- The Biofertilizers can be produced easily on a large scale in the industry.
- No special care is necessary while using biofertilizers.
- They can also be grown by farmers in their own fields.

1.4.3 Advantages to the plant

- Secrete plant growth hormones and anti-metabolites which help in plant growth.
- Act as antagonists and reduce the growth of soil borne plant pathogens by secreting antifungal and antibiotics and help in the control of diseases.
- They help in fixing atmospheric nitrogen in the soil and root nodules of legume crops and make it available to the plant.

1.5 Conclusion

We depend on agriculture for our food and nutrition. However, the soil quality and farming are gradually depleting over time due to various environmental factors. The urgent need to improve the crop quality pushed us towards the addition of chemical fertilizers which can provide the lacking nutrients in the soil. However, in the long run, the soil health was found to degrade over time. The changing climatic conditions and rising demand from the agriculture industry increased the overall cost incurred in the input of raw materials and the quality of products.

It is well-known that microorganisms play a vital role in the agricultural ecosystem. They help in nutrient production, recycling, solubilizing and mobilizing, they have antibiotic activity, as well as help in the degradation of organic wastes. The number of microbes found around a particular agricultural crop are, however, present scantily. Thus, they need to be artificially introduced in the soil environment in the form of artificial cultures or mixed and transferred through suitable carriers. This artificial supply of microbes through culture or inoculum for the plant growth and development is known as biofertilizers. The past two decades saw a rise in the biofertilizer technology in the agricultural sector. The microbial cultures of fungal and bacterial subgroups have now transited to liquid biofertilizers which shows even more durability and higher production. Although the technological advances are made every day, the global warming and changing climatic conditions, demand solution for stress tolerant plants. Apart from this, the availability of the biofertilizers to masses in poorer nations (farmers) is essential. Innovative business strategies, marketing, awareness and research is also continuously needed for impacting the agroindustry globally.

References

Adnan, M., Patel, M., Reddy, M., Khan, S., Alshammari, E., Abdelkareem, A.M., Hadi, S. 2016. Isolation and characterization of effective and efficient plant growth-promoting rhizobacteria from rice rhizosphere of diverse paddy fields of indian soil. ARPN J Agricul Biolog Sci 11:9.

Ahmad, F. Ahmad, I. Khan, M.S. 2005. Indole acetic acid production by the indigenous isolates of Azotobacter and fluorescent Pseudomonas in the presence and absence of tryptophan. Turk J Biol 29:29-34.

Aly, A.H., Edrada-Ebel, R., Wray, V., Müller, W.E., Kozytska, S., Hentschel, U., Proksch, P., Ebel, R. 2008. Bioactive metabolites from the endophytic fungus *Ampelomyces* sp. isolated from the medicinal plant *Urospermum picroides*. Phytochemistry 69(8): 1716-1725. Doi: 10.1016/j.phytochem.2008.02.013

Bal, H.B., Nayak, L., Das, S. Adhya T.K. 2013. Isolation of ACC deaminase producing PGPR from rice rhizosphere and evaluating their plant growth promoting activity under salt stress. Plant Soil 366:93–105. Doi:10.1007/s11104-012-1402-5

Barman, J., Samanta, A., Saha, B., Datta, S. 2016. Mycorrhiza. Reson, 21:1093–1104. Doi: 10.1007/s12045-016-0421-6

Barman, M., Paul, S., Guha, A., Choudhury, P., Sen, J. 2017. Biofertilizer as prospective input for sustainable agriculture in India. Int. J. Curr. Microbiol. App. Sci, 6(11):1177-1186. Doi:10.20546/ijcmas.2017.611.141

Beckers, G.J., Conrath, U. 2007. Priming for stress resistance: from the lab to the field, Current opinion in plant biology 10(4):425-431. Doi:10.1016/j.pbi.2007.06.002

Benhamou, N., le Floch, G., Vallance, J., Gerbore, J., Grizard, D., Rey, P. 2012. *Pythium oligandrum*: an example of opportunistic success. Microbiol, 158(11):2679-2694. Doi:10.1099/mic.0.061457-0

Bharti, N., Barnawal, D., Wasnik, K., Tewari, S.K., Kalra, A. 2016. Co-inoculation of *Dietzia natronolimnaea* and *Glomus intraradices* with vermicompost positively influences *Ocimum basilicum* growth and resident microbial community structure in salt affected low fertility soils. *Appl Soil Ecol*, 100:211-225. Doi:10.1016/j.apsoil.2016.01.003

Bhattacharyya, P.N., Jha, D.K. 2012. Plant growth-promoting rhizobacteria (PGPR): emergence in agriculture. World J Microbiol Biotechnol, 28:1327–1350. Doi:10.1007/s11274-011-0979-9

Bisen, K., Keswani, C., Patel, J., Sarma, B., Singh, H.B. 2016. *Trichoderma* spp.: efficient inducers of systemic resistance in plants. In: D. Choudhary, A. Varma eds, Microbial-mediated Induced Systemic Resistance in Plants. Springer, Singapore, pp. 185-195. Doi: 10.1007/978-981-10-0388-2_12

Brahmaprakash, G., Sahu, P.K. 2012. Biofertilizers for sustainability. *J Ind Inst Sci*, 92:37-62.

Calvo, P., Watts, D.B., Kloepper, J.W., Torbert, H.A. 2017. Effect of microbial based inoculants on nutrient concentrations and early root morphology of corn (*Zea mays*), J Plant Nutr Soil Sci, 180: 56-70. Doi:10.1002/jpln.201500616

Chesti, M.H., Qadri, T., Hamid, A., Qadri, J., Azooz, M., Ahmad P. 2013. Role of Bio-fertilizers in Crop Improvement. In: Hakeem K., Ahmad P., Ozturk M. (eds) Crop Improvement. Springer, Boston, MA. pp. 189-208. Doi:10.1007/978-1-4614-7028-1_5

Clark, D., Klee, H., Dandekar A., 2004. Despite benefits, commercialization of transgenic horticultural crops lags. *California Agri J.*, 58(2). https://escholarship.org/uc/item/3dr9f3dm

Conijn, J., Bindraban, P., Schröder, J., Jongschaap, R. 2018. Can our global food system meet food demand within planetary boundaries? Agric Ecosyst Environ, 251: 244-256. Doi:10.1016/j.agee.2017.06.001

Connor, D.J. 2018. Organic agriculture and food security: a decade of unreason finally implodes. Field Crops Res, 225:128-129. Doi:10.1016/j.fcr.2018.06.008

Dasgan, H.Y., Cetinturk, T., Altuntas, O. 2016. The effects of biofertilisers on soilless organically grown greenhouse tomato. ISHS Acta Horticulturae 1164. III International Symposium on Organic Greenhouse Horticulture, pp. 555-561. Doi:10.17660/ActaHortic.2017.1164.73

de Garcia Salamone, I.E., Hynes, R.K., Nelson, L.M. 2005. Role of Cytokinins in Plant Growth Promotion by Rhizosphere Bacteria. In: Siddiqui Z.A. (eds) PGPR: Biocontrol and Biofertilization. Springer, Dordrecht. 173-195. Doi:10.1007/1-4020-4152-7_6

Deshmukh, R. Khardenavis, A.A., Purohit, H.J. 2016. Diverse metabolic capacities of fungi for bioremediation. Indian J Microbiol, 56:247-264. Doi:10.1007/s12088-016-0584-6

Dey, H. Tayung, K., Bastia, A. 2010. Occurrence of nitrogen-fixing cyanobacteria in local rice fields of Orissa, India. Ecoprint, 17: 77-85. Doi:10.3126/ECO.V17I0.4120

Druzhinina, I.S., Kubicek, C.P. 2007. Environmental and microbial relationships, Springer Science & Business Media. Doi:10.1007/978-3-319-29532-9

García-Fraile, P., Menéndez, E., Rivas, R. 2015. Role of bacterial biofertilizers in agriculture and forestry. AIMS Bioengg, 2(3):183-205. doi: 10.3934/bioeng.2015.3.183.

Gontia-Mishra, I. Sapre, S. Kachare, S. Tiwari, S. 2017. Molecular diversity of 1-aminocyclopropane-1-carboxylate (ACC) deaminase producing PGPR from wheat (*Triticum aestivum* L.) rhizosphere. Plant Soil, 414(1/2):213-227.

Goswami, D. Thakker, J.N. Dhandhukia, P.C. 2016. Portraying mechanics of plant growth promoting rhizobacteria (PGPR): a review, Cogent Food & Agriculture 2(1):1127500. Doi:10.1080/23311932.2015.1127500

Guo, J.-H., Qi, H.-Y., Guo, Y.-H., Ge, H.-L., Gong, L.-Y., Zhang, L.-X., Sun, P.-H. 2004. Biocontrol of tomato wilt by plant growth-promoting rhizobacteria. Biol Control, 29:66-72. Doi:10.1016/S1049-9644(03)00124-5

Gupta, G., Parihar, S.S., Ahirwar, N.K., Snehi, S.K., Singh, V. 2015. Plant growth promoting rhizobacteria (PGPR): current and future prospects for development of sustainable agriculture. J Microb Biochem Technol, 7:096-102.

Hadi, H., Kalantar, A. 2015. Effects of mycorhizal symbiosis, application of super absorbant gel, glycine betain and sugar beet extract on physiological traits and seed yield of castor bean (*Ricinus communis* L.) in drought stress conditions. Iran J Crop Sci, 17(3):236-250.

Hayat, R., Ali, S., Amara, U. Khalid, R. Ahmed, I. 2010. Soil beneficial bacteria and their role in plant growth promotion: a review. Ann Microbiol 60:579–598. Doi:10.1007/s13213-010-0117-1

Heydari, A., Pessarakli, M. 2010. A review on biological control of fungal plant pathogens using microbial antagonists. J Biol Sci, 10(4):273-290. Doi:10.3923/jbs.2010.273.290

Husen, E. 2003. Screening of soil bacteria for plant growth promotion activities in vitro. *Indo J Agricul Sci*, 4:27-31. DOI: http://dx.doi.org/10.21082/ijas.v4n1.2003.p27-31

Ibiene, A., Agogbua, J., Okonko, I., Nwachi, G. 2012. Plant growth promoting rhizobacteria (PGPR) as biofertilizer: Effect on growth of *Lycopersicum esculentus*. J Ame Sci, 8(2):318-324.

Jha, C.K., Saraf, M. 2015. Plant growth promoting rhizobacteria (PGPR): a review. J Agricul Res Develop, 5(2):108-119. Doi:10.13140/RG.2.1.5171.2164

Joo, G.J., Kim, Y.M., Kim, J.T., Rhee, I.K., Kim, J.H., Lee, I.J. 2005. Gibberellins-producing rhizobacteria increase endogenous gibberellins content and promote growth of red peppers. J Microbiol, 43(6):510-515.

Giller, K.E. 2001. Nitrogen fixation in tropical cropping systems, CABI, Wallingford, UK. Doi:10.1079/9780851994178.0000

Kannahi, M., Ramya, R. 2015. Effect of biofertilizer, vermicompost, biocompost and chemical fertilizer on different morphological and phytochemical parameters of *Lycopersicum esculentum* L. Pharm Pharma Sci, 4(9):1460-1469.

Kannaiyan, S. B., William, J. 2001. Inoculant production in developing countries-problems, potentials and success, Netherlands: FAO.

Kawalekar, J.S. 2013. Role of biofertilizers and biopesticides for sustainable agriculture. J Bio Innov 2(3):73-78.

Kaymak, H.Ç., Güvenç, Ý., Yarali, F., Dönmez, M.F. 2009. The effects of bio-priming with PGPR on germination of radish (*Raphanus sativus* L.) seeds under saline conditions. Turk J Agricul Forest, 33(2) 173-179. Doi:10.3906/tar-0806-30

Keswani, C., Singh S.P., Singh H.B. 2013. A superstar in biocontrol enterprise: *Trichoderma* spp, Biotech Today 3(2): 27-30. Doi:10.5958/2322-0996.2014.00005.2

Kiss, L., Russell, J., Szentiványi, O., Xu, X., Jeffries, P. 2004. Biology and biocontrol potential of *Ampelomyces mycoparasites*, natural antagonists of powdery mildew fungi, Biocontrol Science and Technology 14(7):635-651. Doi:10.1080/09583150410001683600

Lambrese, Y., Guiñez, M., Calvente, V., Sansone, G., Cerutti, S., Raba, J., Sanz, M.I. 2018. Production of siderophores by the bacterium *Kosakonia radicincitans* and its application to control of phytopathogenic fungi. Biores Technol Rep 3:82-87.

Li, Y.-z., Zheng, X.-h., Tang, H.-l., Zhu, J.-w., Yang, J.-m. 2003. Increase of beta-1, 3-glucanase and chitinase activities in cotton callus cells treated by salicylic acid and toxin of *Verticillium dahlia*. Acta Botanica Sinica 45(7): 802-808.

Liang, C., Yang, J., Kovács, G.M., Szentiványi, O., Li, B., Xu, X., Kiss, L. 2007. Genetic diversity of *Ampelomyces mycoparasites* isolated from different powdery mildew species in China inferred from analyses of rDNA ITS sequences. *Fungal Diversity*, 24:225-240.

Madhusudhan, L. 2015. Agriculture Role on Indian Economy. Bus Eco J, 6:4. Doi:10.4172/2151-6219.1000176

Maheshwari, D.K., Dheeman, S., Agarwal, M. 2015. Phytohormone-Producing PGPR for Sustainable Agriculture. In: Maheshwari D. (eds) Bacterial Metabolites in Sustainable Agroecosystem. Sustainable Development and Biodiversity, vol 12. Springer, Cham. pp. 159-182. Doi:10.1007/978-3-319-24654-3_7

Manzoor, M., Abbasi, M.K., Sultan, T. 2017. Isolation of phosphate solubilizing bacteria from maize rhizosphere and their potential for rock phosphate solubilization–mineralization and plant growth promotion. Geomicrobiol J, 34:81-95. Doi:10.1080/01490451.2016.1146373

Mazid, M., Khan, T.A. 2014. Future of bio-fertilizers in Indian agriculture: an overview. Int J Agricul Food Res, 3(3):10-23.

Merzaeva, O.V., Shirokikh, I.G. 2010. The production of auxins by the endophytic bacteria of winter rye. Appl Biochem Microbiol, 46:44–50. Doi:10.1134/S0003683810010072

Mishra, U., Pabbi, S. 2004. Cyanobacteria: A potential biofertilizer for rice. Reson 9:6–10. Doi:10.1007/BF02839213

Mohite, B. 2013. Isolation and characterization of indole acetic acid (IAA) producing bacteria from rhizospheric soil and its effect on plant growth. J Soil Sci Plant Nutr, 13(3):638-649. Doi:10.4067/S0718-95162013005000051

Mus, F., Crook, M.B., Garcia, K., Costas, A.G., Geddes, B.A., Kouri, E.D., Paramasivan, P., Ryu, M.-H., Oldroyd, G.E., Poole, P.S. 2016. Symbiotic nitrogen fixation and the challenges to its extension to nonlegumes. Appl Environ Microbiol, 82(13): 3698-3710. Doi: 10.1128/AEM.01055-16

Nagesh, G. 2012. Studies on coffee effluent waste compost on growth and yield of paddy (*Oryza sativa* L.), University of Agricultural Sciences GKVK, Bangalore.

Pal, S., Singh, H., Farooqui, A., Rakshit, A. 2015. Fungal biofertilizers in Indian agriculture: perception, demand and promotion. J Eco-friendly Agri, 10(2):101-113.

Paswan, S.K., Gautam A., Verma P., Rao C.V., Sidhu O.P., Singh A.P., Srivastava S. 2017. The Indian Magical Herb 'Sanjeevni' (*Selaginella bryopteris* L.)-A Promising Anti-inflammatory Phytomedicine for the Treatment of Patients with Inflammatory Skin Diseases. J Pharmacopuncture, 20(2):93-99. Doi: 10.3831/KPI.2017.20.012

Piccoli, P., Bottini, R. 2013. Abiotic Stress Tolerance Induced by Endophytic PGPR. In: Aroca R. (eds) Symbiotic Endophytes. Soil Biology, vol 37. Springer, Berlin, Heidelberg. pp-151-163. Doi:10.1007/978-3-642-39317-4_8

Pindi, P., Satyanarayana, SDV. 2012. Liquid Microbial Consortium-A Potential Tool for Sustainable Soil Health. J Biofertil Biopestici, 3:124. Doi:10.4172/2155-6202.1000124

Prasad, R., Kumar, M., Varma, A. 2015. Role of PGPR in Soil Fertility and Plant Health. In: Egamberdieva D., Shrivastava S., Varma A. (eds) Plant-Growth-Promoting Rhizobacteria (PGPR) and Medicinal Plants. Soil Biology, vol 42. Springer, Cham. pp-247-260. Doi:10.1007/978-3-319-13401-7_12

Prathuangwong, S., Buensanteai, N. 2007. *Bacillus amyloliquefaciens* induced systemic resistance against bacterial pustule pathogen with increased phenols, phenylalanine ammonia lyase, peroxidases and 1, 3-β-glucanases in soybean plants. Acta Phytopathol Entomol Hung, 42(2):321-330. Doi:10.1556/aphyt.42.2007.2.14

Puppo, A., Groten, K., Bastian, F., Carzaniga, R., Soussi, M., Lucas, M.M., De Felipe, M.R., Harrison, J., Vanacker, H., Foyer, C.H. 2005. Legume nodule senescence: roles for redox and hormone signalling in the orchestration of the natural aging process. New Phytol, 165(3):683-701. Doi: 10.1111/j.1469-8137.2004.01285.x

Saadatnia, H. Riahi, H. 2009. Cyanobacteria from paddy fields in Iran as a biofertilizer in rice plants. Plant Soil Environ, 55(5):207-212.

Sahu, D. Priyadarshani, I. Rath, B. 2012. Cyanobacteria–as potential biofertilizer. CIBTech J Microbiol, 1(2-3):20-26.

Santi, C., Bogusz, D., Franche, C. 2013. Biological nitrogen fixation in non-legume plants. Ann Bot, 111(5):743-767. Doi:10.1093/aob/mct048

Saraf, M., Rajkumar, S., Saha, T. 2011. Perspectives of PGPR in Agri-Ecosystems. In: Maheshwari D. (eds) Bacteria in Agrobiology: Crop Ecosystems. Springer, Berlin, Heidelberg. pp- 361-385. Doi:10.1007/978-3-642-18357-7_13

Sarangi, S.K., Maji, B., Singh, S., Sharma, D.K., Burman, D., Mandal, S., Ismail, A.M., Haefele, S.M. 2014. Crop establishment and nutrient management for dry season (boro) rice in coastal areas. Agronomy J, 106(6):2013-2023. Doi:10.2134/agronj14.0182

Satyaprakash M., Nikitha T., Reddi E., Sadhana B., Vani S.S. 2017. Phosphorous and phosphate solubilising bacteria and their role in plant nutrition. Int J Curr Microbiol App Sci 6(4): 2133-2144. Doi:10.20546/ijcmas.2017.604.251

Sayyed, R., Badgujar, M., Sonawane, H., Mhaske, M., Chincholkar, S. 2005. Production of microbial iron chelators (siderophores) by fluorescent Pseudomonads. Ind J Biotechnol, 484-490.

Shah, D.A. Sen, S. Shalini, A. Ghosh, D. Grover, M. Mohapatra, S. 2017. An auxin secreting Pseudomonas putida rhizobacterial strain that negatively impacts water-stress tolerance in Arabidopsis thaliana. Rhizosphere 3:16-19.

Shrivastava, U.P., Kumar, A. 2013. Characterization and optimization of 1-aminocyclopropane-1-carboxylate deaminase (ACCD) activity in different rhizospheric PGPR along with *Microbacterium* sp. strain ECI-12A. Int J Appl Sci Biotechnol, 1:11-15. Doi:10.3126/ijasbt.v1i1.7921

Sindhu S.S., Parmar P., Phour M., Sehrawat A. 2016. Potassium-Solubilizing Microorganisms (KSMs) and Its Effect on Plant Growth Improvement. In: Meena V., Maurya B., Verma J., Meena R. (eds) Potassium Solubilizing Microorganisms for Sustainable Agriculture. Springer, New Delhi. pp. 171-185. Doi:10.1007/978-81-322-2776-2_13

Singh, M. 2018. Biofertilizer: A Pathway Towards Eco-green Environment, Research & Reviews: J of Ecology 4(1):12-18. Doi:10.37591/rrjoe.v4i1.603

Singh, V., Upadhyay, R.S., Sarma, B.K., Singh, H.B. 2016. *Trichoderma asperellum* spore dose depended modulation of plant growth in vegetable crops. Microbiol Res 193:74-86. Doi:10.1016/j.micres.2016.09.002

Soytong, K. 2014. Bio-formulation of *Chaetomium cochliodes* for controlling brown leaf spot of rice. Int J Agricul Technol, 10(2):321-337.

Soytong, K., Ratanacherdchai K. 2005. Application of mycofungicide to control late blight of potato. J Agricul Technol, 1:19-32.

Spadaro, D., Gullino, M.L. 2005. Improving the efficacy of biocontrol agents against soilborne pathogens. Crop Prot, 24(7):601-613. Doi:10.1016/j.cropro.2004.11.003

Tann, H., Soytong, K. 2017. Bioformulations and nano product from *Chaetomium cupreum* CC3003 to control leaf spot of rice var. Sen Pidoa in Cambodia. Int J Plant Biol, 7(1). Doi:10.4081/pb.2016.6413

Tariq, M. Noman, M. Ahmed, T. Hameed, A. Manzoor, N. Zafar, M. 2017. Antagonistic features displayed by Plant Growth Promoting Rhizobacteria (PGPR): A Review. J Plant Sci Phytopathol 1:38-43.

Tejera, N., Lluch, C., Martìnez-Toledo, M.V., Gonzalez-Lopez, J. 2005. Isolation and characterization of Azotobacter and Azospirillum strains from the sugarcane rhizosphere. Plant Soil, 270:223-232. Doi:10.1007/s11104-004-1522-7

Temmink, R.J.M., Harpenslager, S.F., Smolders, A.J.P., Dijk, G., Peters, R.C., Lamers, L.P., Kempen, M.M. 2018. *Azolla* along a phosphorus gradient: biphasic growth response linked to diazotroph traits and phosphorus-induced iron chlorosis. Sci Rep, 8:4451. Doi:10.1038/s41598-018-22760-5

Trabelsi, D., Mhamdi, R. 2013. Microbial inoculants and their impact on soil microbial communities: a review. Biomed Res Int, 2013:863240. Doi:10.1155/2013/863240

Vejan, P., Abdullah, R., Khadiran, T., Ismail, S., Nasrulhaq Boyce, A. 2016. Role of Plant Growth Promoting Rhizobacteria in Agricultural Sustainability-A Review. Molecules, 21(5):573. DOI: 10.3390/molecules21050573

Walia A., Guleria S., Chauhan A., Mehta P. 2017. Endophytic Bacteria: Role in Phosphate Solubilization. In: Maheshwari D., Annapurna K. (eds) Endophytes: Crop Productivity and Protection. Sustainable Development and Biodiversity, vol 16. Springer, Cham. pp. 61-93. Doi:10.1007/978-3-319-66544-3_4

Wani, S.A., Chand, S., Ali, T. 2013. Potential use of *Azotobacter chroococcum* in crop production: an overview. Curr Agricul Res J, 1:35-38. Doi:10.12944/CARJ.1.1.04

Xuan, V.-T. 2018. Rice production, agricultural research, and the environment. In: B.J. Tria Kerkvliet and D.J. Porter eds, Vietnam's rural transformation, Routledge, pp. 185-200. Doi: 10.4324/9780429503269-7

Yang, J., Kloepper, J.W., Ryu, C.-M. 2009. Rhizosphere bacteria help plants tolerate abiotic stress. *Trends Plant Sci*, 14:1-4. Doi:10.1016/j.tplants.2008.10.004

Yanni Youssef, G., Rizk Rizk, Y., El-Fattah Faiza, K. Abd, Squartini, A., Corich, V., Giacomini, A., de Bruijn, F., Rademaker, J., Maya-Flores, J., Ostrom, P., Vega-Hernandez, M., Hollingsworth, R.I., Martinez-Molina, E., Mateos P., Velazquez E., Wopereis J., Triplett E., Umali-Garcia, M., Anarna, J.A., Rolfe, B.G., Ladha J.K., Hill J., Mujoo, R., Ng Perry, K.D.F.B. 2001. The beneficial plant growth-promoting association of *Rhizobium leguminosarum* bv. trifolii with rice roots. Functional Plant Biol, 28:845-870. Doi:10.1071/PP01069

Yazdani, M., Bahmanyar, M.A., Pirdashti, H., Esmaili, M.A. 2009. Effect of phosphate solubilization microorganisms (PSM) and plant growth promoting rhizobacteria (PGPR) on yield and yield components of corn (*Zea mays* L.). World Acad Sci Engg Technol 3:50-52.

Zaidi A., Khan, M.S., Saif, S., Rizvi, A., Ahmed, B., Shahid, M. 2017. Role of nitrogen-fixing plant growth-promoting rhizobacteria in sustainable production of vegetables. In: A. Zaidi, and M. Khan eds, Microbial Strategies for Vegetable Production, Springer, Cham, pp. 49-79. Doi: 10.1007/978-3-319-54401-4_3

2

Plant Growth Promoting Microbes of Rice and Their Application for Sustainable Agriculture

Avishek Banik

School of Biotechnology, Department of Life Sciences, Presidency University Canal Bank Road, DG Block (Newtown), Action Area 1D, Newtown Kolkata-700156, West Bengal, India

Abstract

The concern of global food security is growing day by day as the agriculturally available land is reducing each day with the expansion of industrialization and urbanization. During the course of urban and industrial development, crop production is greatly affected by global warming, contamination of hazardous materials, salinity and drought. To combat against such adversities, scientists have taken several additives measurement like development of transgenic crops, application of indigenous plant growth promoting microbes etc. for sustainable agriculture. Use of plant growth promoting microbes is getting popular day-by-day as they occupy a privilege niche with the autochthonous flora of specific environment. For the expanse of food and shelter, the microbes supply a wide range of plant regulators (e.g. auxins, cytokinins, gibberellins, ABA, phenols, NH_3 etc.), antibiotics, alkaloids, steroids, terpenoids, diterpenes, enzymes (e.g. cellulases, lipases, proteinase, esterases etc.) etc. for overall development of the plant. Apart from growth promotion, it also protects the plants against pathogenic invasion creating a competition with pathogens for nutrition and habitat. In near future, formulation of different plant growth promoting bacteria (like alpha- Azorhizobium, Azospirillum, Ancylobacter, Rhizobium, Bradyrhizobium, Sinorhizobium, Novosphingobium sp., *beta- Burkholderia* sp., *gamma- Acinetobacter, Aeromonas, Azotobacter, Enterobacter, Klebsiella, Pantoea, Pseudomonas, Stenotrophomonas* sp. *Proteobacteria, Bacillus* sp., *Paenibacillus* sp. and *Actinobacteria- Microbacterium* sp. etc.*) and fungus (Aspergillus,*

Trichoderma, Penicillium etc.*) may reduce the use of pesticides and chemical fertilizer to promote greener ways of crop production.*

Keywords: Plant growth promotion; microbes; rice

2.1. Introduction

Rice, *Oryza sativa* L., a member of *Poaceae* family is one of the most important crop and cultivated in more than hundred countries throughout the world (Banik *et al.*, 2016a). Tropical and subtropical climatic countries like China, India, Indonesia, Bangladesh, Vietnam, Thailand, Myanmar, Philippines, Japan etc. are the major contributors to cultivate (Food and Agriculture Organization, United Nations, http://www.fao.org/faostat/en/#data/QC, accessed on 2nd July, 2018) (FAOSTAT, 2013) (**Fig. 2.1**). A large human population of world consume rice as their staple food and use paddy as a fodder (Seck *et al.*, 2012). India, one of the prime rice growing countries having an area ~44.6 million hectares for rice cultivation, distributed among irrigated, rain fed, upland and submerged regions (Soora *et al.*, 2013). Worldwide, enormous importance has been given to this crop because of its promise of food security and huge impact on socioeconomic stability to the day by day increasing population. Thus, *ex* and *in planta* plant growth promoting (PGP) microbes could enhance production of different crops, reduce chemical fertilizer use and there by decrease environmental and biological health hazards (Lin *et al.*, 2010; Lucas *et al.*, 2014; Banik *et al.*, 2018).

2.2. Rice microbiome

The niche specificity of rice microbiome can be classified mainly in three different regions, i.e. rhizosphereric, phyllospheric and endophytic having aerobic and anaerobic members. The microbial community of rice rhizosphere mainly consists of heterotrophs, nitrifying, denitrifying, nitrogen fixing, associative nitrogen fixing, facultative anaerobes and facultative anaerobic nitrogen fixers (Banik *et al.*, 2017).

Frequent and uncontrolled use of chemical nitrogenous fertilizers in rice fields also increases the nitrogen volatilizations which cause many environmental hazards. Recent discoveries on anaerobic oxidation of ammonium (anammox) suggest that, anammox bacteria (Candidatus *Kuenenia*, Candidatus *Brocadia*) play a very critical role in nitrogen loose from the rice rhizosphere (Li *et al.*, 2015).

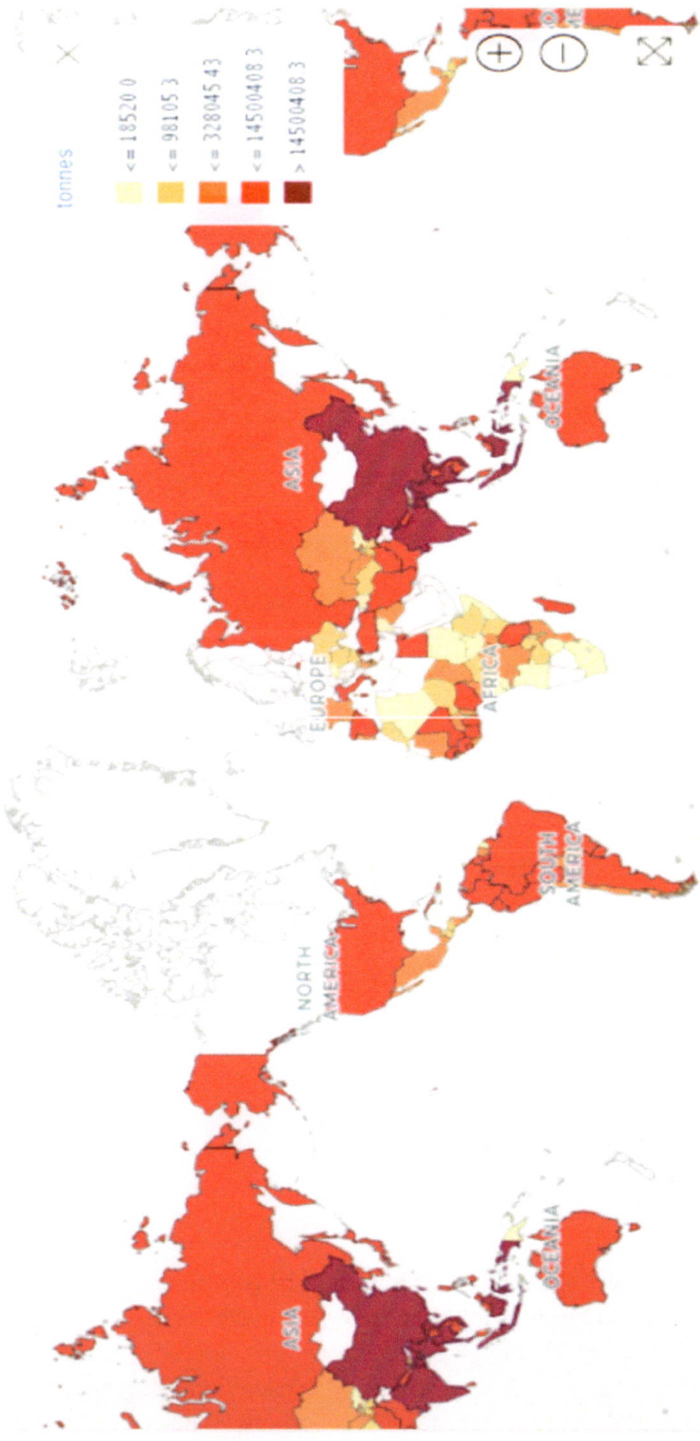

Fig. 2.1: Yield quantities of Rice in world (FAO, average 1994 to 2016, data accessed on 2nd of July, 2018)

2.3. Different plant growth promoting (PGP) traits

Since last decades, exploration of plant growth promoting (PGP) properties of different plant associated bacteria have been investigated and well document (Beneduzi *et al.*, 2008;Nautiyal *et al.*, 2013;Chung *et al.*, 2015;Banik *et al.*, 2016a). However, in many occasions the field application results were not encouraging may be due to their competition with autochthonous microorganism of that specific niche (Mulas *et al.*, 2016).

2.3.1. Important nutrient and its utilization

Plants mines the earth's crust by its root system to construct its cellular macromolecules. Among all the essential elements, the requirement of nitrogen and phosphorus are the most important as they are the major constituent of nitrogenous heterocyclic bases (purine and pyrimidine), amino acids, DNA, chlorophyll, energy-transfer compounds (adenosine triphosphate), phospholipids (lecithin, cephalin etc.), signal transduction etc. (Davidson and Gu, 2012).

2.3.2. Nitrogen fixation

Nitrogen is one of the most abundant (77%) element present in the earth's atmosphere. But, due to nature's irony plant cannot accumulate atmospheric nitrogen to construct cellular building blocks. This is mainly due to the high bond dissociation energy of N=N, which is very energy expensive to break by the living organism (MacKay and Fryzuk, 2004). Thus, plants are totally depending upon diazotrophic microbes for conversion of inert nitrogen to ammonia for construction of amino acids, subsequently cellular macromolecules. The traditional industrial nitrogen fixation by Haber–Bosch process is carried out by generally at >200 atm pressure and <"450°C temperature in presence of iron catalyst which is really an energy demanding process, consuming nearly 1% of fossil fuel of its global demand (Smith, 2002). The assimilation of N_2 by microorganism initiated with reduction of nitrogen to ammonia by nitrogenase system which is assembled by *nif* operon. Generally, the *nif* operon consists of 15- 20 genes, depending upon the microbial families (Temme *et al.*, 2012). Diazotrophically fixed nitrogen, i.e. ammonium is most frequently assimilated through glutamine synthetase (GS) and glutamine oxoglutarate aminotransferase (GOGAT) to form amino acids (Ohyama and Kumazawa, 1980) (Fig. 2.2).

$$N_2 + 8e^- + 16MgATP + 8H^+ \rightarrow 2NH_3 + H_2 + 16MgADP + 16Pi$$ (Ribbe *et al.*, 2013)

A substantial quantity of N_2 has publicized to be combined into crops like rice by biological N_2 fixation. However, selected diazotrophic bacteria such as *Azospirillum* sp., *Azotobacter* sp., *Gluconacetobacter* sp., *Azoarcus* sp., *Burkholderia* sp., *Herbaspirillum* sp., have been studied for significant increase of plants biomass under nitrogen deficient conditions (Banik *et al.*, 2016a; Kumar *et al.*, 2017) (**Table 2.1**).

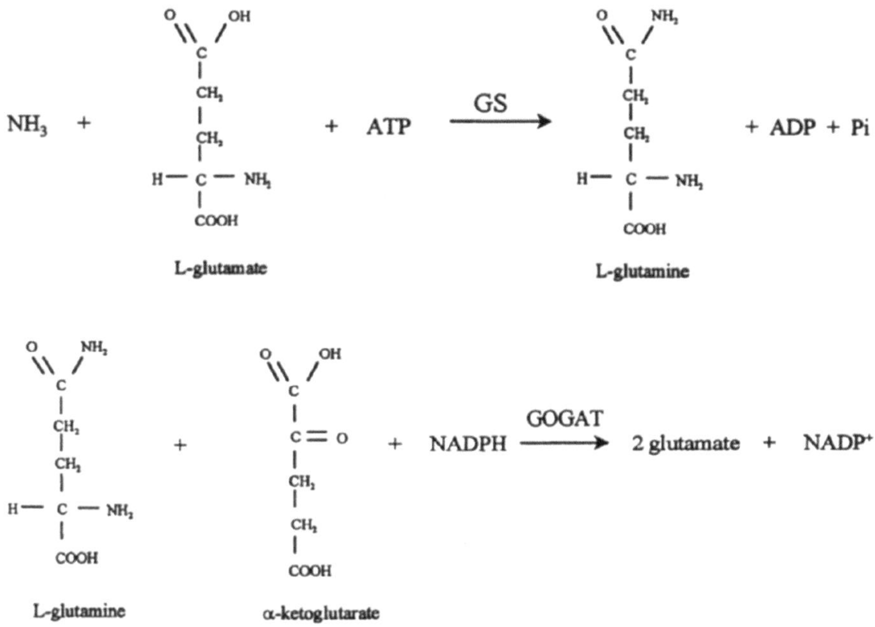

Fig. 2.2: Conversion of biologically fixed ammonia into glutamine and glutamate by GS (glutamine synthetase, EC 6.3.1.2) and GOGAT (glutamine:2-oxoglutarate amidotransferase, EC 1.4.1.13)

2.3.3. Phosphate solubilization

Phosphorus (P), one of the most abundant macroelement present on the earth's crust in both organic and inorganic forms. Although its profuse occurrence, rice ecologies suffer lack of P, as major share present in insoluble form. Rice rhizosphere harbors many bacterial genera like *Pseudomonas*, *Bacillus*, *Enterobacter*, *Azotobacter*, *Burkholderia*, *Microbacterium*, *Rhizobium* sp. etc. and vesicular-arbuscular mycorrhiza (VAM). These microorganisms convert the insoluble P to soluble $H_2PO_4^{-1}$, HPO_4^{-2} by phytase (myo-inositol-hexakisphosphate phosphohydrolase) which assist plants to incorporate them as sugar-phosphate backbone of DNA, RNA, ATP (adenosine triphosphate) and many other signal transducing molecules (Chaiharn and Lumyong, 2009; Panhwar *et al.*, 2012).

2.3.4. Iron chelation (via siderophores)

Iron, one of the most versatile element playing a crucial role in biological system in different forms (Fe^{+} to Fe^{4+}) as a cofactor of enzymes and electron carrier. Microbial siderophores are low molecular weight (H - 700 - 1000 D) compounds, synthesized intracellularly and secreted in the environment to scavenge metals. Till date, more than 500 siderophores have been reported from different organism, each having a distinct structure and recognition properties. Based on their chemical properties, siderophores have been broadly classified into three groups, i.e. carboxylate, catecholate and hydroxamate. Among all of them, enterobactin (belongs from catechol family), a Fe (III) ion siderophore, produced by a member of *Enterobacteriaceae* (*Escherichia coli*) was well investigated by several researchers. Apart from *E. coli*, plant growth promoting microbes like *Streptomyces* (gram positive actinobacteria, produces desferrioxamine), *Pseudomonas* (gammaproteobacterial, produces pyochelin) also possess different groups of siderophores (Liermann *et al.*, 2000, Pahari *et al.*, 2017).

2.3.5. Plant growth regulator

Production of phytohormones is one of the most decisive factor for the establishment of symbiotic plant-microbe interaction. Most of the plant growth promoting bacteria produces common growth regulator such as indole compounds (*Pseudomonas*, *Rhizobium*, *Pantoea*, *Bacillus*, *Azospirillum*, *Azotobacter*, *Enterobacter*, *Novosphingobium*), cytokinins (*Agrobacterium*, *Bacillus*, *Pseudomonas*, *Methylobacterium*, *Klebsiella*, *Xanthomonas* etc.), gibberellins (*Rhizobium meliloti*, *Herbaspirillum*, *Bacillus*, *Acetobacter*, *Pseudomonas*), ACC (1-aminocyclopropane-1-carboxylate) deaminase (*Alcaligens*, *Bacillus*, *Ochrobactrum* sp.) etc. and semisynthetic phytohormones like salicylic acid, jasmonate, nitric oxide, strigolactones etc., which straightly controls plant physiology (Banik *et al.*, 2016a; Defez *et al.*, 2017; Kumar *et al.*, 2017) **(Table 2.1)**.

2.3.6. Disease resistance

PGP bacteria also plays a crucial role for suppression of plant pathogens as they compete with plant pathogenic bacteria for shelter and nutrition. Upon primary exposure to plants, PGP bacteria are able to proliferate and capitalize plant associated niches, resulting blocking of possible plant surface for invasion of the pathogen. Some of the PGP were also reported to produce antimicrobial agents like 2,4-diacetylphloroglucinol, pyrrolnitrin etc. to control the proliferation of the pathogens. Production of iron siderophores also indirectly limits many plant pathogen, as these siderophores chelates iron, which is necessary for pathogens' pathogenesis. Most of the PGP microbes were also found to produce substantial amount of enzymes e.g. β-1,3-glucanase, chitinases etc., which specifically targets the cell was of different pathogenic fungus (fungal cell wall is mainly made up of glucan and chitin). Plant-PGP interaction also evoke many diffusible volatile compounds, which also inhibit growth of many pathogens. Volatile compounds produced by rice rhizospheric strains like *Streptomyces corchorusii* can inhibit growth of *Bipolaris oryzae* (causative agent of brown spot disease), *Curvularia oryzae* (causative agent of leaf spot disease),

Table 2.1: Some important rice associated beneficial bacteria with their niche and phylogenic specification

Niche	Bacterial strain	Phylum	Function	Reference
Rhizosphere soil	*Bacillus methylotrophicus*	Firmicutes	A methanol-utilizing, PGPR	Madhaiyan et al (2010)
	Bacillus amyloliquefaciens	Firmicutes	Salt tolerant *Bacillus*, modulate gene expression, PGPR	Nautiyal et al (2013)
	Alcaligens sp. (P), *Ochrobactrum* sp. (P), *Bacillus* sp. (F)	Proteobacteria (P), Firmicutes (F)	ACC-utilizing bacteria	Bal et al (2013)
	Pseudomonas sp., *Pantoea agglomerans*	Proteobacteria	Suppress rice blast infections	Spence et al (2014)
	Gluconacetobacter diazotrophicus (P), *Pseudomonas stutzeri*(P), *Klebsiella* sp. (P), *Sinorhizobium meliloti* (P), *Enterobacter* sp. (P), *Lysinibacillus* sp. (F), *Bacillus cereus* (F)	Proteobacteria (P), Firmicutes (F)	Diazotrophs of aromatic rice	Kumar et al (2017)
	Bacillus laterosporous MML2270	Firmicutes	Chitinase production	Shanmugaiah et al (2008)
	Pseudomonas aeruginosa	Proteobacteria	Phosphate-solubilizing	Panhwar et al (2012)
	Desulfotomaculum (F), *Desulfobulbus* (P)	Firmicutes (F), Proteobacteria (P)	Sulfate-reducing bacteria	Lin et al (2010)
	Chryseobacterium sp., *Pseudomonas* sp.,	Bacteroidetes (B), Proteobacteria (P)	Confers high protection against biotic and abiotic stress inducing systemic resistance	Lucas et al (2014)
Endophytes and epiphytes	*Ancylobacter*, *Azorhizobium*, *Azospirillum*, *Rhizobium*, *Bradyrhizobium*, *Sinorhizobium*, *Novosphingobium*, *Acinetobacter*, *Aeromonas*, *Azotobacter*,	Proteobacteria	Nitrogen fixing polyvalent PGP	Banik et al (2016a)

(Contd.)

	Enterobacter, *Klebsiella*, *Pantoea*, *Pseudomonas*, *Stenotrophomonas*			
	Stenotrophomonas maltophilia, *Ochrobactrum* sp., *Pseudomonas oryzihabitans*, *Rhizobium radiobacter*	Proteobacteria	Seed born endophyte, PGP	Hardoim et al (2012)
	Streptomyces sporocinereus	Actinobacteria	Pathogen inhibition	Zeng et al (2018)
	Enterobacter cloacae, *Klebsiella variicola*	Proteobacteria	IAA overproduction improves nitrogen-fixing apparatus of endophytic bacteria	Defez et al (2017)
	Enterobacter sp. and *Kosakonia* sp.	Proteobacteria	ACC deaminase production, increase saline resistance, regulation of plant ethylene synthesis	Liu et al (2017)
	Pseudomonas aeruginosa (P), *Sphingobacterium siyangensis* (P), *Stenotrophomonas pavanii* (P), *Bacillus megaterium* (A), *Curtobacterium plantarum*(A)	Proteobacteria (P), Firmicutes (F), Actinobacteria (A)	Enhanced degradation of chlorpyrifos	Feng et al (2017)
	Streptomyces endus	Actinobacteria	Biocontrol of rice blast disease	Xu et al (2017)
	Xanthomonas oryzae	Proteobacteria	Biocontrol of bacterial blight	Yousefi et al (2018)
	Bacillus amyloliquefaciens	Firmicutes	Seed born endophyte, produces gibberellins and regulates endogenous phytohormones	Shahzad et al (2016)
Phyllosphere and Phylloplane	*Methylobacterium* sp.	Proteobacteria	Growth promotion and induction of systemic resistance	Madhaiyan et al (2004)
	Methylobacterium phyllosphaerae	Proteobacteria	Methylotroph and pigment production	Madhaiyan et al (2009)
	Pseudomonas sp.	Proteobacteria	Biocontrol of *Rhizoctonia solani*	Akter et al (2016)
	Pantoea, *Burkholderia*, *Klebsiella*, *Enterobacter*	Proteobacteria	Nitrogen fixing polyvalent PGP	Banik et al (2016a)
	Enterobacter sp.	Proteobacteria	PGP	Nutaratat et al (2017)

Fusarium oxysporum (causative agent of root rot disease), *Pyricularia oryzae* (causative agent of blast disease), *Rhizoctonia solani* (causative agent of sheath blight disease), *Rhizoctonia oryzae-sativae* (causative agent of aggregate sheath blight disease) (Spence *et al.*, 2014; Akter *et al.*, 2016; Tamreihao *et al.*, 2016; Zeng *et al.*, 2018) **(Table 2.1)**.

2.4. Microbial distribution at different rice niches

2.4.1. Rhizosphere and rhizoplane

Rhizosphere is the region of soil, situated at the very close vicinity of plant roots and influenced by nutrient exchange between surrounding and plant roots. During the growth of plants, rice roots secrete different amino acid (mainly alanine, glycine, histidine, proline, valine etc.), sugars (arabinose, mannose, glucose, glucuronic acid, galactose etc.) and different secondary metabolites (organic acid, vitamins, enzymes, nucleoside, ions, volatiles etc.) known as root exudates. These exudates activate microbial chemotaxis towards root vicinity and aid to select specific microbial groups for early colonization in the rhizosphere (Bacilio-Jiménez *et al.*, 2003) (**Fig. 2.3**). Globally rice is cultivated in flooded soil, resulting oxic and anoxic zones at root surrounding areas. Depending upon the availability of oxygen, the colonization patterns of microflora varies from aerobic, facultative anaerobic to strictly anaerobic metabolism. The anaerobic microbiome of rice has been identified as the major causative agents to emit around 15–20% of global anthropogenic CH_4 from the paddy fields. Rice methanogens such as *Methanobacterium*, *Methanosaeta*, *Methanobrevibacter*, *Methanoregula*, *Methanocella*, *Methanosarcina*, *Methanosphaerula* etc. use H_2 + CO_2 or acetate as a substrate produced from fermentation of soil organic matters (Lee *et al.*, 2015). The members of aerobic, facultative aerobic and micro aerophilic such as *Azotobacter*, *Azospirillum*, *Agrobacterium*, *Pseudomonas*, *Burkholderia*, *Herbaspirillum*, *Azoarcus*, *Rhizobium*, *Flavobacterium*, *Bacillus* etc. are mostly popular for their PGP attributes. The root surface area is known as the rhizoplane which is one of the important niche of biofilm from microbes to communicate with plant and other microbes present in the rhizosphere. Studies conducted on rice rhizoplane suggest that the members of *Proteobacteria* predominates with lesser abundance of *Actinobacteria* and *Acidobacteria* (Madhaiyan *et al.*, 2010; Bal *et al.*, 2013; Nayak *et al.*, 2020 Lucas *et al.*, 2014; Ueda *et al.*, 2016) **(Table 2.1)**.

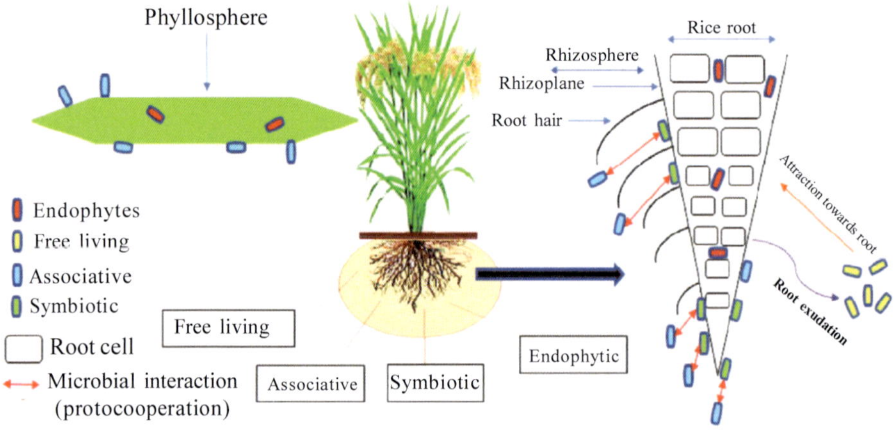

Fig. 2.3: Distribution and interaction of rice associated microbes at their specific niche

2.4.2. Endophytes

The application of plant growth promoting endophytic bacteria for overall growth of rice cultivation is getting popular day by day as endophytes successfully colonize different plant parts such as root, stem, leaf and seed. Different rhizospheric members of genus like *Azotobacter*, *Gluconobacter*, *Pseudomonas*, *Burkholderia*, *Herbaspirillum*, *Azoarcus*, *Azospirillum*, *Enterobacter*, *Erwinia*, *Klebsiella* etc. were also reported for successful endophytic colonization in different parts of rice. The purpose of endophytic adaptation of these microbes have been hypothesized by several scientific groups. The anticipated endophytic occupancy inside the plants may be due to comparatively lesser than rhizosphere, requirement of specific nutrition for proliferation at a particular niche, phytonic position acts as a shield from different environmental stress (UV, salt, desiccation, flood etc.) etc. The entry points of most of the endophytes are either cracks of lateral roots for root and stem endophytes (after entry from roots they travel to stem and leaf) or stomata for leaf endophytes (Banik *et al.*, 2016b). Different researchers have isolated different strains of *Azospirillum*, *Azotobacter*, *Herbaspirillum*, *Serratia*, *Klebsiella*, *Azoarcus*, *Gluconobacter*, *Novosphingobium*, *Azorhizobium*, *Enterobacter* from rice root, *Methylobacterium*, *Flavobacterium*, *Herbaspirillum*, *Rhodopseudomonas*, *Cytophagales* from rice stem, *Bacillus*, *Curtobacterium*, *Aurantimonas*, *Pantoea*, *Burkholderia*, *Klebsiella* from leaf and seed of rice plants (Hardoim *et al.*, 2012; Banik *et al.*, 2016a; Shahzad *et al.*, 2016) **(Table 2.1)**.

2.4.3. Phyllosphere and Phylloplane

The of the surface of the leaf as associated areas known as the phyllosphere which harbors mainly non-pathogenic bacteria, belonging from *Gammaproteobacteria*, *Alphaproteobacteria*, *Actinobacteria*, *Enterobacteriaceae* and *Bacteroidetes*. The microbial community at phyllosphere and phylloplane varies greatly as because it comes under the direct exposure of several biotic and a biotic factor. The leaf surface area of the phyllosphere is nearly covers the double than the surface of lands, where microbes mostly colonized as aggregates and enters through the stomata (Vorholt, 2012). Investigation on phyllosphere microorganism of Indian rice cultivars suggested that the dominance of different species *Pantoea*, *Exiguobacterium*, *Stenotrophomonas*, *Burkholderia*, *Klebsiella*, *Enterobacter* and *Bacillus* through culture dependent approach (Banik *et al.*, 2016a; Venkatachalam *et al.*, 2016; Thapa *et al.*, 2017) **(Table 2.1)**.

2.5. Futuristic approaches for application of PGP

Plants may look like solitary individual, but the ground beneath of it demonstrates a different story. They communicate with each other by using a group of microbes which proliferate as endophytically or association with the plants. These microbial network helps the plants to share their resource with each other, which is known as Wood Wide Web (WWW). This network performers as older plant provides nutrients to younger plants, aiding them for grater probabilities of survival. Here the role of microbes become important as plants uses them as a vehicle to send the signaling molecule to the neighboring plants for pathogen attack so that neighbors can raise their defense against that particular pathogen (Gilbert and Johnson, 2017).

Due to upcoming challenges like global warming, drought, salinity, contamination of metals due to industrialization etc. the role of PGP microbes would be immense in terms of crop production (Banik *et al.*, 2018). However, the most challenging part to use the PGP microbes is the optimization and reduction of variation between laboratory, green house and filed trails. Considering the presence autochromes microorganism of that specific niche, polyvalent consortia of microbes should be the chosen for application as PGP traits do not work individualistically in all environment.

Futuristic research such as tracking the plant life-cycle from rice microbiome opening new horizons in rice microbe interactions. Edwards *et al* (2018) demonstrates that how roots associated rice microbiome influences health and nutrition of the plants. They demonstrated the extremely dynamic nature of roots microbiome during plant's vegetative phase. However, it settles down after the flowering phase. The microbial community among members of archaea and bacteria were conserve independent of site variation predicts features of rice plant age by modeling using a random forest approach.

References

Akter S, Kadir J, Juraimi AS (2016) In vitro evaluation of *Pseudomonas* bacterial isolates from rice phylloplane for biocontrol of *Rhizoctonia solani* and plant growth promoting traits. J Environ Biol 37:597.

Bacilio-Jiménez M, Aguilar-Flores S, Ventura-Zapata E *et al* (2003) Chemical characterization of root exudates from rice (*Oryza* sativa) and their effects on the chemotactic response of endophytic bacteria. Plant Soil 249: 271-277.

Bal HB, Nayak L, Das S *et al* (2013) Isolation of ACC deaminase producing PGPR from rice rhizosphere and evaluating their plant growth promoting activity under salt stress. Plant soil 366: 93-105.

Banik A, Mukhopadhaya SK, Dangar TK (2016a) Characterization of N2-fixing plant growth promoting endophytic and epiphytic bacterial community of Indian cultivated and wild rice (*Oryza* spp.) genotypes. Planta243: 799-812

Banik A, Mukhopadhaya SK, Sahana A *et al* (2016b) Fluorescence resonance energy transfer (FRET)-based technique for tracking of endophytic bacteria in rice roots. BiolFertil Soils52: 277-282

Banik A, Pandya P, Patel B *et al* (2018) Characterization of halotolerant, pigmented, plant growth promoting bacteria of groundnut rhizosphere and its in-vitro evaluation of plant-microbe protocooperation to withstand salinity and metal stress. Sci Total Environ 630:231-42

Banik A., Kumar U, Mukhopadhyay SK *et al* (2017) Dynamics of endophytic and epiphytic bacterial communities of Indian cultivated and wild rice (*Oryza* spp.) genotypes. Ecol Gen Genom3:7-17. Doi: 10.1016/j.egg.2017.06.001

Beneduzi A, Peres D, Vargas LK *et al* (2008) Evaluation of genetic diversity and plant growth promoting activities of nitrogen-fixing bacilli isolated from rice fields in South Brazil. Appl Soil Ecol39: 311-320

Chaiharn M, Lumyong S (2009) Phosphate solubilization potential and stress tolerance of rhizobacteria from rice soil in Northern Thailand. World J Microbiol biotechno, 25:305-314. doi:10.1007/s11274-008-9892-2

Chung EJ, Hossain, MT, Khan A, *et al* (2015) *Bacillus oryzicola* sp. nov., an endophytic bacterium isolated from the roots of rice with antimicrobial, plant growth promoting, and systemic resistance inducing activities in rice. Plant Pathol J 31(2): 152. doi: 10.5423/PPJ.OA.12.2014.0136

Davidson D, Gu FX (2012) Materials for sustained and controlled release of nutrients and molecules to support plant growth. J Agric Food Chem 60: 870-876

Defez R, Andreozzi A, Bianco C (2017) The overproduction of indole-3-acetic acid (IAA) in endophytes upregulates nitrogen fixation in both bacterial cultures and inoculated rice plants. MicrobEcol 74: 441-452

Edwards JA, Santos-Medellín CM, Liechty ZS *et al* (2018) Compositional shifts in root-associated bacterial and archaeal microbiota track the plant life cycle in field-grown rice. PLoSbiolo, 16:2003862

FAOSTAT, Food and Agricultural Organization of the United Nations, Statistics Division. 2013

Feng F, Ge J, Li Y *et al* (2017) Enhanced degradation of chlorpyrifos in rice (*Oryza sativa* L.) by five strains of endophytic bacteria and their plant growth promotional ability. Chemosphere 184: 505-513. doi: 10.1016/j.chemosphere.2017.05.178.

Gilbert L, Johnson D (2017) Plant–Plant Communication Through Common Mycorrhizal Networks. In Advances in Botanical Research. Academic Press 82: 83-97

Hardoim PR, Hardoim CC, Van Overbeek LS *et al* (2012) Dynamics of seed-borne rice endophytes on early plant growth stages. PLoS One 7:30438

Kumar U, Panneerselvam P, Govindasamy V *et al* (2017) Long-term aromatic rice cultivation effect on frequency and diversity of diazotrophs in its rhizosphere. EcolEng 101: 227-236.

Lee HJ, Jeong SE, Kim PJ *et al* (2015) High resolution depth distribution of Bacteria, Archaea, methanotrophs, and methanogens in the bulk and rhizosphere soils of a flooded rice paddy. Front Microbiol6:639

Li H, Yang X, Zhang Z *et al* (2015) Nitrogen loss by anaerobic oxidation of ammonium in rice rhizosphere. ISME J9: 2059–2067

Liermann LJ, Kalinowski BE, Brantley SL *et al* (2000) Role of bacterial siderophores in dissolution of hornblende. Geochimica et Cosmochimica Acta, 64: 587-602

Lin H, Shi J, Chen X *et al* (2010) Effects of lead upon the actions of sulfate-reducing bacteria in the rice rhizosphere. Soil Biol Biochem 42:1038-1044.

Liu Y, Cao L, Tan H *et al* (2017) Surface display of ACC deaminase on endophytic *Enterobacteriaceae* strains to increase saline resistance of host rice sprouts by regulating plant ethylene synthesis. Microb Cell Fact 16: 214

Lucas JA, García-Cristobal J, Bonilla A *et al* (2014) Beneficial rhizobacteria from rice rhizosphere confers high protection against biotic and abiotic stress inducing systemic resistance in rice seedlings. Plant Physiol Biochem 82: 44-53

MacKay BA, Fryzuk MD (2004) Dinitrogen coordination chemistry: on the biomimetic borderlands. Chem rev, 104(2): 385-402. Doi:10.1021/cr020610c

Madhaiyan M, Poonguzhali S, Kwon SW *et al* (2009) *Methylobacterium phyllosphaerae* sp. nov., a pink-pigmented, facultative methylotroph from the phyllosphere of rice. Int J SystEvolMicrobiol 59: 22-27.

Madhaiyan M, Poonguzhali S, Kwon SW *et al* (2010) *Bacillus methylotrophicus* sp. nov., a methanol-utilizing, plant-growth-promoting bacterium isolated from rice rhizosphere soil. Int J Syst Evol Microbiol 60: 2490-2495

Madhaiyan M, Poonguzhali S, Senthilkumar M (2004) Growth promotion and induction of systemic resistance in rice cultivar Co-47 (*Oryza sativa* L.) by *Methylobacterium* spp. Botanical Bulletin of Academia Sinica, 45.

Mulas D, Díaz-Alcántara CA, Mulas R *et al* (2016) Inoculants based in autochthonous microorganisms, a strategy to optimize agronomic performance of biofertilizers. In MBR González and J Gonzalez-Lopez eds: Beneficial Plant-microbial Interactions: Ecology and Applications, CRC Press, pg:300-328.

Nautiyal CS, Srivastava S, Chauhan PS *et al* (2013) Plant growth-promoting bacteria *Bacillus amyloliquefaciens* NBRISN13 modulates gene expression profile of leaf and rhizosphere community in rice during salt stress. Plant PhysiolBiochem 66: 1-9.

Nayak SK, Nayak S, Patra JK (2020) Rhizobacteria and its biofilm for sustainable agriculture: A concise review. In MK Yadav and BP Singh eds. New and Future Developments in Microbial Biotechnology and Bioengineering, Elsevier, Netherlands, pp- 165-175. Doi:10.1016/B978-0-444-64279-0.00013-X

Nutaratat P, Apitchaya M, Nantana S (2017) High-yield production of indole-3-acetic acid by *Enterobacter* sp. DMKU-RP206, a rice phyllosphere bacterium that possesses plant growth-promoting traits. 3 Biotech 7: 305. doi: 10.1007/s13205-017-0937-9.

Ohyama T, Kumazawa K (1980) Nitrogen assimilation in soybean nodules: I. The role of GS/GOGAT system in the assimilation of ammonia produced by N2-fixation. Soil Science and Plant Nutrition, 26(1), pp.109-115.

Pahari A., Pradhan A., Nayak S.K., B.B. Mishra (2017) Bacterial Siderophore as a Plant Growth Promoter. In J.K. Patra *et al.* (eds.), Microbial Biotechnology, Springer, Singapore, pp-163-180. Doi:10.1007/978-981-10-6847-8_7

Panhwar QA, Othman R, Rahman ZA *et al* (2012) Isolation and characterization of phosphate-solubilizing bacteria from aerobic rice. Afr J Biotechnol 11: 2711-2719.

Ribbe MW, Hu Y, Hodgson KO *et al* (2013) Biosynthesis of nitrogenase metalloclusters. Chem rev 114: 4063-4080.

Seck PA, Diagne A, Mohanty S *et al* (2012) Crops that feed the world 7: Rice. Food security4: 7-24.

Shahzad R, Waqas M, Khan AL *et al* (2016) Seed-borne endophytic *Bacillus amyloliquefaciens* RWL-1 produces gibberellins and regulates endogenous phytohormones of *Oryza sativa*. Plant Physiol Biochem 106: 236-243

Shanmugaiah V, Mathivanan N, Balasubramanian N *et al* (2008) Optimization of cultural conditions for production of chitinase by *Bacillus laterosporous* MML2270 isolated from rice rhizosphere soil. Afr J Biotechnol 7: 15

Smith BE (2002) Nitrogen reveals its inner secrets. Science 297:1654–1655.

Soora NK, Aggarwal PK, Saxena R *et al* (2013) An assessment of regional vulnerability of rice to climate change in India. Climatic change, 118: 683-699.

Spence C, Alff E, Johnson C *et al* (2014) Natural rice rhizospheric microbes suppress rice blast infections. BMC plant biol 14:130.

Tamreihao K, Ningthoujam DS, Nimaichand S *et al* (2016) Biocontrol and plant growth promoting activities of a *Streptomyces corchorusii* strain UCR3-16 and preparation of powder formulation for application as biofertilizer agents for rice plant. Microbiol res192: 260-270.

Temme K, Zhao D, Voigt CA (2012) Refactoring the nitrogen fixation gene cluster from *Klebsiella oxytoca*. Proc Natl Acad Sci U S A 109: 7085-7090

Thapa S, Prasanna R, Ranjan K *et al* (2017) Nutrients and host attributes modulate the abundance and functional traits of phyllosphere microbiome in rice. Microbiol res 204: 55-64

Ueda Y, Frindte K, Knief C *et al* (2016) Effects of elevated tropospheric ozone concentration on the bacterial community in the phyllosphere and rhizoplane of rice. PloS one: 0163178. Doi:10.1371/journal.pone.0163178

Venkatachalam S, Ranjan K, Prasanna R *et al* (2016) Diversity and functional traits of culturable microbiome members, including cyanobacteria in the rice phyllosphere. Plant Biol18: 627-637.

Vorholt JA (2012) Microbial life in the phyllosphere. Nat Rev Microbiol 9: 2059–2067

Xu T, Li Y, Zeng X (2017) Isolation and evaluation of endophytic *Streptomyces endus* OsiSh 2 with potential application for biocontrol of rice blast disease. J Sci Food Agric 97: 1149-1157. doi: 10.1002/jsfa.7841.

Yousefi H, Hassanzadeh N, Behboudi K *et al* (2018) Identification and determination of characteristics of endophytes from rice plants and their role in biocontrol of bacterial blight caused by *Xanthomonas oryzae* pv. oryzae. Hellenic Plant Protect J, 11: 19-33

Zeng J, Xu T, Cao L *et al* (2018) The Role of Iron Competition in the Antagonistic Action of the Rice Endophyte *Streptomyces sporocinereus* OsiSh-2 Against the Pathogen *Magnaporthe oryzae*. MicrobEcol 1-9. doi: 10.1007/s00248-018-1189-x

3

Role of Biofertilizer in Sustainable Agriculture: Enhancing Soil Fertility Plant Tolerance and Crop Productivity

Priyanka Verma[1]*, Dheer Singh[2], Ishwar Prasad Pathania[3] and Komal Aggarwal[4]

[1]*Department of Microbiology, Eternal University, Sirmaur-173101, India*
[2]*Department of Microbiology, Bundelkhand University, Jhansi-284128, India*
[3]*Department of Biotechnology, Sharda University, Greater Noida-201306, India*
[4]*Department of Biotechnology, Gautam Buddha University Greater Noida-201310, India*

Abstract

Agriculture is one of the most important factors contributing to the economic growth of India. Out of the 329 million hectares of India's geographical area, about 114 million hectares are under cultivation. There are 17 essential elements required for proper plant growth. Of the mineral elements, the primary macronutrients (nitrogen, phosphorous, and potassium) are needed in the greatest quantities and are most likely to be in short supply in agricultural soils. Secondary macronutrients are needed in smaller quantities, and are typically found in sufficient quantities in agricultural soil, and therefore do not often limit crop growth. Micronutrients, or trace nutrients, are needed in very small amounts and can be toxic to plants in excess. Silicon (Si) and sodium (Na) are sometimes considered essential plant nutrients, due to their ubiquitous presence in soils.

The economy of India thrives on agriculture, the most practiced occupation in the country. Agricultural fertilizers are essential to enhance proper growth of plants and crop yield. Recently, farmers have been using chemical fertilizers for quicker and better yield. But these fertilizers endanger ecosystems, soil, plants, human and animal lives. In contrast, naturally grown biofertilizers not only give a better yield, but are also harmless to humans. This chapter aims to study the biofertilizers play a key role for enhancing economic development of sustainable agriculture (EDSA) in comparison to chemical fertilizers. Biofertilizers improve the crop yields, plant resistance, soil health, profit to farmers, and the benefit to cost ratio of biofertilizers were higher than chemical fertilizer, proved that the biofertilizers would lead to better sustainable economic development for the farmers and their country.

Keywords: Agriculture, Biofertilizers, Economic development, Plant Tolerance, Soil fertility

3.1 Introduction

The growing population and consumption, and reduction in available land and other productive units are placing unprecedented pressure on the current agriculture and natural resources to meet the increasing food demand. Providing food for human under sustainable systems having a significant challenge in the developing world and is highly critical for alleviating poverty. To circumvent this challenge, farmers tended to overuse certain inputs such as chemical, agricultural inputs, which in turn have already started deteriorating soil-plant-microbes environmental system. To meet the world's future food security and sustainability needs, food production must grow substantially while agriculture's environmental impact must decrease dramatically at the same time. Global food production must increase by 70% till 2050 to fulfill the increasing demand. In order to meet this challenging target, an average annual increase in cereal production of 43 million metric tons per year is required. A biofertilizers is an organic product containing a specific microorganism in concentrated form, which is derived either from the plant roots or from the soil of root zone (Chen, 2006; Gupta and Sen, 2013). These bioinputs, or bioinoculants, improve plant growth and yield. Biofertilizers are the products comprising living cells of different types of microorganisms that have the ability to mobilize nutritionally important elements from non-usable forms through biological stress. Biofertilizers can be defined as substances that contain living microorganisms that colonize in the rhizosphere, or the interior of the plant. These promote growth by increasing the availability of primary nutrients and/or growth stimulus to the target crop when applied to seed, plant surfaces, or soil (Muraleedharan *et al.*, 2010). Biofertilizers have shown great potential as supplementary, renewable and environmental friendly sources of plant nutrients and are an important component of Integrated Nutrient Management (INM) and Integrated Plant Nutrition System (IPNS) (Raghuwanshi, 2012). The production of biofertilizers in the country is 10,000 mt/annum and the production capacity is 18,000 mt/annum. The average annual consumption of biofertilizers in the country is around 64g/ha.

Sustainable agriculture is vital in today's world as it offers the potential to meet our agricultural needs, something that conventional agriculture fails to do. This type of agriculture uses a special farming technique wherein the environmental resources can be fully utilized and at the same time ensuring that no harm was done to it. Thus, the technique is environment friendly and ensures safe and healthy agricultural products. Microbial populations are instrumental to fundamental processes that drive stability and productivity of agro-ecosystems. Several investigations addressed at improving understanding of the diversity, dynamics and importance of soil microbial communities and their beneficial and cooperative roles in agricultural productivity.

For many years, soil microbiologists and microbial ecologists differentiated soil microorganisms as 'beneficial' or 'harmful' depending how they affect soil quality, crop growth and yield. Beneficial microorganisms are those that fix atmospheric N_2, decompose organic wastes and residues, detoxify pesticides, suppress plant diseases and soil-borne pathogens, enhance nutrient cycling and produce bioactive compounds such as vitamins, hormones and enzymes that stimulate plant growth which is the recent interest in eco-friendly and sustainable agricultural practices (Harish *et al.*, 2009a; Kavino *et al.*, 2010). Biofertilizers and biopesticide containing efficient microorganisms, improve plant growth in many ways compared to synthetic fertilizers, insecticides and pesticides by way of enhancing crop growth and thus help in sustainability of environment and crop productivity. The rhizospheric soils contain diverse type of efficient microbes with beneficial effects on crop productivity. The plant growth promoting rhizobacteria (PGPR) and cyanobacteria are often rhizospheric microbes and produce bioactive substances to promote plant growth and/or protect them against pathogens (Glick, 1995; Harish, *et al.*, 2009b).

3.2 Importance of sustainable agriculture

Sustainable agriculture is a broad based concept rather than a specific methodology. It encompasses advances in agricultural management practices and technology, and the growing recognition indicates that the conventional agriculture that developed post World War-II, will not be able to meet the needs of the growing population in the 21st Century. Conventional agriculture is facing either reduced production or increased costs, or both. Farming monocultures, such as wheat fields, repeated on the same land results in the depletion of topsoil, soil vitality, groundwater purity and beneficial microbe, insect life, making the crop plants vulnerable to parasites and pathogens.

An ever-increasing amount of fertilizers and pesticides as well as the energy requirements for tilling to aerate soils and increasing irrigation costs are of prime concern. While conventional methods enabled large increases in crop yields, thus high profits only initially, have failed to be considered as the ideal approach for future.

3.3 Efficient soil microbes for sustainable agriculture and environment

Low agricultural production efficiency is closely related to a poor coordination of energy conversion which, in turn, is influenced by crop physiological factors, the environment, and other biological factors including soil microbes. The soil and rhizosphere microflora can accelerate the growth of plants and enhance their resistance to pathogens and harmful insects by producing bioactive metabolites. Such microorganisms maintain growth of plants, and thus have primary effects on both soil and crop quality.

A wide range of benefits are possible depending on their predominance and activity at any one time. However, there is a growing consensus that it is feasible to obtain maximum economically agronomic yield of high quality at higher net returns, without the use of artificial fertilizers, herbicides, insecticides and pesticides. Until recently, this was not thought to be the very likely possibility using conventional agricultural practices. However, it is important to recognize that the best soil and agricultural management practices to attain a more sustainable and green agriculture will also enhance the growth, number and activities of efficient soil microflora that, in turn, can enhance the growth, yield and agriculture quality. In particular, healthy living soil with improved quality is the very foundation of a future sustainable agriculture **(Fig. 3.1)**.

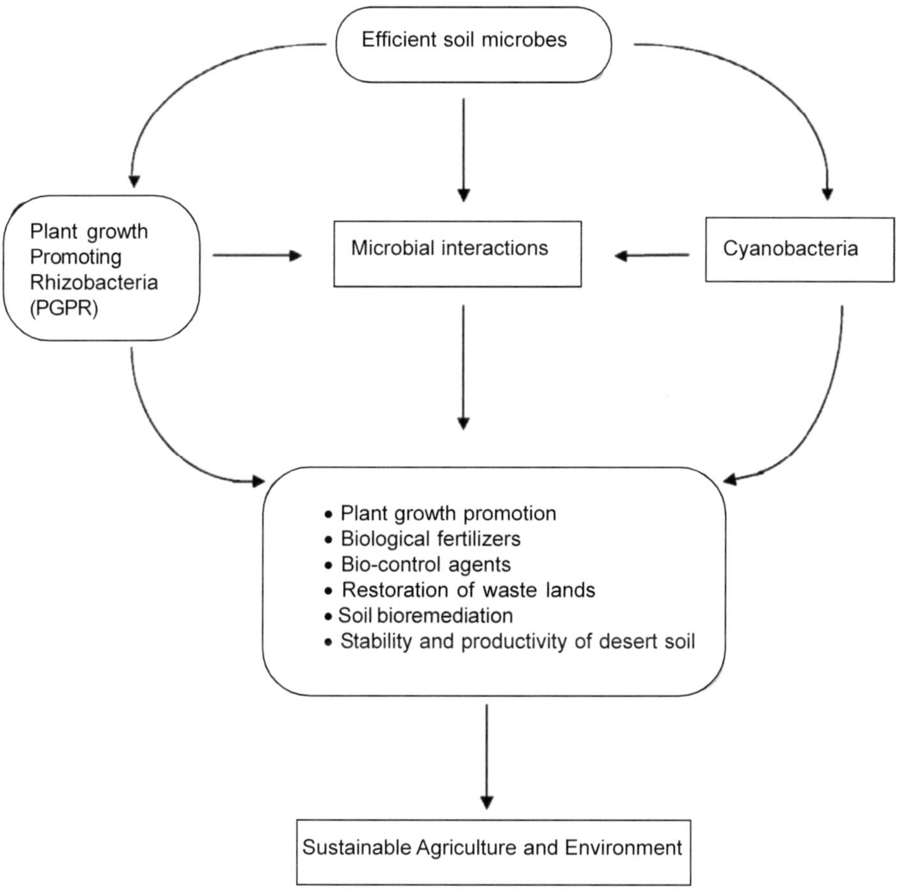

Fig. 3.1: Microbes play a key role in sustainable agriculture and environment (Adopted from Singh et al., 2011)

3.4 Impact of biofertilizers on crop productivity

Biofertilizers is a substance which contains living microorganism which, when applied to seed, plant surfaces, or soil, colonizes the rhizosphere or the interior of the plant and promotes growth by increasing the supply or availability of primary nutrients to the host plant **(Table 3.1).** Very often microorganisms are not as efficient in natural surroundings as one would expect them to be and therefore artificially multiplied cultures of efficient selected microorganisms play a vital role in accelerating the microbial processes in soil. Use of biofertilizers is one of the important components of INM, as they are cost effective and renewable source of plant nutrients to supplement the chemical fertilizers for sustainable agriculture. Several microorganisms and their association with crop plants are being exploited in the production of biofertilizers **(Table 3.2).** A number of microorganisms are considered as beneficial for agriculture and used as biofertilizers viz. *Rhizobium*, *Azotobacter*, *Azospirillum*, *Cyanobacteria*, *Azolla*, are phosphate and potassium solubilizing microorganisms. In addition, silicate solubilizing bacteria and plant growth promoting rhizobacteria are also presently available as liquid biofertilizers (Parewa *et al*., 2014; Suman *et al*., 2016a; Verma *et al*., 2014; Zhang and Kong, 2014).

Table 3.1: Essential plant nutrient elements and their primary form utilized by plants (Parikh and James, 2012)

S.N.	Essential plant element		Symbol	Primary form
1.	Non-Mineral Elements			
		Carbon	C	CO_2 (g)
		Hydrogen	H	H_2O (l), H^+
		Oxygen	O	H_2O (l), O_2(g)
2.	Mineral Elements			
i	Primary Macronutrients	Nitrogen	N	NH_4^+, NO_3^-
		Phosphorus	P	HPO_4^{2-}, H_2PO_4-
		Potassium	K	K^+
		Calcium	Ca	Ca_2^+
ii	Secondary Macronutrients	Magnesium	Mg	Mg_2^+
		Sulfur	S	SO_4^{2-}
		Iron	Fe	Fe_3^+, Fe_2^+
		Manganese	Mn	Mn^{2+}
		Zinc	Zn	Zn^{2+}
iii	Micronutrients	Copper	Cu	Cu^{2+}
		Boron	B	$B(OH)_3$
		Molybdenum	Mo	MoO_4^{2-}
		Chlorine	Cl	Cl^-
		Nickel	Ni	Ni^{2+}

3.5 Biofertilizers

Indiscriminate use of synthetic fertilizers has led to the pollution and contamination of the soil, has polluted water basins, destroyed microorganisms and friendly insects, making the crop more prone to diseases along with decline in soil fertility. Demand is much higher than the availability. It is estimated that by 2020, to achieve the targeted production of 321 million tons of food grain, the requirement of nutrient will be 28.8 million tons, while their availability will be only 21.6 million tones being a deficit of about 7.2 million tones. Depleting feedstocks/ fossilfuels (energy crisis) and increasing cost of fertilizers which become unaffordable for small and marginal farmers and is a major concern now. Depleting soil fertility due to significant gap between nutrient removal and supplies along with the environmental hazards of the fertilizers are threats to sustainable agriculture and productivity.

Table 3.2. Various crop diseases. regulate by plant growth promoting microorganisms (PGPM) as biocontrol agents

Sl.No.	Disease	PGPM	References
1.	*Cucumber Mosaic Cucumovirus* (CMV) of tomato (*Lycopersicon esculentum*)	*Bacillus pumilus* strain SE34, *Kluyvera cryocrescens* strain IN114, *B. amyloliquefaciens* strain IN937a, and *B. subtilis* strain IN937b.	(Zehnder et al., 2001)
2.	Tomato mottle virus	*B. amyloliquefaciens* 937a, *B. subtilis* 937b, and *B. pumilus* SE34	(Murphy et al., 2000)
3.	Bacterial wilt disease incucumber (*Cucumis sativus*)	*B. pumilus* strain INR7 F	(Zehnder et al., 2001)
4.	Sheath blight disease and leaffolder insect in rice (*Oryza sativa*)	*Pseudomonas fluorescens* based bioformulation	(Commare et al., 2002)
5.	Blue mold disease of tobacco(*Nicotiana*)	*B. pumilus* strain SE34	(Zhang et al. and Zhang et al. 2002)
6.	Downy mildew in pearl millet(*Pennisetum glaucum*)	*B. subtilis* strain GBO3, *B. pumilus* strain INR7 and *B.pumilus* strain T	(Raj et al., 2003)
7.	CMV in cucumber	*B. subtilis* strain IN937a	(Jetiyanon et al., 2003)
8.	Foliar diseases of tomato	*B. cereus* strains B101R, B212R, and A068R	(Silva et al., 2004)
9.	Blight of bell pepper (*Capsicum annuum*)	*Bacillus* strains BB11 and FH17	(Jiang et al., 2006)
10.	Saline resistance in groundnut (*Arachis hypogea*)	*Pseudomonas fluorescens*	(Saravanakumar and Samiyappan, 2007)
11.	Maize (*Zea mays*) rot	*Burkholderia* strains MBf21 and MBf15	(Hernández-Rodríguez et al., 2008)
12.	Soil borne pathogen of cucumber and pepper (*Piper*)	*B. subtilis* ME488	(Chung et al., 2008)
13.	Significantly reduce the Banana Bunchy Top Virus (BBTV) incidence	*P. fluorescens* strain CHA0 + chitin bioformulations	(Kavino et al., 2008)
14.	Rice blast Naureen	*Bacillus* sp., and *Azospirillum* strains SPS2, WBPS1 and Z2–7	(Zakira, 2009)
15.	Rice Sheath rot (*Sarcocladium oryzae*)	Fluorescent *Pseudomonas* sp.	(Saravanakumar et al., 2009)
16.	Blight of squash	*Bacillus* strain	(Zhang et al., 2010)

Besides above facts, the longterm use of biofertilizers is economical, eco-friendly, more efficient, productive and accessible to marginal and small farmers over chemical fertilizers (Venkataraman and Shanmugasundaram, 1992).

3.5.1 Plant growth promoting rhizobacteria as biological fertilizers

In general, huge bulk artificial fertilizes is applied to replenish soil N and P which resultant in high cost and environmental risk. Most of P compounds are insoluble and unavailable to plants. N_2-fixing and P-solubilizing bacteria (PSB) are important for crop plants as they increase N and P uptake and play a crucial role as PGPR in the biofertilization (Zahir *et al.*, 2004; Zaidi and Khan, 2006). Thus, the application of such microbes as environment friendly biofertilizers may contribute to minimize the use of expensive phosphatic fertilizers. Phosphorus biofertilizers increase the availability of accumulated P (by solubilization), efficiency of biological N_2-fixation and the availability of K, Zn, etc., due to production of plant growth promoting substances (Kucey *et al.*, 1989). Inoculation of N_2 fixing and PSB in synergy was more effective than the monoculture for providing a more balanced nutrition to crops such as sorghum, barley, black gram, soybean and wheat (Abd-Alla *et al.*, 2001; Alagawadi and Gaur, 1992; Galal, 2003; Tanwar *et al.*, 2003). It also facilitates the essential plant nutrient elements and their primary form utilized by plants (Parikh and James, 2012) **Fig 3.2**.

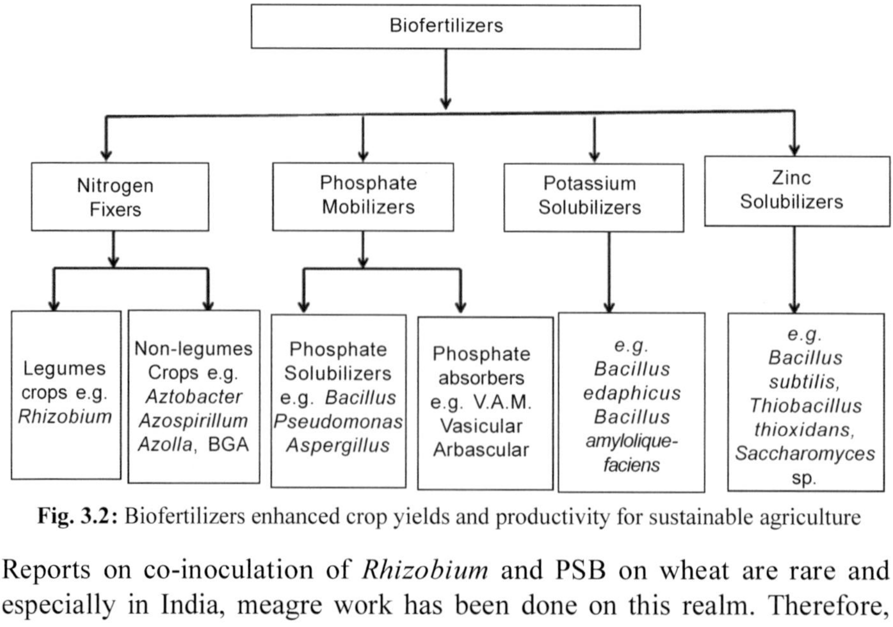

Fig. 3.2: Biofertilizers enhanced crop yields and productivity for sustainable agriculture

Reports on co-inoculation of *Rhizobium* and PSB on wheat are rare and especially in India, meagre work has been done on this realm. Therefore, extensive investigations to explore the effect of single and dual inoculations of

N_2-fixing and P-solubilizing bacterial species on yields of various crops are required urgently. More recent findings indicated that the treatment of arable soils with PGPR inoculations significantly increases agronomic yields (Harish *et al.*, 2009a). The PGPR strains *Pseudomonas alcaligenes* PsA15, *Bacillus polymyxa* BcP26 and *Mycobacterium phlei* MbP18 have the pronounced stimulatory effects on plant growth and uptake of N, P and K of maize in nutrient-deficient calcisol soils (Egamberdiyeva, 2007). The enhancement in various agronomic yields by PGPR was because of the production of growth promoting phytohormones, phosphate mobilization, production of siderophore and antibiotics, inhibition of plant ethylene synthesis and induction of plant systemic resistances to pathogens (Cakmakçi *et al.*, 2006; Han and Lee, 2006; Kohler *et al.*, 2006; Turan *et al.*, 2005; Zahir *et al.*, 2004; Zaidi and Khan, 2006). Recently, (Kavino *et al.*, 2010) reported that *P. fluorescens* CHA0 in combination with chitin increased growth, leaf nutrient contents and yield of banana plants under perennial cropping systems thus suggesting that in view of the environmental problems, due to excessive use and high production costs of fertilizers, PGPR may represent the potential soil microflora to be deployed for sustainable and eco-friendly agriculture (Suman *et al.*, 2016b).

In the broadest sense, plant growth promoting rhizobacteria include the N_2-fixing rhizobacteria that colonize the rhizosphere, providing N to plants in addition to the well characterized legume rhizobia symbionts. Regardless of the mechanism(s) of plant growth promotion, PGPR must colonize the rhizosphere around the roots, the rhizoplane (root surface) or the root itself (within root tissue). PGPR can affect plant growth either indirectly or directly; indirect promotion of plant growth is affected when PGPR lessen or antagonize the deleterious effects of one or more phytopathogens; while direct route by PGPR involves either providing plants with the compounds synthesized by the bacterium or facilitating the uptake of certain nutrients from the environment (Glick, 1995, Nayak *et al.*, 2017). PGPR regulation of various growth parameters/yields of crop/fruit plants have been listed in **Table 3.3.**

Among the diverse bacteria identified as PGPR, the Bacilli and Pseudomonads are the predominant ones (Podile and Krishna Kishore, 2007). PGPR exert a direct effect on plant growth by production of phytohormones, solubilization of inorganic phosphates, increased iron nutrition through iron-chelating siderophores and the volatile compounds that affect the plant signaling pathways. Additionally, by antibiosis, competition for space and nutrients and induction of systemic resistance in plants against a broad-spectrum of root and foliar pathogens, PGPR reduce the populations of root pathogens and other deleterious microorganisms in the rhizosphere, thus benefiting the plant growth (Verma *et al*, 2017a).

Table 3.3: Plant growth promoting microorganisms regulating various plant growth parameters to improve crop yield

Sl.No.	Promotes plant growth	Microorganisms	References
1.	Direct growth promotion of canola and lettuce	*Rhizobium leguminosarum*	(Noel et al., 1996)
2.	Early developments of canola seedlings	*Pseudomonas putida* G 12–2	(Glick et al., 1997)
3.	Growth of wheat and maize plants	*Azospirillum brasilense* and *A. irakense* strains	(Dobbelaere et al., 2002)
4.	Growth of pearl millet	*P. fluorescens* strain	(Raj et al., 2003)
5.	Growth stimulation of tomato plant	*P. putida* strain	(Gravel et al., 2007)
6.	Growth and productivity of canola	*Azotobacter* and *Azospirillum* strains	(Yasari and Patwardhan, 2007)
7.	Enhance uptake of N, P and K by maize crop in nutrient deficient calcisol soil	*P. alcaligenes* PsA15, *Bacillus polymyxa* BcP26 and *Mycobacterium phlei* MbP18,	(Egamberdiyeva, 2007)
8.	Stimulates growth and yield of chick pea (*Cicer arietinum)*	*Pseudomonas*, *Azotobacter* and *Azospirillum* strains	(Rokhzadi et al., 2008)
9.	Improve the yield and phosphorus uptake in wheat	*R. leguminismarum* (Thal-8/SK8) and *Pseudomonas* sp.strain 54RB	(Afzal and Bano, 2008)
10.	Improves seed germination, seedling growth and yield of maize	*P. putida* strains R-168 and DSM-291; *P. fluorescens* strains R-98 and DSM-50090; *A. brasilense* DSM-1691 and *A. lipoferum* DSM-1690	(Gholami et al., 2009)
11.	Seed germination, growth parameters of maize seedling in green house and also grain yield of field grown maize	*P. putida* strain R-168,	(Gholami et al., 2009)

3.5.2 Potential characteristic features of some biofertilizers

3.5.2.1 Nitrogen fixers

3.5.2.1.1 *Rhizobium*

Rhizobium belongs to family *Rhizobiaceae*, symbiotic in nature, fix nitrogen 50-100 kg/ha with legumes only. It is useful for pulse legumes like chickpea, red-gram, pea, lentil, black gram, etc., oil-seed legumes like soybean and groundnut and forage legumes like berseem and lucerne. Successful nodulation of leguminous crops by *Rhizobium* largely depends on the availability of compatible strain for a particular legume. It colonizes the roots of specific legumes to form tumor like growths called root nodules, which acts as factories of ammonia production. *Rhizobium* has ability to fix atmospheric nitrogen in symbiotic association with legumes and certain non-legumes like *Parasponia*. *Rhizobium* population in the soil depends on the presence of legume crops in the field. In absence of legumes, the population decreases. Artificial seed inoculation is often needed to restore the population of effective strains of the *Rhizobium* near the rhizosphere to hasten N-fixation. Each legume requires a specific species of *Rhizobium* to form effective nodules. Many legumes may be modulated by diverse strains of *Rhizobia*, but growth is enhanced only when nodules are produced by effective strains of *Rhizobia*. It is thus extremely important to match microsymbionts prudently for maximum nitrogen fixation. A strain of *Rhizobia* that nodulates and fixes a large amount of nitrogen in association with one legume species may also do the same in association with certain other legume species. This must be verified by testing. Leguminous plants that demonstrate this tendency to respond similarly to particular strains of *Rhizobia* are considered "effectiveness" group (Wani and Lee, 2002).

3.5.2.1.2 *Azospirillum*

Azospirillum belongs to family *Spirilaceae*, heterotrophic and associative in nature. In addition to their nitrogen fixing ability of about 20-40 kg/ha, they also produce growth regulating substances. Although there are many species under this genus like, *A. amazonense*, *A. halopraeferens, A. brasilense*, but, worldwide distribution and benefits of inoculation have been proved mainly with the *A. lipoferum* and *A. brasilense.* The *Azospirillum* form associative symbiosis with many plants particularly with those having the C4-dicarboxyliac pathway of photosynthesis (Hatch and Slack pathway), because they grow and fix nitrogen on salts of organic acids such as malic, aspartic acid (Arun, 2007). Thus, it is mainly recommended for maize, sugarcane, sorghum, pearl millet etc.

3.5.2.1.3 *Azotobacter*

Azotobacter belongs to family *Azotobacteriaceae*, aerobic, free living, and heterotrophic in nature. *Azotobacters* are present in neutral or alkaline soils and *A. chroococcum* is the most commonly occurring species in arable soils. *A. vinelandii, A. beijerinckii, A. insignis* and *A. macrocytogenes* are other reported species. The number of *Azotobacter* rarely exceeds of 10^4 to 10^5 g^{-1} of soil due to lack of organic matter and presence of antagonistic microorganisms in soil. The bacterium produces antifungal antibiotics which inhibits the growth of several pathogenic fungi in the root region thereby preventing seedling mortality to a certain extent (Subba Roa, 2001; Tippannavar and Reddy, 1993). The isolated culture of *Azotobacter* fixes about 10 mg nitrogen g^{-1} of carbon source under *in vitro* conditions. *Azotobacter* also to known to synthesize biologically active growth promoting substances such as vitamins of B group, indole acetic acid (IAA) and gibberellins. Many strains of *Azotobacter* also exhibited fungi static properties against plant pathogens such as *Fusarium, Alternaria* and *Helminthosporium*. The population of *Azotobacter* is generally low in the rhizosphere of the crop plants and in uncultivated soils. The *Azotobacter* colonizing the roots not only remains on the root surface but also a sizable proportion of them penetrates into the root tissues and lives in harmony with the plants. They do not, however, produce any visible nodules or out growth on root tissue. The occurrence of this organism has been reported from the rhizosphere of a number of crop plants such as rice, maize, sugarcane, bajra, vegetables and plantation crops (Arun, 2007).

3.5.2.1.4 *Blue Green Algae (Cyanobacteria)*

These belongs to eight different families, phototrophic in nature and produce Auxin, Indole acetic acid and Gibberllic acid, fix 20-30 kg N/ha in submerged rice fields as they are abundant in paddy, so also referred as 'paddy organisms'. N is the key input required in large quantities for low land rice production. Soil N and BNF by associated organisms are major sources of N for low land rice. The 50-60% N requirement is met through the combination of mineralization of soil organic N and BNF by free living and rice plant associated bacteria (Roger and Ladha, 1992). To achieve food security through sustainable agriculture, the requirement for fixed nitrogen must be increasingly met by BNF rather than by industrial nitrogen fixation. Most N fixing BGA are filamentous, consisting of chain of vegetative cells including specialized cells called heterocyst which function as micro BGA forms symbiotic association capable of fixing nitrogen with fungi, liverworts, ferns and flowering plants, but the most common symbiotic association has been found between a free floating aquatic fern, the *Azolla* and *Anabaena azollae* (BGA).

Azolla contains 4-5% N on dry basis and 0.2-0.4% on wet basis and can be the potential source of organic manure and nitrogen in rice production. The important factor in using *Azolla* as biofertilizer for rice crop is its quick decomposition in the soil and efficient availability of its nitrogen to rice plants (Kannaiyan, 2002). Besides N-fixation, these biofertilizers or biomanures also contribute significant amounts of P, K, S, Zn, Fe, Mb and other micronutrient. The fern forms a green mat over water with a branched stem, deeply bilobed leaves and roots. The dorsal fleshy lobe of the leaf contains the algal symbiont within the central cavity. *Azolla* can be applied as green manure by incorporating in the fields prior to rice planting. The most common species occurring in India is *A. pinnata* and same can be propagated on commercial scale by vegetative means. It may yield on average about 1.5 kg per square meter in a week. India has recently introduced some species of *Azolla* for their large biomass production, which are *A. caroliniana*, *A. microphylla*, *A. filiculoides* and *A. mexicana.*

3.5.2.2 Phosphate Solubilizers

Several reports have examined the ability of different bacterial species to solubilize insoluble inorganic phosphate compounds, such as tricalcium phosphate, dicalcium phosphate, hydroxyapatite, and rock phosphate. Among the bacterial genera with this capacity are *Pseudomonas*, *Bacillus*, *Rhizobium*, *Burkholderia*, *Achromobacter*, *Agrobacterium*, *Micrococcus*, *Aereobacter*, *Flavobacterium* and *Erwinia.* There are considerable populations of phosphate solubilizing bacteria in soil and in plant rhizospheres (Kasiamdari *et al*., 2002). These include both aerobic and anaerobic strains, with a prevalence of aerobic strains in submerged soils. The soil bacteria belonging to the genera *Pseudomonas* and *Bacillus* and fungi are more common. The major microbiological means by which insoluble-P compounds are mobilized is by the production of organic acids, accompanied by acidification of the medium. The organic and inorganic acids convert tricalcium phosphate to di- and- monobasic phosphates with the net result of an enhanced availability of the element to the plant. The type of organic acid produced and their amounts differ with different organisms. Tri- and di-carboxylic acids are more effective as compared to mono basic and aromatic acids. Aliphatic acids are also found to be more effective in P solubilization compared to phenolic, citric and fumaric acids. The analysis of culture filtrates of PSMs has shown the presence of number of organic acids including citric, fumaric, lactic, 2-ketogluconic, gluconic, glyoxylic and ketobutyric acids.

3.5.2.3 Phosphate absorbers

Mycorrhiza (an ancient symbiosis in organic agriculture)

The term Mycorrhiza denotes "fungus roots". It is a symbiotic association between host plants and certain group of fungi at the root system, in which the fungal partner is benefited by obtaining its carbon requirements from the photosynthatase of the host and the host in turn is benefited by obtaining the much needed nutrients especially phosphorus, calcium, copper, zinc etc., which are otherwise inaccessible to it, with the help of the fine absorbing hyphae of the fungus. These fungi are associated with majority of agricultural crops, except with those crops/plants belonging to families of *Chenopodiaceae*, *Amaranthaceae*, *Caryophyllaceae*, *Polygonaceae*, *Brassicaceae*, *Commelinaceae*, *Juncaceae* and *Cyperaceae*. They are ubiquitous in geographic distribution occurring with plants growing in artic, temperate and tropical regions alike. VAM occur over a broad ecological range from aquatic to desert environments (Mosse *et al*., 1981). Of 150 species of fungi that have been described in order *Glomales* of class *Zygomycetes*, only small proportions are presumed to be *mycorrhizal.* There are six *Zygomycetes*, only small proportions are genera of fungi that contain species, which are known to produce *Arbuscular mycorrhizal fungi* (AMF) with plants. Two of these genera, *Glomus* and *Sclerocytis*, produce chlamydospores only. Four genera form spores that are similar to azygospores: *Gigaspora*, *Scutellospora*, *Acaulospora* and *Entrophospora.* The oldest and most prevalent of these associations are the arbuscular mycorrhizal (AM) symbioses that first evolved 400 million years ago, coinciding with the appearance of the first land plants. Crop domestication, in comparison, is a relatively recent event, beginning 10,000 years ago (Sawers *et al*., 2008).

3.5.2.4 Potassium solubilizers

Potassium (K) is the third major essential macronutrient for plant growth. The concentrations of soluble potassium in the soil are usually very low and more than 90% of potassium in the soil exists in the form of insoluble rocks and silicate minerals. The K is one of the major plant macronutrients influencing plant growth, development and grain quality, its plays a key role in the synthesis of cells, enzymes, proteins, starch, cellulose, and vitamins. Moreover, K not only participates in nutrient transportation and uptake, but also confers resistance to abiotic and biotic stresses, leading to enhanced production of quality crops and provides resistance to plant diseases. The K is absorbed by plants in large amount than any other mineral element except nitrogen (N) and, in some cases, calcium (Ca). Chemical or synthetic K fertilizers are the largest available sources of K in the rhizosphere; therefore, larger amounts of K fertilizers can be used to promote the availability of K for plant uptake. The concentration of K in straw and grain serves as an indicator whether the K status of crop is deficient or sufficient. However, K uptake by aboveground parts of plants is assimilated mainly into the straw but not into the grain (Verma *et al.*, 2017b).

Two potassium (K)-bearing minerals, Nanjing feldspar and Suzhou illite, were used to investigate K mobilization by the wild-type strain NBT of *Bacillus edaphicus* (Sheng and He, 2006). Solubilization of K from its sources by the wild-type NBT and the MPs^{++} mutant resulted mostly from the action of organic acids and capsular polysaccharides. Oxalic acid seemed to be a more active agent for the solubilization of Nanjing feldspar. Bacterial inoculation also resulted in significantly higher N, P, and K contents of plant components. The bacteria were able to survive in the wheat rhizosphere soils after root inoculation. Verma *et al.* (2015), evaluated forty one endophytic bacteria were isolated from surface-sterilized roots and culms of wheat *var.* HS507, growing in NW Indian Himalayas. These bacteria were screened *in vitro* for multifarious plant growth promoting attributes such as solubilization of phosphorus, potassium, zinc; production of indole acetic acids, hydrogen cyanide, gibberellic acid, siderophore and activities of nitrogen fixation, ACC deaminase and biocontrol against *Rhizoctonia solani* and *Macrophomina phaseolina* at low temperature (4°C). One isolate IARI-HHS2-30, showed appreciable level of potassium solubilization was further characterized *in vivo* at control condition of low temperature. Based on 16S rDNA sequence analysis, this isolate was identified as *Bacillus amyloliquefaciens* assigned accession number KF054757.

3.5.2.5 Zinc solubilizers

The nitrogen fixers like *Rhizobium, Azospirillum, Azotobacter*, BGA and Phosphate solubilizing bacteria like *B. magaterium*, *Pseudomonas striata*, and phosphate mobilizing *Mycorrhiza* have been widely accepted as biofertilizers (Subba Roa, 2001). However these supply only major nutrients but a host of microorganism that can transform micronutrients are there in soil that can be used as bio-fertilizers to supply micronutrients like zinc, iron, copper etc., zinc being utmost important is found in the earth's crust to the tune of 0.008 per cent but more than 50 per cent of Indian soils exhibit deficiency of zinc with content must below the critical level of 1.5 ppm of available zinc (Katyal *et al.*, 1994). The plant constraints in absorbing zinc from the soil are overcome by external application of soluble zinc sulphate ($ZnSO_4$). But the fate of applied zinc in the submerged soil conditions is pathetic and only 1-4% of total available zinc is utilized by the crop and 75% of applied zinc is transformed into different mineral fractions (Zn-fixation) which are not available for plant absorption (crystalline iron oxide bound and residual zinc). There appears to be two main mechanisms of zinc fixation, one operates in acidic soils and is closely related with cat ion exchange and other operates in alkaline conditions where fixation takes by means of chemisorptions, (chemisorptions of zinc on calcium carbonate formed a solid-solution of $ZnCaCO_3$), and by complexation by organic ligands (Alloway, 2008).

The zinc can be solubilized by microorganisms viz., *B. subtilis*, *Thiobacillus thioxidans* and *Saccharomyces* sp. These microorganisms can be used as biofertilizers for solubilization of fixed micronutrients like zinc (Azooz and Ahmad, 2013). The results have shown that a *Bacillus* sp. (Zn solubilizing bacteria) can be used as biofertilizer for zinc or in soils where native zinc is higher or in conjunction with insoluble cheaper zinc compounds like zinc oxide (ZnO), zinc carbonate ($ZnCO_3$) and zinc sulphide (ZnS) instead of costly zinc sulphate (Mahdi *et al.*, 2010).

3.6 Inoculation of Biofertilizers

AM fungi have the potential to reduce damage caused by soil-borne pathogenic fungi, nematodes, and bacteria. Meta-analysis showed that AM fungi generally decreased the effects of fungal pathogens. A variety of mechanisms have been proposed to explain the protective role of mycorrhizal fungi. A major mechanism is nutritional, because plants with a good phosphorus status are less sensitive to pathogen damage. Non-nutritional mechanisms are also important, because mycorrhizal and nonmycorrhizal plants with the same internal phosphorus concentration may still be differentially affected by pathogens. Such non-nutritional mechanisms include activation of plant defense systems, changes in exudation patterns and concomitant changes in mycorrhizosphere populations, increased lignification of cell walls, and competition for space for colonization and infection sites (Kasiamdari *et al.*, 2002). It is also reported that increased production and activity of phenolic and phytoalexin compounds with due to AM-inoculation considerably increases the defense mechanism there by imparts the resistance to plants.

3.7 Mechanisms of biological fertilizers to increase crop yield

Bowen and Rovira (1999) reviewed the biology of the rhizosphere and its management to improve plant growth, summarizing their interest in this area from an agronomic point of view. Their review commenced with the increases in growth when tomatoes were inoculated with *Azotobacter* (Brown and Littna, 1964). Similar results were obtained by Rovira (1965) for wheat following inoculation with *Azotobacter chroococcum*, *Clostridium pasteurianum* and *Bacillus polymyxa*. Bowen and Rovira (1999) likened the widespread failure of inoculation with *Azotobacter* to increase yields in field experiments in Russia (only one-third were successful) to the study by Reuter *et al*., (1995) of responses of crop yields to fertilizer P. In 580 field experiments conducted in southern Australia, where soil tests predicted a positive crop response to the application of superphosphate in 30% of the cases, only 10% gave a positive yield response. It was possible to improve the predictability of the test if other soil properties were considered as extra information. The lesson to be drawn from the similar failures with inoculation of field crops is the need to also. Plant growth promoting microorganisms regulating various plant growth parameters to improve crop yield listed in **Table 3.3**.

3.8 Improve Soil Fertility Utilized Biofertilizers

Whereas the role of mycorrhizal associations in enhancing nutrient uptake will mainly be relevant in lower input agroecosystems, the mycorrhizal role in maintaining soil structure is important in all ecosystems (Ryan and Graham, 2002). Formation and maintenance of soil structure will be influenced by soil properties, root architecture and management practices. The use of machines and fertilizers are considered to be responsible for soil degradation. The specific adsorption of P by functional groups can affect the charge balance and cause dispersion of particles (Lima *et al*., 2000). Soil aggregation is one component of soil structure. Mycorrhizal fungi contribute to soil structure by (1) growth of external hyphae into the soil to create a skeletal structure that holds soil particles together; (2) creation by external hyphae of conditions that are conducive for the formation of micro-aggregates; (3) enmeshment of micro aggregates by external hyphae and roots to form macro aggregates; and (4) directly tapping carbon resources of the plant to the soils (Miller and Jastrow, 1990). This direct access will influence the formation of soil aggregates, because soil carbon is crucial to form organic materials necessary to cement soil particles.

Hyphae of AM fungi may be more important in this regard than hyphae of saprotrophic fungi due to their longer residence time in soil, because fungivorous soil fauna prefers hyphae of the latter over those of AM fungi (Gange, 2000; Klironomos and Kendrick, 1996). In addition, AM fungi produce glomalin (12-45 mg/cm^3), a specific soil protein, whose biochemical nature is still unknown. Glomalin is quantified by measuring several glomalin related soil protein (GRSP) pools (Rillig, 2004). Glomalin has a longer residence time in soil than hyphae, allowing for a long persistent contribution to soil aggregate stabilization. The residence time for hyphae is considered to vary from days to months (Staddon, 2003) and for glomalin from 6 to 42 years (Rillig *et al*., 2001). Steinberg and Rillig (2003) demonstrated that even under relatively favorable conditions for decomposition, 40% of AM fungal hyphae and 75% of total glomalin could be extracted from the soil 150 days after being separated from their host. Glomalin is considered to stably glue hyphae to soil. The mechanism is the formation of a 'sticky' string-bag of hyphae which leads to the stability of aggregates.

3.9 Economic Development in Agriculture

The economy of India thrives on agriculture, the most practiced occupation in the country. Agricultural fertilizers are essential to enhance proper growth and crop yield. Recently, farmers have been using chemical fertilizers for quicker and better yield. But these fertilizers endanger ecosystems, soil, plants, and human and animal lives. In contrast, naturally grown biofertilizers not only give a better yield, but are also harmless to humans.

Agriculture is one of the most important factors contributing to the economic growth of India. Out of the 329 million hectares of India's geographical area, about 114 million hectares are under cultivation (Raghuwanshi, 2012). In order to reap a better harvest, farmers inoculate the soil with fertilizers. Fertilizers come in two types—they are either chemical- or biofertilizer. Increasingly high inputs of chemical fertilizers during last 150 years have not only left soils degraded, polluted and less productive but have also posed severe health and environmental hazards. Organic farming methods (such as the use of biofertilizers) would solve these issues and make the ecosystem healthier (Shukla *et al.*, 2016). The current global market for organically raised agricultural products is valued at around US$ 30 billion with a growth rate of around 8 percent. Nearly 22 million hectares of land are now cultivated organically. Organic cultivation represents less than 1 percent of the world's conventional agricultural production and about 9 percent of the total agricultural area.

3.10 Conclusion

In conclusion, the present book chapter revealed about the ecological significance of biofertilizers and role of microbes in plant growth. The modern agriculture is mostly dependent on chemical fertilizers which can be replaced by eco-friendly PGP microbial consortium or biofertilizers. Application of high doses of chemical fertilizers may temporarily help to increase crop production. However, this may turn into bitter and highly regrettable consequences where soil fertility will be depleted or become acidic and devoid of macro and micro nutrients for crops to grow and microorganisms to proliferate. Thus, it is absolutely necessary to awake timely and be able to use eco-friendly inputs such as beneficial plant growth promoting microflora to save our 'currency', the soil and its constituents.

Developing suitable alternate formulations viz., liquid inoculants / granular formulations for all bioinoculants to carrier-based inoculants. Standardizing the media, method of inoculation etc., for the new formulations is also the need of the hour. In addition employing microbiologists in production units to monitor the production, developing cold storage facilities in production centers, technical training for the production unit and updating the quality control units with present state of art facilities and futuristic developments in the manufacturing units with organizational training to the unit extension workers and farmers pave the way to popularize the technology.

Acknowledgement: The authors are grateful to the Department of Microbiology, Akal college of Basic Science, Eternal University, Himachal Pradesh and Department of Biotechnology (DBT), Ministry of Science and Technology for providing the facilities and financial support.

References

Abd-Alla, Mohamed Hemida, Omar, Shukry Ahmed, & Omar, SA. (2001). Survival of *rhizobia/ bradyrhizobia* and a rock-phosphate-solubilizing fungus *Aspergillus niger* on various carriers from some agro-industrial wastes and their effects on nodulation and growth of faba bean and soybean. J Plant Nutrition, 24(2), 261-272.

Afzal, Aftab, & Bano, Asghari. (2008). *Rhizobium* and phosphate solubilizing bacteria improve the yield and phosphorus uptake in wheat (*Triticum aestivum*). Int J Agric Biol, 10(1), 85-88.

Alagawadi, A Rl, & Gaur, AC. (1992). Inoculation of *azospirillum brasilense* and phosphate-solubilizing bacteria on yield of *sorghum [sorghum bicolor* (l.) moench] in dry land. Tropical Agriculture (Trinidad and Tobago).

Alloway, BJ. (2008). Zinc in soils and crop nutrition. International zinc association, brussels. International Fertilizer Industry Association, Paris.

Arun, KS. (2007). Bio-fertilizers for sustainable agriculture. Mechanism of p solubilization. Agribios publishers, 196, 197.

Azooz, Mohamed Mahgoub, & Ahmad, Parvaiz. (2013). Role of bio-fertilizers in crop improvement. Crop Improvement: New Approaches and Modern Techniques, 189.

Bowen, GD, & Rovira, AD. (1999). The rhizosphere and its management to improve plant growth Adv in Agronomy (Vol. 66, pp. 1-102): Elsevier.

Brown, Donald D, & Littna, Elizabeth. (1964). Variations in the synthesis of stable rna's during oogenesis and development of xenopus laevis. Journal of Molecular Biology, 8(5), 688-695, IN10.

Cakmakçi, R., Dönmez, F., Aydýn, A., and Sahin, F. (2006). Growth promotion of plants by plant growth-promoting rhizobacteria under greenhouse and two different field soil conditions. Soil Biology and Biochemistry, 38(6), 1482-1487.

Chen, Jen-Hshuan. (2006). The combined use of chemical and organic fertilizers and/or biofertilizer for crop growth and soil fertility. Paper presented at the International workshop on sustained management of the soil-rhizosphere system for efficient crop production and fertilizer use.

Chung, Soohee, Kong, Hyesuk, Buyer, Jeffrey S, Lakshman, Dilip K, Lydon, John, Kim, Sang-Dal, and Roberts, Daniel P. (2008). Isolation and partial characterization of bacillus subtilis me488 for suppression of soilborne pathogens of cucumber and pepper. App Microbiol Biotechnol, 80(1), 115-123.

Commare, R Radja, Nandakumar, R, Kandan, A, Suresh, S, Bharathi, M, Raguchander, T, & Samiyappan, R. (2002). Pseudomonas fluorescens based bio-formulation for the management of sheath blight disease and leaffolder insect in rice. Crop Protection, 21(8), 671-677.

Dobbelaere, Sofie, Croonenborghs, Anja, Thys, Amber, Ptacek, David, Okon, Yaacov, and Vanderleyden, Jos. (2002). Effect of inoculation with wild type azospirillum brasilense and a. Irakense strains on development and nitrogen uptake of spring wheat and grain maize. Biolo Ferti Soils, 36(4), 284-297.

Egamberdiyeva, Dilfuza. (2007). The effect of plant growth promoting bacteria on growth and nutrient uptake of maize in two different soils. Applied Soil Ecology, 36(2-3), 184-189.

Galal, YGM. (2003). Assessment of nitrogen availability to wheat (*Triticum aestivum* l.) from inorganic and organic n sources as affected by azospirillum brasilense and rhizobium leguminosarum inoculation. Egypt J Microbiol, 38, 57-73.

Gange, A. (2000). Arbuscular mycorrhizal fungi, collembola and plant growth. Trends in Ecology & Evolution, 15(9), 369-372.

Gholami, A, Shahsavani, S, & Nezarat, S. (2009). The effect of plant growth promoting rhizobacteria (pgpr) on germination, seedling growth and yield of maize. Int J Biol Life Sci, 5(1), 35-40.

Glick, Bernard R. (1995). The enhancement of plant growth by free-living bacteria. Canad J Microbiol, 41(2), 109-117.

Glick, Bernard R, Liu, Changping, Ghosh, Sibdas, and Dumbroff, Erwin B. (1997). Early development of canola seedlings in the presence of the plant growth-promoting rhizobacterium pseudomonas putida gr12-2. Soil Biology and Biochemistry, 29(8), 1233-1239.

Gravel, Valerie, Antoun, Hani, and Tweddell, Russell J. (2007). Growth stimulation and fruit yield improvement of greenhouse tomato plants by inoculation with pseudomonas putida or trichoderma atroviride: Possible role of indole acetic acid (iaa). Soil Biol Biochem, 39(8), 1968-1977.

Gupta, A, and Sen, S. (2013). Role of biofertilisers and biopesticides for sustainable agriculture, scholar. Google. Com.

Han, Hyo-Shim, & Lee, KD. (2006). Effect of co-inoculation with phosphate and potassium solubilizing bacteria on mineral uptake and growth of pepper and cucumber. Plant soil Environ, 52(3), 130.

Harish, S, Kavino, M, Kumar, N, Balasubramanian, P, and Samiyappan, R. (2009a). Induction of defense-related proteins by mixtures of plant growth promoting endophytic bacteria against banana bunchy top virus. Biol Cont, 51(1), 16-25.

Harish, S, Kavino, M, Kumar, N, & Samiyappan, R. (2009b). Differential expression of pathogenesis-related proteins and defense enzymes in banana: Interaction between endophytic bacteria, banana bunchy top virus and pentalonia nigronervosa. Biocon Science Technol, 19(8), 843-857.

Hernández-Rodríguez, A, Heydrich-Pérez, M, Acebo-Guerrero, Y, Velazquez-Del Valle, Miguel Gerardo, & Hernandez-Lauzardo, Ana Niurka. (2008). Antagonistic activity of cuban native rhizobacteria against fusarium verticillioides (sacc.) nirenb. In maize (*zea mays* l.). App Soil Ecol, 39(2), 180-186.

Jetiyanon, Kanchalee, Fowler, William D, and Kloepper, Joseph W. (2003). Broad-spectrum protection against several pathogens by pgpr mixtures under field conditions in thailand. Plant Disease, 87(11), 1390-1394.

Jiang, Zhi-Qiang, Guo, Ya-Hui, Li, Shi-Mo, Qi, Hong-Ying, & Guo, Jian-Hua. (2006). Evaluation of biocontrol efficiency of different bacillus preparations and field application methods against phytophthora blight of bell pepper. Biol Control, 36(2), 216-223.

Kannaiyan, S. (2002). Biotechnology of biofertilizers: Alpha Science Int'l Ltd.

Kasiamdari, RS, Smith, SE, Smith, FA, and Scott, ES. (2002). Influence of the mycorrhizal fungus, glomus coronatum, and soil phosphorus on infection and disease caused by binucleate rhizoctonia and rhizoctonia solani on mung bean (*vigna radiata*). Plant and Soil, 238(2), 235-244.

Katyal, JC, Venkateswarlu, B, & Das, SK. (1994). Biofertilisers for nutrient supplementation in dryland agriculture potentials and problems. Fertiliser News, 39, 27-27.

Kavino, M, Harish, S, Kumar, N, Saravanakumar, D, and Samiyappan, R. (2010). Effect of chitinolytic pgpr on growth, yield and physiological attributes of banana (*Musa* spp.) under field conditions. App Soil Ecol, 45(2), 71-77.

Kavino, M., Harish, S., Kumar, N., Saravanakumar, D., and Samiyappan, R. (2008). Induction of systemic resistance in banana (*musa* spp.) against banana bunchy top virus (bbtv) by combining chitin with root-colonizing pseudomonas fluorescens strain cha0. Euro J Plant Pathol, 120(4), 353-362.

Klironomos, J. N, & Kendrick, W B. (1996). Palatability of microfungi to soil arthropods in relation to the functioning of arbuscular mycorrhizae. Biol Fert of Soils, 21(1-2), 43-52.

Kohler, J, Caravaca, F, Carrasco, L, & Roldán, A. (2006). Contribution of pseudomonas mendocina and glomus intraradices to aggregate stabilization and promotion of biological fertility in rhizosphere soil of lettuce plants under field conditions. Soil Use and Management, 22(3), 298-304.

Kucey, RMN, Janzen, HH, and Leggett, ME. (1989). Microbially mediated increases in plant-available phosphorus Adv agro (Vol. 42, pp. 199-228): Elsevier.

Lima, J. M., Anderson, S. J. and Curi, N. (2000). Phosphate-induced clay dispersion as related to aggregate size and composition in hapludoxs.

Mahdi, SS, Dar, SA, Ahmad, S, & Hassan, GI. (2010). Zinc availability–a major issue in agriculture. Res J Agric Sci, 3(3), 78-79.

Miller, RM, & Jastrow, JD. (1990). Hierarchy of root and mycorrhizal fungal interactions with soil aggregation. Soil Biol Biochem, 22(5), 579-584.

Mosse, B, Stribley, DP, & LeTacon, F. (1981). Ecology of mycorrhizae and mycorrhizal fungi Advances in microbial ecology (pp. 137-210): Springer.

Muraleedharan, Hari, Seshadri, S, & Perumal, K. (2010). Biofertilizer (phosphobacteria). Shri AMM Murugappa Chettiar Research Centre, Taramani, Chennai, 600(113), 1-16.

Murphy, John F, Zehnder, Geoffrey W, Schuster, David J, Sikora, Edward J, Polston, Jane E, & Kloepper, Joseph W. (2000). Plant growth-promoting rhizobacterial mediated protection in tomato against tomato mottle virus. Plant Disease, 84(7), 779-784.

Nayak S.K., Nayak S., Mishra B.B. (2017) Antimycotic Role of Soil Bacillus sp. Against Rice Pathogens: A Biocontrol Prospective. In: Patra J., Vishnuprasad C., Das G. (eds) Microbial Biotechnology. Springer, Singapore, pp29-60. Doi:10.1007/978-981-10-6847-8_2

Noel, Tanya C, Sheng, C, Yost, CK, Pharis, RP, & Hynes, MF. (1996). Rhizobium leguminosarum as a plant growth-promoting rhizobacterium: Direct growth promotion of canola and lettuce. Can J Microbiol, 42(3), 279-283.

Parewa, Hanuman Prasad, Yadav, J, Rakshit, A, Meena, VS, & Karthikeyan, N. (2014). Plant growth promoting rhizobacteria enhance growth and nutrient uptake of crops. Agric Sustain Dev, 2(2), 101-116.

Parikh, SJ, and James, BR. (2012). Soil: The foundation of agriculture. Nature Education Knowledge, 3(10), 2.

Podile, A. R, and Krishna Kishore G. (2007). Plant growth-promoting rhizobacteria Plant-associated bacteria (pp. 195-230): Springer.

Raghuwanshi, R. (2012). Opportunities and challenges to sustainable agriculture in india. NeBIO, 3(2), 78-86.

Raj, S Niranjan, Deepak, SA, Basavaraju, P, Shetty, H Shekar, Reddy, MS, & Kloepper, Joseph W. (2003). Comparative performance of formulations of plant growth promoting rhizobacteria in growth promotion and suppression of downy mildew in pearl millet. Crop Protection, 22(4), 579-588.

Reuter, DJ, Dyson, CB, Elliott, DE, Lewis, DC, & Rudd, CL. (1995). An appraisal of soil phosphorus testing data for crops and pastures in south australia. Aust J Exp Agri, 35(7), 979-995.

Steinberg P. and Rillig M.C. (2003). Differential decomposition of arbuscular mycorrhizal fungal hyphae and glomalin. Soil Biology and Biochemistry, 35(1), 191-194.

Rillig, Matthias C. (2004). Arbuscular mycorrhizae and terrestrial ecosystem processes. Ecology Letters, 7(8), 740-754.

Rillig, Matthias C, Wright, Sara F, Nichols, Kristine A, Schmidt, Walter F, & Torn, Margaret S. (2001). Large contribution of arbuscular mycorrhizal fungi to soil carbon pools in tropical forest soils. Plant and Soil, 233(2), 167-177.

Roger, Pierre-Armand, & Ladha, JK. (1992). Biological n 2 fixation in wetland rice fields: Estimation and contribution to nitrogen balance Biological nitrogen fixation for sustainable agriculture (pp. 41-55): Springer.

Rokhzadi, Asad, Asgharzadeh, Ahmad, Darvish, Farrokh, Nour-Mohammadi, G, & Majidi, Eslam. (2008). Influence of plant growth-promoting rhizobacteria on dry matter accumulation and yield of chickpea (*cicer arietinum* l.) under field conditions. American-Eurasian J Agri and Environ Sci, 3(2), 253-257.

Rovira, AD. (1965). Interactions between plant roots and soil microorganisms. Annual Reviews in Microbiology, 19(1), 241-266.

Ryan, Megan H, & Graham, James H. (2002). Is there a role for arbuscular mycorrhizal fungi in production agriculture? Plant and soil, 244(1-2), 263-271.

Saravanakumar, D, & Samiyappan, R. (2007). Acc deaminase from pseudomonas fluorescens mediated saline resistance in groundnut (*arachis hypogea*) plants. J Applied Microbiol, 102(5), 1283-1292.

Saravanakumar, Duraisamy, Lavanya, Nallathambi, Muthumeena, Kannappan, Raguchander, Thiruvengadam, & Samiyappan, Ramasamy. (2009). Fluorescent pseudomonad mixtures mediate disease resistance in rice plants against sheath rot (*sarocladium oryzae*) disease. Biocontrol, 54(2), 273.

Sawers, Ruairidh JH, Gutjahr, Caroline, & Paszkowski, Uta. (2008). Cereal mycorrhiza: An ancient symbiosis in modern agriculture. Trends Plant Sci, 13(2), 93-97.

Sheng, Xia Fang, & He, Lin Yan. (2006). Solubilization of potassium-bearing minerals by a wild-type strain of bacillus edaphicus and its mutants and increased potassium uptake by wheat. Canad J of Microbiol, 52(1), 66-72.

Shukla, Livleen, Suman, Archna, Yadav, AN, Verma, P, & Saxena, Anil Kumar. (2016). Syntrophic microbial system for ex-situ degradation of paddy straw at low temperature under controlled and natural environment. J App Biol Biotech, 4(2), 30-37.

Silva, Harllen Sandro Alves, da Silva Romeiro, Reginaldo, Macagnan, Dirceu, de Almeida Halfeld-Vieira, Bernardo, Pereira, Maria Cristina Baracat, & Mounteer, Ann. (2004). Rhizobacterial induction of systemic resistance in tomato plants: Non-specific protection and increase in enzyme activities. Biological Control, 29(2), 288-295.

Singh, J. S., Pandey V. C. and Singh D.P.(2011). Efficient soil microorganisms: A new dimension for sustainable agriculture and environmental development. Agriculture, Ecosystems & Environment 140(3-4), 339-353.

Staddon. (2003). Rapid turnover of hyphae of mycorrhizal fungi determined by ams microanalysis of 14c. Science, 300(5622), 1138-1140.

Subba Roa, NS. (2001). An appraisal of biofertilizers in india. The biotechnology of biofertilizers. ed.) S. Kannaiyan, Narosa Pub. House, Delhi, India.

Suman, Archna, Verma, Priyanka, Yadav, Ajar Nath, Srinivasamurthy, R, Singh, Anupama, & Prasanna, Radha. (2016a). Development of hydrogel based bio-inoculant formulations and their impact on plant biometric parameters of wheat (*Triticum aestivum* l.). Int. J. Curr. Microbiol. App. Sci, 5(3), 890-901.

Suman, Archna, Yadav, Ajar Nath, and Verma, P. (2016b). Endophytic microbes in crops: Diversity and beneficial impact for sustainable agriculture Microbial inoculants in sustainable agricultural productivity (pp. 117-143): Springer.

Tanwar, SPS, Sharma, GL, and Chahar, MS. (2003). Effect of phosphorus and biofertilizers on yield, nutrient content and uptake by black gram [vigna mungo (l.) hepper]. Legume Research-An International Journal, 26(1), 39-41.

Tippannavar, CM, and Reddy, TKR. (1993). Seed treatment of wheat (*Triticum aesativum* l.) on the survival of seed borne azotobacter chroococcum. Karnataka Journal of Agricultural Sciences, 6, 310-312.

Turan, A, Kaya, G, Karamanlioðlu, B, Pamukcu, Z, & Apfel, CC. (2005). Effect of oral gabapentin on postoperative epidural analgesia. British Journal of Anaesthesia, 96(2), 242-246.

Venkataraman, GS, and Shanmugasundaram, S. (1992). Algal biofertilizer technology for rice. DBT Centre for BGA Biofertilizer, Madurai, Kamraj University (TN), India.

Verma, J. P., Yadav, J., Tiwari, K. N., and Jaiswal, D. K. (2014). Evaluation of plant growth promoting activities of microbial strains and their effect on growth and yield of chickpea (*cicer arietinum* l.) in India. Soil Biology and Biochemistry, 70, 33-37.

Verma, Priyanka, Yadav, Ajar Nath, Shukla, Livleen, Saxena, Anil Kumar, and Suman, Archna. (2015). Alleviation of cold stress in wheat seedlings by bacillus amyloliquefaciens IARI-HHS2-30, an endophytic psychrotolerant k-solubilizing bacterium from nw indian himalayas. Natl J Life Sci, 12(2), 105-110.

Verma, P. (2013). Elucidating the diversity and plant growth promoting attributes of wheat (Triticum aestivum) associated acidotolerant bacteria from southern hills zone of India. Natl J Life Sci, 10(2), 219-226.

Verma, P. (2014). Evaluating the diversity and phylogeny of plant growth promoting bacteria associated with wheat (Triticum aestivum) growing in central zone of india. Int J Curr Microbiol Appl Sci, 3(5), 432-447.

Verma, P. (2015). Assessment of genetic diversity and plant growth promoting attributes of psychrotolerant bacteria allied with wheat (*Triticum aestivum*) from the northern hills zone of India. Ann Microbiol, 65(4), 1885-1899. DOI: 10.1007/s13213-014-1027-4

Verma, P. (2016a). Appraisal of diversity and functional attributes of thermotolerant wheat associated bacteria from the peninsular zone of India. Saudi J Biolo Sci.

Verma, P. Yadav AN, Khannam KS, Kumar S, Saxena AK, Suman A. (2016). Molecular diversity and multifarious plant growth promoting attributes of bacilli associated with wheat (*Triticum aestivum* l.) rhizosphere from six diverse agro ecological zones of india. Journal of Basic Microbiology, 56(1), 44-58. DOI: 10.1002/jobm.201500459.

Verma, Priyanka, Yadav, Ajar Nath, Kumar, Vinod, Singh, Dhananjaya Pratap, and Saxena, Anil Kumar. (2017a). Beneficial plant-microbes interactions: Biodiversity of microbes from diverse extreme environments and its impact for crop improvement Plant-microbe interactions in agro-ecological perspectives (pp. 543-580): Springer

Verma, Priyanka, Yadav, Ajar Nath, Khannam, Kazy Sufia, Saxena, Anil Kumar, and Suman, Archna. (2017b). Potassium-solubilizing microbes: Diversity, distribution, and role in plant growth promotion Microorganisms for green revolution (pp. 125-149): Springer.

Wani, SP, and Lee, KK. (2002). Population dynamics of nitrogen fixing bacteria associated with pearl millet (*P. Americanum l.*). Biotechnology of nitrogen fixation in the tropics. University of Pertanian, Malaysia, 21-30.

Yasari, Esmaeil, and Patwardhan, AM. (2007). Effects of (azotobacter and azospirillum) inoculants and chemical fertilizers on growth and productivity of canola (*brassica napus* l.). Asian J. Plant Sci, 6(1), 77-82.

Zahir, Zahir A, Arshad, Muhammad, and Frankenberger, William T. (2004). Plant growth promoting rhizobacteria: Applications and perspectives in agriculture. Advances in Agronomy, 81, 98-169.

Zaidi, Almas, and Khan, Mohammad Saghir. (2006). Co-inoculation effects of phosphate solubilizing microorganisms and glomus fasciculatum on green gram-bradyrhizobium symbiosis. Turkish J Agricul forestry, 30(3), 223-230.

Zakira, Fareed. (2009). Africa's new path: Paul kagame charts a way forward. Newsweek, July, 18.

Zehnder, Geoffrey W, Murphy, John F, Sikora, Edward J, and Kloepper, Joseph W. (2001). Application of rhizobacteria for induced resistance. European Journal of Plant Pathology, 107(1), 39-50.

Zhang, Chengsheng, and Kong, Fanyu. (2014). Isolation and identification of potassium-solubilizing bacteria from tobacco rhizospheric soil and their effect on tobacco plants. Appl Soil Ecol, 82, 18-25.

Zhang, Shouan, Moyne, Anne-Laure, Reddy, MS, and Kloepper, Joseph W. (2002). The role of salicylic acid in induced systemic resistance elicited by plant growth-promoting rhizobacteria against blue mold of tobacco. Biological Control, 25(3), 288-296.

Zhang, Shouan, White, Thomas L, Martinez, Miriam C, McInroy, John A, Kloepper, Joseph W, and Klassen, Waldemar. (2010). Evaluation of plant growth-promoting rhizobacteria for control of phytophthora blight on squash under greenhouse conditions. Biological Control, 53(1), 129-135.

4

Role of Plant Growth Promoting Rhizobacteria in Mitigating Salt Stress

Priyanka Chandra*, Awtar Singh, Madhu Choudhary and R. K. Yadav

ICAR-Central Soil Salinity Research Institute, Karnal-132001, Haryana, India

Abstract

Soil salinity is one of the major environmental stresses affecting crop productivity and soil health drastically, particularly in arid and semi-arid areas. Most of the vegetable crops being salt sensitive, grow poorly in saline and alkaline soils. Soil salinity affects crops mainly due to the osmotic stress and excessive accumulation of toxic ions (Na^+, Cl^-, and SO_4^{2-}), whereas alkali soils are generally characterized by poor physical conditions due to high concentrations of bicarbonate (HCO_3^-) and carbonate (CO_3^{2-}) as well as high exchangeable Na^+. Excessive concentrations of sodium chloride (NaCl) in soil or irrigation water can induce several morphological, physiological, and biochemical responses in plants leading to stunted growth and yield. This is due to osmotic (i.e., water deficit) and ionic (i.e., Na^+ and Cl^-) effects on nutrient uptake/translocation and metabolic processes such as nitrogen assimilation, photosynthesis, and protein synthesis. Plant growth promoting microbes (PGPMs) are considered as promising tools to overcome the limitations of salinity on crop growth and productivity. Many PGPMs can enhance tolerance of crops to salt stress by increasing nutrient uptake through better solubilization of nutrients like phosphorus and zinc, and also by production of small peptides, volatiles and metabolites with hormone activities such as indole-3-acetic acid or auxin analogs in the rhizosphere. This chapter discusses the benefits of PGPMs in plant adaptations to salt stress conditions.

Keywords: Plant growth promoting microbes, salt stress, salinity, ACC deaminase, nutrient uptake

4.1 Introduction

Soil salinity and sodicity are the most serious abiotic stresses which drastically reduce crop growth, crop productivity and soil health worldwide. Soil salinity affects plant growth and development in many ways such as osmotic effects, nutritional imbalance or nutritional disorders and/or specific ion toxicity like sodium and chloride. In sodic soils, excess of exchangable sodium disperses the soil structure which leads to poor hydraulic conductivity infilteration rate and show high impedance for crop root growth (Abrol *et al*., 1988).

Therefore, the development of salt-tolerant plants is a much-desired scientific goal. However, efforts have resulted in limited success, and only a few major genetic determinants of salt tolerance have been identified. Interestingly, as an alternative to breeding and genetic manipulation, plant salt tolerance can also be improved by use of salt-tolerant beneficial microorganisms. Plant growth promoting rhizobacteria (PGPR) inhabiting the plant rhizosphere have been investigated for their potential to alleviate salt stress (Dimkpa *et al*., 2009; Yang *et al*., 2009).

PGPR refers to the beneficial free living soil bacteria that colonize the rhizosphere of plants and have ability to promote growth of plants by direct and/or indirect mechanisms (Kloepper *et al*., 1980, 1989). PGPR can influence the plant root activities in rhizospheric environment (Chandra and Chandra, 2016). Biogeochemical interactions also take place between soil microbes and plants, because rhizospheric microorganisms provide a critical link between plant and soil (Kumar and Verma, 2017). Indeed rhizobacteria counteract osmotic stress and help to improve plant growth. This chapter reviews the benefits of plant-growth-promoting rhizobacteria for plants growing in salt affected soils. Some of the salient mechanisms of PGPRs are highlighted as below:

(1) Plants treated with rhizobacteria have better root and shoot growth, nutrient uptake, hydration, chlorophyll content, and resistance to diseases.

(2) Stress tolerance can be explained by nutrient mobilization and biocontrol of phytopathogens in the rhizosphere and by production of phytohormones and 1-aminocyclopropane-1-carboxylate deaminase.

(3) Rhizobacteria favour the recycling of plant nutrients in the rhizosphere.

(4) Rhizobacteria favour osmolyte accumulation in plants.

(5) Plants inoculated with rhizobacteria have higher K^+ ion concentration and, in turn, a higher K^+/Na^+ ratio that favours salinity tolerance.

(6) Rhizobacteria induce synthesis of antioxidative enzymes in plants that degrade reactive oxygen species generated upon salt shock.

4.2 Salt affected soils

Salt affected soils comprise of saline and sodic soils which differs each other mainly on their origin, physico-chemical properties and constraints to the plant growth. They are identified on the basis of EC, ESP and SAR determination, and further classified on the basis of USDA classification described in **Table 4.1.**

Salinity primarily results from accumulation of salts over long period of time, in the soil or groundwater, which is generally caused by two natural processes. Firstly, weathering of parent materials breaking down rocks and release of various soluble salts. These mainly include chlorides of sodium, calcium sulphates and magnesium, to a lesser extent, and carbonates, with sodium corbonates as predominant soluble salt. Secondary salinization occurs due to accumulation of dissolved salts in the soil water to an extent which could inhibit plant growth.

Table 4.1: Characterstics of saline, saline alkali and alkali soils on the basis of USDA classification

Characterstics	Saline soils	Saline-alkali soils	Alkali soils
Electrical conductivity (dS m^{-1})	> 4	>4	<4
pHs	<8.5	>8.5	>8.5
Exchangeable Sodium Percent (ESP)	<15	>15	>15

4.2.1 Saline soils

Saline soil have white salt encrustation on their surface. They have predominantley chlorides and sulphates of sodium, calcium and magnessium. Electrical conductivity (ECe) of soil saturation paste extract of these soils is more than 4 dS m^{-1}at 25°C. The exchangeable sodium percentage of saline soil is less than 15, and the pH is less than 8.5.

4.2.2 Alkali soils

Alkali /sodic soil contains excess of salts capable of alkaline hydrolysis such as sodium carbonate, sodium bicorbonate and sodium silicate. They also have sufficient exchangable sodium which results into poor soil physical conditions which is affecting plant growth. The electrical conductivity of the saturation paste extract is less than 4 dS m^{-1}at 25°C, exchangeable sodium percentage is greater than 15 and the pHs varies between 8.5-10. The pH of the calcareous alkali soils is highly related to their ESP.

4.2.3 Saline-Alkali soils

These soils have electrical conductivity of the saturation paste extract more than 4 dS m^{-1} and the exchangable sodium percentage greater than 15 while pH range is higher than 8.5. These soils form as a results of combined process of salinization and alkalinization. As long as excess soluble salts are present, these soils exhibits the properties of saline soils. On leaching of excess soluble salts downwards, the properties of these soils will become like that of alkali soils. On leaching of excess soluble salts, the soil becomes strongly alkaline (pHs above 8.5). Consequently soils particles disperse, and the soil becomes unfavourable for movement of water and for tillage (USDA, 1954).

Based on the world soil map, the regional approximation of the saline and sodic soils is between 397 and 434 million hectares, which comprise nearly 15 per cent of the global land area (FAO 1997; Mandal *et al.*, 2009). In India, salt affected area is 6.73 million ha, out of which saline soils occupy 2.95 million ha and sodic soils occupied 3.78 million ha (Mandal *et al.*, 2009). It is assumed that if the soil degradation continues at the present rate, India will have around 11.7 million ha area under soil salinity and alkalinity in 2025 (Chaudhari *et al.*, 2013).

In arid and semiarid regions of Mediterranean ecosystems, presence of excess salts in soil is a very serious agro ecological problem (Yaish and Kumar, 2015). According to Ashraf and Harris (2004) potential of yield loss is about 20% due to salinity. Salt affected soils are mainly confined to the arid and semi-arid and subhumid (dry) regions and also in the coastal areas in India. The salt deposits comprise of sodium carbonate, sulphate and chloride along with calcium and magnesium. These soils vary in nature from saline to nonsaline sodic. In coastal regions, saline soils are the most predominant. They have high soluble salts (ECe>4 dS m^{-1}) of chloride and sulphate of sodium, calcium and magnesium, low ESP with pHs value less than 8.2. The classification of salt affected soils according to Indian system is presented in **Table 4.2**.

Table 4.2: Indian system of classification of salt affected soils (Adopted from Arora *et al.*, 2017).

Soil characterstics	Saline soils	Alkali soils
pHs	<8.2	>8.2
ESP	<15	>15
ECe	>4 dS m^{-1}	Variable, mostly <4 dS m^{-1}
Nature of soluble salts	Neutral,mostly Cl^-,SO_4^{2-}, HCO_3^-, may be present but CO_3^{2-} is absent	Capable of alkaline hydrolysis, preponderance of HCO_3^- and CO_3^{2-} of Na^+

4.3 Adverse effect of salt stress on crops

Salinity affects plant growth and yield in many of crops in different ways. Crops such as cereals (rice and maize), forages (clover) or horticultural crops (potatoes and tomatoes) are very much susceptible to excessive concentration of salts, either dissolved in irrigation water or present in soil (rhizosphere) solution. Soil salinity has been reported to reduce yields, nodulation and the total nitrogen content in legume plants (Singleton and Bohlool, 1984). El-Fouly *et al.*(2001)

found that the dry weight of different plant organs of tomato were reduced in response to the increase of NaCl level in the root growth medium.

Under salt stress conditions, all major physiological activities of plants, such as photosynthesis, protein synthesis, energy and lipid metabolism are adversely affected. Photosynthetic activity gets reduced with increasing osmotic stress which results in partial closure of stomata. Plants also get affected by membrane destabilization and nutrient imbalance due to accumulation of salts in soil which blocks the nutrient uptake (Shrivastava and Kumar, 2015).

Salt stress also leades to decrease in cell growth and development, reduced leaf area and chlorophyll content, accelerated defoliation and senescence in plants. An increase in the uptake of Na^+ or decrease in the uptake of Ca^{2+} and K^+ in leaves, thereby leading to nutritional imbalances. Accumulation of excess Na^+also causes metabolic disorders whereas low Na^+ and high K^+ or Ca^{2+} are required for optimum function. Excess sodium and chloride affects the plant enzymes which causes cell swelling and thus reduction in energy production and other physiological changes. Uptake and accumulation of Cl^- disrupt photosynthetic function through inhibition of nitrate reductase activity. Under excessive Na^+ and Cl^- concentrations in soil, competitive interactions among other nutrient ions (K^+, NO^{3-} and $H_2PO^-_4$) for binding sites take place which lead to transport of proteins in root cells, and thereafter translocation, deposition and partitioning within the plant tissues (Xu *et al.*, 2000).

Once the capacity of cells to store salts is exhausted, salts get accumulated in the intercellular space leading to cell dehydration and death.Under saline conditions, osmotic pressure in the rhizosphere exceeds that of root cells which influence the water and nutrient uptake. Almost all micro- and macronutrient contents decrease in the roots and shoots with increase in the NaCl concentration. Salama *et al.* (1996) reported limited uptake of nutrients in different organs of wheat plants when there was high salt concentration in the root growth medium. The plants with perturbed nutrients relations are more susceptible to invasion of different pathogenic microorganisms and physiological dysfunctions, where as their edible parts have markedly less economic and nutritional value due to reduced fruit size and shelf life, non-uniform fruitshape, decreased vitamin content, etc (Carmen and Roberto, 2011).

The primary salinity effects give rise to oxidative stress which is characterised by accumulation of reactive oxygen species (ROS). ROS is very much harmful to biomembranes, proteins, nucleic acids and enzymes. To protect against oxidative stress, plant cells produce both antioxidant enzymes and non-enzymatic antioxidants, and the modulation of antioxidant enzyme activity and concentrations are frequently used as indicators of oxidative stress in plants (Mayak *et al.*, 2004).

4.4 Salt stress mitigation by plant growth-promoting rhizobacteria (PGPR)

Plant growth-promoting rhizobacteria (PGPR) are the heterogeneous group of bacteria that are found in the rhizosphere, at root surfaces and in association with roots. PGPR have the potential to improve the extent or quality of plant growth directly and/or indirectly. They are important in stimulating plant growth because of their positive effects on soil conditions and nutrient availability and thereby plant growth and yields (Nehra and Choudhary, 2015). In last few decades a large array of bacteria including species of *Pseudomonas, Azospirillum, Azotobacter, Klebsiella, Enterobacter, Alcaligenes, Arthrobacter, Burkholderia, Bacillus* and *Serratia* have been reported to enhance plant growth (Kloepper *et al.*, 1989; Gupta *et al.*, 2015). They are known to colonize the rhizosphere of wheat, potato, maize, grasses, pea, sugarcane, cotton, and cucumber and significantly improve the growth and yield of crops (Chandra and Chandra 2017; Nadeem *et al.*, 2006; Nehra *et al.*, 2016). Bacterial inoculants increase plant growth and germination rate, improve seedling emergence, cascading external stress factors and protect plants from diseases (Siddiqui and Shaukat, 2002)(**Table 4.3**).

4.5 Benefits of plant growth-promoting rhizobacteria (PGPR)

Plant growth promoting rhizobacteria (PGPR) improve plant growth and yield by direct and indirect mechanisms. Direct mechanisms of plant growth promotion by PGPR can be demonstrated by absence of plant pathogens or other harmful rhizosphere microorganisms, while indirect mechanisms involve the ability of PGPR to reduce the deleterious effects of plant pathogens on crop yield (**Figure 4.1**). The direct growth promotion by PGPR entails either providing the plant with plant growth promoting substances that are synthesized by the bacterium or facilitating the uptake of certain plant nutrients from the environment. They directly enhance plant growth by a variety of mechanisms such as: fixation of atmospheric nitrogen that is transferred to the plant, production of siderophores that chelate iron and make it available to the plant root, solubilization of minerals such as phosphorus, and synthesis of phytohormones (Glick, 1995). Direct enhancement of mineral uptake due to increases in specific ion fluxes at the root surface in the presence of PGPR has also been reported (Bertrand *et al.*, 2000). PGPRs manipulate rhizosphere through one or more of these mechanisms.

The indirect promotion of plant growth occurs when PGPR prevent deleterious effects of one or more phytopathogenic microorganisms. The exact mechanisms by which PGPR promote plant growth are not fully understood, but considered to include (i) the ability to produce or change the concentration of plant growth regulators like indoleacetic acid, gibberellic acid, cytokinins and ethylene (ii)

Table 4.3: Plant growth promoting rhizobacteria demonstrating mitigation of salt stress.

PGPR	Mechanism to mitigate salt stress	Crops	References
Bacillus amyloliquefaciens, Bacillus insolitus, Microbacterium sp., *P. syringae*	Inhibited influx of sodium due to formation of rhizosheath by exopolysaccharide	Wheat	Ashraf *et al.* (2004)
Azospirillum lipoferum	Indoleactetic acid (IAA) and GA_3 production	Maize	Lucangeli and Bottini (1996)
Pseudomonas fluorescens	Production of ACC deaminase	Groundnut	Saravanakumar and Samiyappan (2007)
Bradyrhizobium japonicum	Indole 3-acetic acid production	Chick pea	Bano *et al.*, (2010)
Pseudomonas putida	ACC deaminase activity	Canola	Cheng *et al.*, (2007)
Bacillus pumilus	Phosphate solubilization	Tomato,bhindi, gladiolus and banana	Damodaran *et al.,* (2013a, 2013b, 2013c, 2014)
Pseudomonas syringae, Pseudomonas fluorescens, Rhizobium phaseoli	ACC deaminase activity	Moongbean	Ahmad *et al.* (2012)
Bacillus subtilis, Arthrobacter	Proline accumulation	Wheat	Upadhyay *et al.,* 2011
Bacillus thuringiensis	Production of IAA and Siderophore	Tomato, bhindi, Gladiolus, banana	Damodaran *et al.*, (2013b, 2013c, 2014)
Azotobacter chroococcum	Increase in polyphenol as well as K^+/Na^+ ratio	Maize	Rojas-Tapias *et al.*(2012)
Pseudomonas aurantiaca, Pseudomonas extremorientalis	Auxin production	Wheat	Egamberdieva and Kucharova (2009)
Pseudomonas chlororaphis	IAA production	Cucumber, Tomato	Egamberdieva (2012)
Bacillus megaterium	Phosphorus solubilization	Tomato	Chookietwattana and Maneewan (2012)
Bacillus subtilis	Increase K^+/Na^+ ratio	White clover	Han *et al.*, (2014)
Azotobacter vinelandii	Nitrogen fixation	Wheat	Aly *et al.*, (2012)
Pseudomonas aeruginosa	Exopolysaccharide	Sunflower	Tewari and Arora (2014)
Bacillus subtilis	Phosphate solubilization and auxin production	Rice	Kannan *et al.*, 2015
Pseudomonas sp.	ACC deaminase activity	Tomato	Ali *et al.* (2014)
A. lipoferum, A. brasilense,	Nitrogen fixation	Kallar grass	Malik et al. (1997)

(Contd.)

Azoarcus, Pseudomonas sp.		(*Leptochloa fusca*) and rice	
*Serratia marcescens*CDP 13	ACC deaminase activity, phosphate solubilization, productionof siderophore, indole acetic acid production, nitrogen fixation	Wheat (*Triticum aestivum* L.)	Singh and Jha, 2016
B. megaterium	Regulation of plasma membrane aquaporins	Maize (*Zea maize* L.)	Marulanda et al. (2010)
Azospirillum sp.	Production of phytohormones and osmoprotectants	Durum wheat (*Triticum durum*)	Nabti et al. (2010)
P. putida	Production of phytohormones	Cotton (*Gossypium hirsutum*)	Yao et al. (2010)
Pseudomonas pseudoalcaligenes and *Bacillus pumilus*	Reduce lipid peroxidation andsuperoxide dismutase activity	Rice	Jha and Subramanian, 2014
Acinetobacter sp. and *Pseudomonas* sp..	Production of ACC deaminase and IAA	Barley and oats	Chang et al. (2014)
Streptomyces sp. strain PGPA39	ACC deaminase activity and IAA production7 and phosphate solubilization	'Micro tom' tomato	Palaniyandi et al. (2014)
Enterobacter sp. UPMR18	Increased antioxidant enzyme activities (SOD, APX, and CAT) and upregulation of ROS pathway genes (CAT, APX, GR, and DHAR)	Okra	Habib et al., 2016
Dietzia natronolimnaea STR1	Modulation of an ABA "signaling cascade,	Wheat	Bharti et al., 2016
Burkholderia cepacia SE4, *Promicromonospora* sp. SE188 and *Acinetobacter calcoaceticus*	Increased water potential and decreased electrolyte leakage	Cucumber (*Cucumis sativus*)	Kang et al., 2014
Bacillus pumilus, Arthrobacter sp., *B. aquimaris, B. cereus,* and *Pseudomonas mendocina*	Solubilization and mineralization of nutrients, siderophore-production , phytohormones production	Wheat	Upadhyay and Singh, 2015
P. fluorescens, P. aeruginosa, P. stutzeri	Solubilize phosphate, and produce phytohormones, siderophores and 1-aminocyclopropane-1-carboxylic acid (ACC) deaminase enzyme.	Tomato	Tank and Saraf (2010)

(Contd.)

Arthrobacter protophormiae and *Dietzia natronolimnaea*	Enhancing IAA content, reducing ABA/ACC content, modulating expression of a regulatory component(CTR1) of ethylene signaling pathway and DREB2 transcription factor.	Wheat	Barnawal et al., 2017
Planococcus rifietoensis	Phytohormones production and nutrient solubilization	Wheat (*Triticum aestivum*)	Rajput et al., 2013
Halomonas sp.	Nitrogen fixation ability and ACC-deaminase activity	Salicornia plants	Mapelli et al., 2013
Pseudomonas sp.	ACC deaminase activity	Canola	Akhgar et al., 2014
Alcaligens sp., *Bacillus* sp. and *Ochrobactrum* sp.	ACC deaminase activity	Rice	Bal et al., 2013
Brevibacterium epidermidis, Brevibacterium iodinum, Planococcus rifietoensis, Exiguobacterium acetylicum, Arthrobacter nicotianae, Zhihengliuella alba, Micrococcus yunnanensis, Oceanimonas smirnovii, Bacillus licheniformis, Bacillus stratosphericus, Bacillus aryabhattai, Bacillus stratosphericus, Corynebacterium variabile, Bacillus aryabhattai	ACC deaminase activity	Canola	Siddikee et al., 2010
Serratia sp. and *Rhizobium* sp.	Enhanced mineral uptake and production of antioxidants	Lettuce (*Lactuca sativa* L.)	Han and Lee, 2005b
Pantoea agglomerans	Exopolysaccharide production	Wheat	Amellal et al., 1999
Bacillus subtilis, Bacillus atrophaeus, Bacillus spharicus	Indole-3-acetic acid and cytokinins, N_2-fixation ability, phosphate-solubilizing capacity, and antimicrobial substance production	Strawberry (*Fragaria ananassa*)	Karlidag et al., 2013
Staphylococcus haemolyticus and *Bacillus subtilis*	Osmolytes proline, glycine betaine and choline accumulation	Chickpea	Qurashi and Sabri, 2013

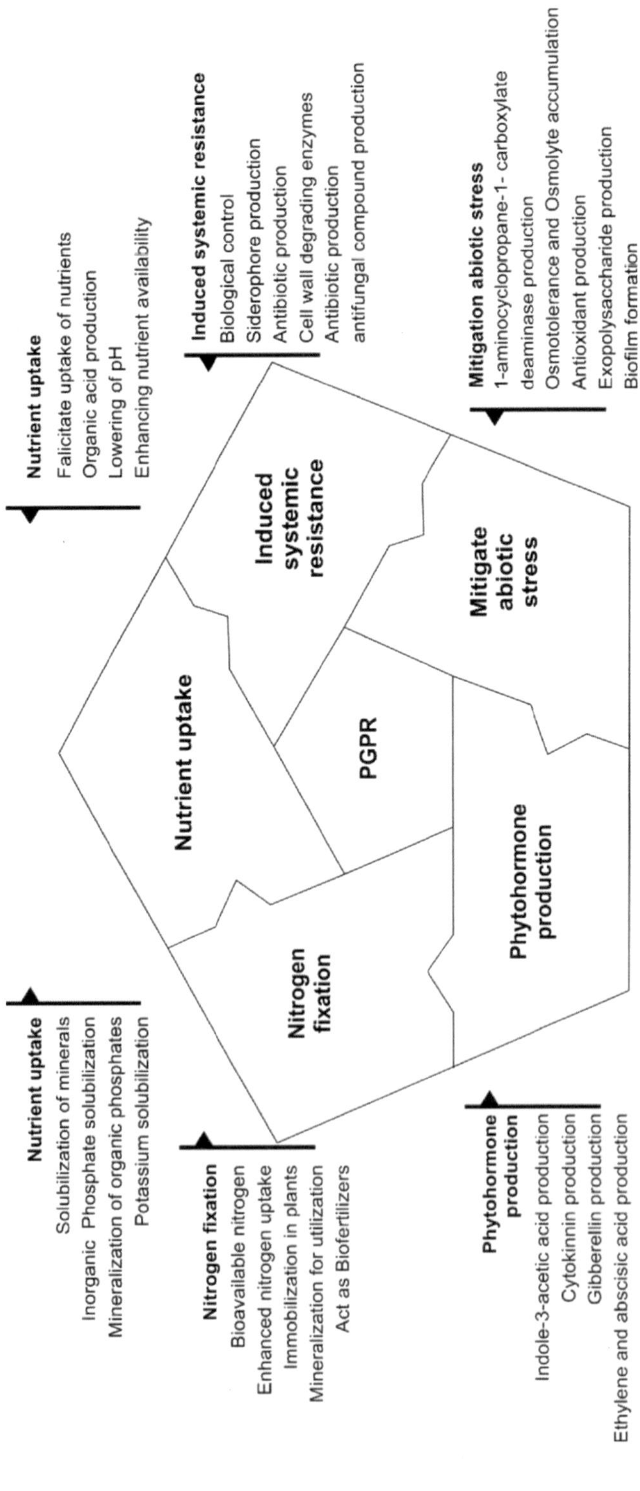

Fig. 4.1: Benefits of Plant Growth Promoting Rhizobacteria (PGPR)

asymbiotic N_2 fixation, (iii) antagonism against phytopathogenic microorganisms by production of siderophores and cyanide, (iv) solubilization of mineral phosphates and other nutrients and (v) the ability to synthesize anti-fungal metabolites. These metabolites include antibiotics, fungal cell wall-lysing enzymes, which suppress the growth of fungal pathogens. The most popular bacteria studied and exploited as biocontrol agent include the species of *Pseudomonas* and *Bacillus*. Some PGPR may promote plant growth indirectly by affecting symbiotic N_2 fixation, nodulation or nodule occupancy. However, role of cyanide producing bacteria is contradictory as it may be associated with deleterious as well as beneficial rhizobacteria (Beneduzi *et al*., 2012; Gupta *et al*., 2015).

4.6 Mechanism of mitigating salt stress

PGPRs regulate osmotic balance and ion homeostasis through modulation of phytohormone status, gene expression, protein function, and metabolite synthesis in plants (Paul and Lade, 2014). As a result, improved antioxidant activity, osmolyte accumulation, proton transport machinery, salt compartmentalization, and nutrient status etc. help to reduce osmotic stress and ion toxicity. Furthermore, in addition to indole-3- acetic acid and 1-aminocyclopropane-1-carboxylic acid deaminase biosynthesis, other extracellular secretions of the rhizobacteria function as signaling molecules and elicit stress responsive pathways (**Figure 4.2**). PGPRs mitigate salt stress in crops through following mechanims:

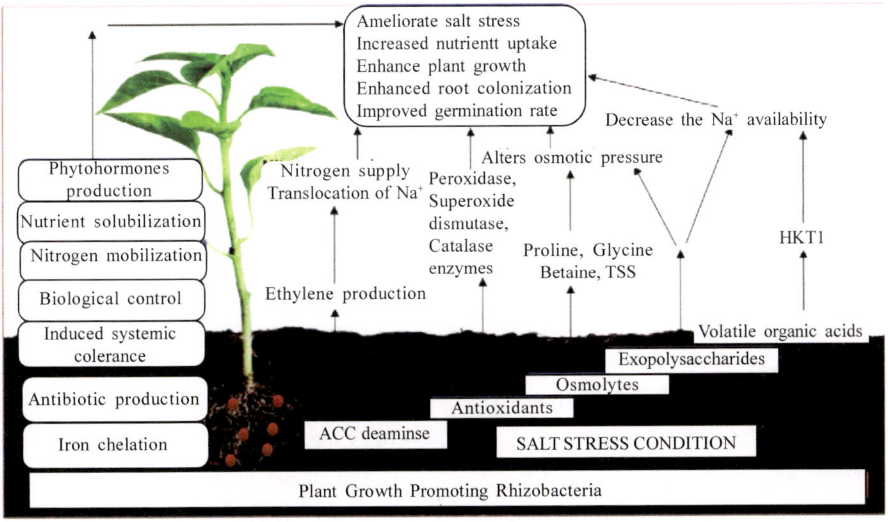

Fig. 4.2: Mechanisms of Plant Growth Promoting Rhizobacteria (PGPR) to ameliorate salt stress

4.6.1 Production of ACC deaminase

Ethylene is a plant growth regulator which is involved in various physiological responses. However, it is regarded as a stress hormone because it is synthesized at a rapid rate under stress. Ethylene decreases seed germination and root development and eventually hinders plant growth. Under salinity stress, levels of 1-aminocyclopropane-1-carboxylate (ACC) increase in the plants which leading to increase in concentration of ethylene which damages the plant cells (Botella *et al.*, 2000). Many rhizobacteria have potential to produce ACC deaminase enzymes which can convert ACC into α-ketobutyrate and ammonium, thereby lowering levels of ethylene and protect the plants from salinity stress and enhance the plant growth (Akhgar *et al.*, 2014).

In the presence of 1-aminocyclopropane-1-carboxylate deaminase producing bacteria, 1-aminocyclopropane-1-carboxylate produced in plant under salinity stress is sequestered and degraded by bacterial cells to supply nitrogen and energy facilitating plant growth under the stress.The effectiveness of rhizobacteria which possess ACC deaminase for enhancing salt tolerance and subsequently the growth of rice, tomato and various other crops under salt-stress conditions have been well proved (Bal *et al.*, 2013; Mayak *et al.*, 2004; Paul and Lade, 2014). The halotolerant strains of *Bacillus*, *Halomonas*, *Exiguobacterium*, *Oceanimonas*, *Corynebacterium*, *Brevibacterium*, *Planococcus*, *Zhihengliuella*, *Arthrobacter* and *Micrococcus* have been observed to possess ACC deaminase (Siddikee *et al.*, 2010; Paul and Lade, 2014).

Application of PGPR with ACC deaminase as a soil conditioner resulted in enhanced seed germination, chlorophyll content, and growth of okra plants under salinity stress by maintaining low stress ethylene levels and increasing the ROS-scavenging enzymes. In this study, two PGPR *B. megaterium* UPMR2 and *Enterobacter* sp. UPMR18 that possessed characteristics of ACC deaminase activity demonstrated their effectiveness in inducing salt tolerance and consequent improvement in the growth of okra plants under salt stress (Habib *et al.*, 2016).

In yet another study, *Pseudomonas fluorescens* strain TDK1 possessing ACC deaminase activity enhanced the salinity tolerance in groundnut plants, which in turn resulted in increased yield when compared with the groundnuts treated with *Pseudomonas* isolates not having ACC deaminase activity. The promising role of ACC deaminase from *P. fluorescens* strain TDK1 in alleviating saline stress has been concluded in groundnut plants (Saravanakumar and Samiyappan, 2007). Several studies have suggested possible productive use of ACC deaminase from microbial strains against various biotic and abiotic stresses in plants wherever ACC accumulated as precursor for ethylene biosynthesis. Cheng *et al*. (2007) have also confirmed that ACC deaminase bacteria conferred salt tolerance in plants by lowering the synthesis of salt-induced stress ethylene and promoted the growth of canola in saline environment. Nadeem *et al*., (2006) have also observed that inoculation of maize with ACC deaminase producing PGPR had effect on growth under salt stress.

4.6.2 Facilitation of nutrient uptake

Under salt stress, crop performance is mainly adversely affected due to induced nutritional disorders. Nitrogen, in one form or another, accounts for about 80 % of the total mineral nutrients absorbed by plants, and inadequate nitrogen uptake often proves a growth-limiting nutritional stress (He and Dijkstra, 2014). Studies indicate that salinity reduces N uptake/accumulation (Fageria *et al*., 2011). In addition, it is also known that salt stress causes reduction in P accumulation in plants, which develop P-deficiency symptoms. Salinity reduces phosphate uptake/accumulation in crops grown in soils primarily by reducing phosphate availability. Besides, maintenance of adequate levels of K^+ is essential for plant survival in saline habitats, and sodium-induced K^+ deficiency has been implicated in various crops (Qadir *et al*., 2009).

PGPRs have been proved to be vital for circulation of plant nutrients in many ways, thereby reducing the need for chemical fertilizers. Apart from fixing N_2, many strains of PGPR can affect plant growth directly by solubilising inorganic phosphate, improving nutrient uptake and mineralizing organic phosphate. Plants generally take phosphorus from soil solution in the form of $H_2PO_4^-$ and HOP42-. Large deposits of phosphorus are present in soils, but majority of this is unavailable to plants. After application of phosphate fertilizer to soil, it's readily unavailable to plants because of its fixation and precipitation with some cations like iron and aluminium in acid soil and calcium and magnesium in saline soils, resulting in low use efficiency of phosphatic fertilizers (Khan *et al*., 2010).

PGPR, which mobilize and/or solubilize nutrients, increase water use efficiency and stimulate root growth, could produce these positive effects via phytohormones and enzymatic lowering of plant ethylene concentrations. In addition, several physiological, enzymatic and biochemical changes in plants, after inoculation with PGPR, have been suggested to help alleviate salt or drought stress (Tank and Saraf, 2010; Dimkpa *et al*., 2009; Barnawal *et al*., 2017). Inoculation with isolate of *B. pumilus* resulted in significantly higher accumulation of N at both 60 and 90 days. The P content of wheat plants was significantly increased after inoculation with *B. pumilus*, *Arthrobacter* sp., *B. cereus*, *B. aquimaris* and *B. subtilis* as compared to controls, because these strains can solubilize phosphate (Upadhyay *et al*., 2009). Egamberdiyeva and Hoflich (2003) reported that PGPR increased uptake of N, P and K in wheat under drought and salinity stress, concentrations of N, P and K in lettuce increased 5, 70 and 50%, respectively when inoculated with *Bacillus* sp. (Vivas *et al*., 2003). Treatment with PGPR strains can alleviate effects of potentially toxic ions and thus improve plant growth (Yue *et al*., 2007).

Furthermore, it has been suggested that phosphate nutrition is a limiting factor for the growth of salt-stressed plants. In a study, two of the PGPR strains (*M. oleivorans* and *R. massiliae*) were able to solubilize phosphate and, thus improved phosphate nutrition, which partially accounted for the ability of the PGPR-treated plants to overcome salinity stress. Indeed, many previous studies have reported that phosphate-solubilizing organisms are associated with increased plant phosphate content (Hahm *et al*., 2017). Ammonia production is as important as nitrogen fixation, because ammonia is taken up by plants as nitrogen source for the nitrogen containing biomolecules. The halotolerant and halophilic bacteria are found to show ammonia production, phosphate solubilization and nitrogen fixation activities *in vitro* (Paul and Lade, 2014). Under salt stress, these bacteria significantly increased the root and shoot length and total fresh weight of the plants. The growth rates of the plants inoculated with bacterial strains ranged from 62.2% to 78.1% (Mapelli *et al*., 2013). A recent study from Southern Tunisia reported that 20 *Halomonas* strains having tolerance to a wide set of abiotic stresses and possessing PGP activities such as IAA production, phosphate solubilization and potential of nitrogen fixation (Mapelli *et al*., 2013). Rajput *et al*. (2013) showed that the halophilic bacteria *Planococcus rifietoensis*, has IAA producing activity, phosphate solubilizing activity and ACC deaminase activity that enhance the growth and yield of *T. aestivum* under salinity stress. In the study carried out by Siddike *et al*., (2010) halotolerant strains of *Oceanobacillus* sp., *Halomonas* sp., *Exiguobacterium* sp., *Zhihengliuella* sp. and several *Bacillus* sp. were found to have IAA production, nitrogen fixation, phosphate solubilization and ammonia production abilities.

Iron is one of the essential micronutrient for proper growth and development of plants. After oxygen, silica and aluminium, iron is the fourth one which found abundantly in earth crust(5%). This element can exist in soil in ferrous (Fe^{2+}) and ferric (Fe^{3+}) form, but plants assimilate iron as Fe^{2+}, while Fe^{3+} is directly unavailable for plants, because it often forms insoluble oxides or hydroxides which limit bioavailability (Zuo and Zhang 2011). Plants acquire iron by acidification of the rhizosphere, which is helpful for reducing Fe^{3+} into Fe^{2+}.In addition, plants release low molecular weight phytosiderophores for solubilization and binding of iron, which is transported into root cells through membrane proteins (Altomare and Tringovska, 2011; Guerinot, 2010). However these mechanisms are not enough, when plants grow in alkaline conditions (Radzki *et al*., 2013).

Therefore, PGPR is a good alternative for improving iron nutrition in these soils by releasing siderophore compounds, which have tendency of chelating iron with high affinity and in a reversible manner (Pii *et al.*, 2014; Saha *et al.*, 2016). These siderophores 'also create competition with the phytopathogens, which makes the entry of phytopathogens prohibited in plants. Previous studies have also shown that PGPRs release siderophores in rhizosphere and siderophore-producing bacteria influence the plant uptake of various metals, including Fe, Zn, and Cu. PGPR strains *M. oleivorans*, *B. iodinum*, and *R. massiliae* was also observed to produce siderophore (Hahm *et al.*, 2017), which suggests that PGPR and other microorganisms can also affect plant stress tolerance by influencing the bioavailability of metal ions required by their host plants (Dimkpa *et al.* 2009). The study carried out by Hahm *et al.*, (2017) also revealed that PGPR inoculation enhanced the chlorophyll content and relative water content in leaves of pepper. Similar results were reported by Mayak *et al* (2004), who observed that the fresh and dry weight of tomato got improved by inoculation with *Achromobacter piechaudii*. Another study showed that inoculation with PGPR, including *Bacillus subtilis*, *Bacillus atrophaeus*, *Bacillus sphaericus*, *Staphylococcus kloosii*, and *Kocuria erythromyxa*, increased both the shoot and root weights (fresh and dry) of strawberry plants growing under high-salinity conditions (Karlidag *et al.*, 2013). Han and Lee (2005b) also reported that PGPR, such as *Serratia* and *Rhizobium* species, enhance the growth, nutrient uptake, and chlorophyll content of lettuce grown under different levels of soil salinity.

4.6.3 Exopolysaccharide production

Polysaccharides have an extreme diverse structure; therefore, it plays varied role in nature and may get altered during stress conditions (Ozturk and Aslim, 2009). Exopolysaccharides(EPS) play an important role in plant protection by creating boundry around the cell (Caiola *et al.,* 1996; Arora *et al.*, 2010). To reduce the adverse effect of salinity, EPS production by salt tolerant bacteria is a significant approach to assist metabolism and can be helpful against osmotic stress (Qurashi and Sabri, 2012). EPS-producing plant growth promoting rhizobacteria can decrease the Na^+ availability for plant uptake in the rhizosphere and consequently prevent the salt stress in plants growing under saline environments (Upadhyay *et al.*, 2011). The PGPR which produces EPS play important role in rhizosheaths formation. The rhizosheath is active site for biogeochemical processes and regulate the ion and water flow across the plant roots (Amellal *et al.*, 1999; Awad *et al.*, 2012). EPS-producing PGPRs enhance the soil structure by promoting soil aggregation, which results in water retention and increased provision of nutrients to plants. EPS can also alleviate plant salt stress by binding Na^+; this binding is due to the hydroxyl, sulfhydryl, carboxyl and phosphoryl functional groups characteristic of bacterial EPS (Watanabe *et al.*, 2003; Nunkaew *et al.*, 2015). *Aeromonas hydrophila/caviae*, *Bacillus* sp., *Planococcus rifietoensis*, *Halomonas variabilis*, *Burkholderia*, *Enterobacter*, *Microbacterium*, and *Paenibacillus* are some of the PGPRs that produce EPS and facilitate biofilm formation (Upadhyay *et al.*, 2011; Qurashi and Sabri, 2012; Etesami and Beattie, 2018, Nayak *et al.*, 2020).

4.6.4 Plant growth hormones production

Production of phytohormone like auxin and cytokinins is reduced under salt stress in plants (Etesami and Beattie, 2018). PGPR have ability to produce indole acetic acid, gibberellins and cytokinins, which are responsible for increasing length of root, surface area of root, number of root tips, and all these factors contribute to enhance water and nutrient uptake, ultimately improvement in plant growth under salt stress environment (Etesami and Beattie, 2017; Etesami, 2018). Increase in indole acetic acid content in wheat seedlings, tomato and cucumber plants, when inoculated with *P. chlororaphis*, enhanced the water conductivity under salinity conditions (Egamberdieva and Kucharova, 2009; Egamberdieva, 2011, 2012). Another important phytohormone is cytokinin (CK) which plays important role in cell devision, cell enlargement, and shoot growth (O'Brien and Benková, 2013). It has been observed that 90% of rhizospheric microorganism have capacity to release CK, when cultured *in vitro* (Pallai *et al.*, 2012; Kapoor and Kaur, 2016). The *Platycladus orientalis* plants inoculated with a CK-producing PGPR strain *B. subtilis* had increased CK levels in the shoots and were more resistant to drought (Liu *et al.*, 2013).

Moreover, both *M. oleivorans* and *R. massiliae* are also able to produce IAA. It has been suggested that the depressive effect of salinity on plant growth is related to reductions in endogenous levels of hormones. Therefore, the application of additional natural phytohormones,such as bacterial auxins, could positively affect plant development under high-salinity conditions (Hahm *et al.*, 2017; Ahmad *et al.*, 2012; Zaidi *et al.*, 2004). Orhan (2016) reported that *Z. halotolerans*, *S. succinus*, *B. gibsonii*, *O. oncorhynchi*, *Halomonas* sp. and *Thalassobacillus* sp. are able to produce IAA, significantly improve growth in *Triticum aestivum* under salt stress (200mM NaCl). A study by Desale *et al.* (2014) revealed that *Halomonas* sp. and *Halobacillus* sp. have IAA production potential under salinity stress. Halophilic bacterial strain *Promicromonospora* sp., isolated from agricultural field soil in Republic of Korea showed phosphate solubilizing and gibberellin-producing ability (Kang *et al.*, 2012).

4.6.5 Production of antioxidant enzymes

Production of reactive oxygen species in plants under salt stress, damage the lipids, protein and nucleic acids. In saline conditions, reactive oxygen species formation is favoured because of inhibition of photosynthetic electron chain due to the limiting of photosynthesis (Tripathy and Oelmüller, 2012). Antioxidants play important role in reducing the oxidative damage (Arora and Chandra, 2011). Under salinity conditions, antioxidative enzymes activities (guaicol peroxidase, catalase and superoxide dismutase) increase in plants, these enzymes significantly increase the tolerance in plants against salt stress (Mittova *et al.*, 2002, 2003). Application of PGPRs increase the plant defence-associated enzymes like, peroxidase, superoxide dismutase,catalase, phenylalanine ammonia-lyase, lipoxygenase, polyphenol oxidase and phenolics in plants (Nautiyal *et al.*, 2008; Chakraborty *et al.*, 2013). Antioxidative enzymes, which are induced by PGPR, are believed to increase tolerance in plants against salinity also by eliminating hydrogen peroxide from roots under salinity-stress (Kim *et al.*, 2005). But, few data are present regarding mechanisms involved in plant antioxidative protection by PGPR (Gouda *et al.*, 2018).

Salt stress induces the formation of reactive oxygen species (ROS), which can cause severe oxidative damage to plants. Antioxidant systems play an important role in protecting plants from oxidative stress and involve a variety of antioxidant enzymes, including superoxidedismutase (SOD), dehydroascorbatereductase (DHAR), glutathione reductase (GR), APX, CAT, and GPX. In plant systems, enzymes and redox metabolites act in synergy to detoxify ROS. For example, both APX and GPX catalyze the conversion of H_2O_2 to water, and CAT converts H_2O_2 to oxygen and water. In a study, the activities of antioxidant enzymes (APX, CAT, and GPX) in the leaf extracts of PGPR-inoculated pepper plants were significantly more than those observed for the non inoculated control plants, regardless of growing conditions (i.e., normal or saline). Gururani *et al.* (2012) also reported that the activities of ROS scavenging enzymes, such as APX, CAT, DHAR, GR, and SOD, were enhanced in PGPR-inoculated potato plants exposed to various stressors (salt, drought, and heavy metals). Moreover, increased SOD, APX, and CAT activities were also observed in salt-stressed okra plants treated with the PGPR *Enterobacter* sp. UPMR18 (Habib *et al.*, 2016).

4.6.6 Accumulation of osmolytes

For preventing the adverse effect of salt stress on growth and yield of plant, functioning of photosynthetic structure and water homeostasis maintenance are very important. Enhancement in concentration of compatible organic solutes (proline and glycine betaine) in plants under salinity stress (Serraj and Sinclair, 2002) has significant contribution in protecting the plants from salt stress.

Proline accumulation is an adaptive response by plants to both general stress and salinity, since it mediates osmotic adjustment at the cellular level, thereby protecting intracellular macromolecules from dehydration, and also because it serves as a hydroxyl radical scavenger. Increased total soluble sugar (TSS) content is another important defence strategy for plants facing salinity stress, as these compounds helps in maintaining the high leaf water potential in plants. Proline and TSS also reduces the salt stress in plants by altering osmotic pressure, and stabilizing complex II electron transport and enzymes like ribulosebisphosphate carboxylase/oxygenase (Makela *et al.*, 2000). PGPRs have capacity to augment plant tolerance against stress by enhancing osmolyte content in plants. In a study Hahm *et al.* (2017) found that the contents of both proline and TSS were enhanced in the PGPR-inoculated pepper plants under saline conditions. Therefore, it is likely that the PGPR strains promoted plant growth under salinity stress by enhancing metabolic defence strategies. Increase in proline accumulation in soyabean plants was found, when inoculated with PGPR strains which reduced salinity stress and improved the growth (Han and Lee, 2005a). Increase in proline content and total soluble sugar in wheat plants, which is treated with PGPR; significantly increase their osmotolerance (Upadhyay *et al.*, 2012). According to a report by Qurashi and Sabri (2013) endogenous osmolytes such as proline, glycine betaine and choline are accumulated in moderately halophilic bacterial strains (*S. haemolyticus* and *B. subtilis*) isolated from saline rhizosphere of chickpea. These osmolytes improve the growth of bacterial strains and plants by alleviating salt stress. A more recent study has revealed that inoculation of two halophilic bacteria (*V. marismortui* and *T. halophilus*) to tomato seeds improved the stem growth compared to the uninoculated control. In addition, these halophilic bacteria produce halotolerant and thermotolerant chitinases that help in decompositing chitin-based organic matters. Kavamura *et al.* (2013) have also reported that a *Virgibacillus* strain produces exopolysaccharide and has the ability to grow in a medium with reduced water availability.

4.7 Conclusion

Plant growth and crop productivity is adversely affected by salt stress on very large scale, and further the salt stressed areas are increasing across the world. It is a major concern for researchers that how to treat these soils. These salt affected soils also have negative impact on microbial activity which is a biological indicator of soil health. So, evolving eco-friendly techniques is needed for sustainable management of these soils. Application of PGPRs, is an important alternative for management of salt affected soils, and it is now widely accepted and put in practice. PGPRs that live in rhizosphere have ability to enhance plant growth and reduce stress by various machanisms such as phosphorus solubilization, IAA production, symbiotic N_2 fixation, siderophore production, and ACC deaminase activity. PGPRs alleviate salt stress in plants by improving osmolyte content in plants, increasing K^+ content and maintaining a high K^+/Na^+ ratio and scavenging ROS produced by plants with bacterial antioxidative enzymes etc. Currently, numbers of PGPR inoculants have been commercialized which can enhance crop growth or alleviate salt stress through atleast one mechanism. For future research in this direction it is very crucial to understand how these PGPR inoculants interact with plant systems in enhancing growth of plants and increasing salt stress tolerance. Rhizobacteria can play a significant role in conferring resistance to plants against salt stresses and for improving crop yield.

References

Abrol, I.P., Yadav, J.S.P., Massoud, F.I. (1988) Salt-affected soils and their management. Food and Agriculture Organization of the United Nations, Soils Bull. 39, Rome, Italy.

Ahmad, M., Zahir, Z.A., Asghar, H.N., Arshad, M. 2012. The combined application of rhizobial strains and plant growth promoting rhizobacteria improves growth and productivity of mung bean (*Vigna radiata* L.) under salt-stressed conditions. Ann. Microbiol. 62: 1321-1330.

Akhgar, A. R., Arzanlou, M., Bakker, P. A. H. M. and Hamidpour, M. (2014). Characterization of 1-Aminocyclopropane-1-Carboxylate (ACC) Deaminase-Containing Pseudomonas spp. in the Rhizosphere of Salt-Stressed Canola. Pedosphere 24(4): 461–468.

Ali, S., Charles, T.C., Glick, B.R., (2014). Amelioration of high salinity stress damage by plant growth promoting bacterial endophytes that contain ACC deaminase. Plant Physiol. Biochem. 80, 160–167.

Altomare, C., Tringovska, I. (2011) Beneficial soil microorganisms, an ecological alternative for soil fertility management. In: Lichtfouse E (ed) Genetics, biofuels and local farming systems. Sustainable agriculture reviews, vol 7. Springer, Netherlands, pp 161–214.

Aly, M. M., Tork, S., AlGarni, S. and Nawar L (2012) Production of lipase from genetically improved. Afr J Microbiol Res 6(6):1125-1137. DOI: 10.5897/AJMR11.1123

Amellal, N., F. Bartoli, G. Villemin, A. Talouizte and T. Heulin, (1999). Effect of inoculation of EPS-producing *Pantoea agglomerans* on wheat rhizosphere soil aggregation. Plant and Soil, 211: 93-101.

Arora M, Kaushik A., Rani N and Kaushik C.P. (2010). Effect of cyanobacterial exopolysaccharides on salt stress alleviation and seed germination, Journal of Environmental Biology. 31(5) 701-704

Arora, D. S. and Chandra, P. (2011). *In vitro* antioxidant potential of some soil fungi: screening of functional compounds and their purification from *Penicillium citrinum*. Applied Biochemistry and Biotechnology165:639-651

Arora, S. (2017). Diagnostic Properties and Constraints of Salt-Affected Soils. In: Arora S., Singh A., Singh Y. (eds) Bioremediation of Salt Affected Soils: An Indian Perspective. Springer, Cham, pp 41–52.

Ashraf, M., and Harris, P.J.C. (2004). Potential biochemical indicators of salinity tolerance in plants. Plant Sci 166: 3–16.

Ashraf, M., Hasnain, S., Berge, O., Mahmood, T. (2004) Inoculating wheat seedlings with exopolysaccharide-producing bacteria restricts sodium uptake and stimulates plant growth under salt stress. Biol Fertil Soils 40(3):157–62. doi:10.1007/s00374-004-0766-y

Awad, N. M., Turky, A. S., Abdelhamid M. T., and Attia M. (2012). Ameliorate of Environmental Salt Stress on the Growth of *Zea mays* L. Plants By Exopolysaccharides Producing Bacteria. Journal of Applied Sciences Research, 8(4): 2033-2044.

Bal, H.B., Nayak, L., Das, S., Adhya, T.K. (2013). Isolation of ACC deaminase producing PGPR from rice rhizosphere and evaluating their plant growth promoting activity under salt stress. Plant Soil 366(1–2):93–105. doi:10.1007/s11104-012-1402-5

Bano, A., Batool, R., and Dazzo, F. (2010) Adaptation of chickpea to desiccation stress is enhanced by symbiotic rhizobia. Symbiosis 50:129–133.

Barnawal, D., Bharti, N., Pandey, S.S., Pandey, A., Chanotiya, C.S. and Kalra, A. (2017). Plant growth-promoting rhizobacteria enhance wheat salt and drought stress tolerance by altering endogenous phytohormone levels and *TaCTR1/TaDREB2* expression. Physiologia Plantarum 161: 502–514. 2017.

Beneduzi, A., Ambrosini, A., and Passaglia, L.M.P. (2012). Plant growth-promoting rhizobacteria (PGPR): Their potential as antagonists and biocontrol agents. Genetics and Molecular Biology.35(4 Suppl):1044-1051.

Bertrand, H., Plassard C., Pinochet, X., Toraine B., Normand P. and Cleyet-Marel J. C. (2000). Stimulation of the ionic transport system in *Brassica napus* by a plant growth-promoting rhizobacterium (*Achromobacter* sp.). Can. J. Microbiol., 46: 229-236.

Bharti, N., Pandey, S. S., Barnawal, D., Patel, V. K., and Kalra, A. (2016). Plant growth promoting rhizobacteria Dietzia natronolimnaea modulates the expression of stress responsive genes providing protection of wheat from salinity stress. Sci. Rep. 6:34768. doi: 10.1038/srep34768.

Botella, M.A., del Amor FM, Amoros A, Serrano M, Martinez V, Cerda A (2000). Polyamine, ethylene and other physico-chemical parameters in tomato (*Lycopersicon esculentum*) fruits as affected by salinity. Physiol Plant 109(4):428–34. doi:10.1034/j.1399-3054.2000.100409.x

Caiola, M.G., D. Billi and E.I. Friedmann (1996). Effect of desiccation on envelopes of the cyanobacterium *Chroococcidiopsis* sp. (*Chroococcales*). Eur. J. Phycol., 31, 97-105.

Carmen, B., and Roberto, D. (2011). Soil bacteria support and protect plants against abiotic stresses. In: Shan er, A. (Ed.), Abiotic Stress in Plantsmechanisms and Adaptations. Pub. InTech, pp. 143–170.

Chakraborty, N., Ghosh, R., Ghosh, S, Narula, K., Tayal, R, Datta, A.,Chakraborty, S. (2013). Reduction of oxalate levels in tomato fruit and consequentmetabolic remodeling following overexpression of afungal oxalate decarboxylase1[W]. Plant Physiol 162(1):364–78. doi:10.1104/pp.112.209197

Chandra, P. and Chandra, A. (2016). Elucidation and phylogenetic analysis of 16S rRNA in plant growth promoting Streptomyces sp. isolated from wheat-sugarcane cropping system. Journal in Wheat Research

Chandra, P. and Chandra, A. (2017). Plant growth promoting attributes of rhizospheric bacteria and their potential to improve growth of the sugarcane plants. pp: 41-44. In: Arora Sanjay, Singh S.R., Pathak. J., Yadav, P.P.S., Lal Bahadur, and Varshney, P. (Eds.) Souvenir: National Seminar on "Healthy Soil for Healthy Life" December 5, 2017 ISSS Lucknow Chapter, Lucknow.

Chang, P., Gerhardt, K.E., Huang, Xiao-Dong, Yu, Xiao-Ming, Glick, B.R., Gerwing, P.D., Greenberg, B.M. (2014). Plant growth promoting bacteria facilitate the growth of barley and oats in salt impacted soil: implications for phytoremediation of saline soils. Int. J. Phytorem. 16 (11), 1133–1147.

Chaudhari, S.K., Chinchmalatpure, A.R. and Sharma, D.K. (2013). Climate Change Impact on Salt-affected Soils and Their Crop Productivity. CSSRI Karnal/Technical Manual/2013/4. Central Soil Salinity Research Institute, Karnal 132001, Haryana, India

Cheng, Z., Park, E., and Glick, B.R. (2007). 1-Aminocyclopropane-1-carboxylate (ACC) deaminase from *Pseudomonas putida* UW4 facilitates the growth of canola in the presence of salt. Can J Microbiol 53(7): 912-918. https://doi.org/10.1139/W07-050

Chookietwattana, K., Maneewan, K. (2012) Screening of efficient halotolerant phosphate solubilising bacterium and its effect on promoting plant growth under saline conditions. World Appl Sci J 16(8):1110–1117.

Damodaran, T., Mishra, V.K., Sharma, D.K., Jha, S.K., Verma, C.L., Rai, R.B., Kannan, R., Nayak, A.K., and Dhama, K. (2013a). Management of sub soil sodicity for sustainable banana production in sodic soil – an approach. Int J Curr Res 5:1930–1934.

Damodaran, T., Rai, R.B., Sharma, D.K., Mishra, V.K., and Jha, S.K. (2013b). Rhizosphere engineering: An Approach for sustainable vegetable production in sodic soils. National symposium on abiotic and biotic stress management in vegetable crops, North America. http://conference.isvs.org.in/Index.php/conf_Doc/ncab/paper/view/382

Damodaran, T., Vijaya, S., Rai, R.B., Sharma, D.K., Mishra, V.K., Jha, S.K., and Kannan, R. (2013c). Isolation of rhizospheric bacteria by natural selection and screening for PGPR and salt tolerance traits. Afr J Microbiol 7(44):5082–5089.

Damodaran, T., Rai, R.B., Kannan, R., Pandey, B.K., Sharma, D.K., Misra,V.K.., Vijayalaxmi, S., and Jha SK (2014). Rhizosphere and endophytic bacteria for induction of salt tolerance in gladiolus grown in sodic soils. J Plant Interact 9(1):577–584.

Desale, P., Patel, B., Singh, S., Malhotra, A., Nawani, N. (2014). Plant growth promoting properties of *Halobacillus* sp. and *Halomonas* sp. in presence of salinity and heavy metals. J Basic Microbiol.54:781–791.37.

Dimkpa, C., Weinand, T., and Asch, F. (2009). Plant-rhizobacteria interactions alleviate abiotic stress conditions. Plant Cell Environ 32: 1682-1694.

Egamberdieva, D. (2011) Survival of*pseudomonas extremorientalis*TSAU20 and *P. Chlororaphis* TSAU13 in the rhizosphere of commonbean (*Phaseolus vulgaris*) under saline conditions. Plant SoilEnviron 57(3):122–7.

Egamberdieva, D. (2012) *Pseudomonas chlororaphis*: a salt-tolerant bacterialinoculant for plant growth stimulation under saline soil conditions.Acta Physiol Plant 34(2):751–56. doi:10.1007/s11738-011-0875-9.

Egamberdieva, D., Kucharova, Z. (2009) Selection for root colonisingbacteria stimulating wheat growth in saline soils. Biol Fertil Soils45(6):563–71. doi:10.1007/s00374-009-0366-y.

Egamberdiyeva D., Hoflich G. (2003) Influence of growth-promoting bacteria on the growth of wheat in different soils and temperatures. Soil Biology and Biochemistry, 35, 973–978.

El-Fouly, M.M., Zeinab, M.M., Zeinab, A.S. *et al* (2001) Micronutrient sprays as a tool to increase tolerance of faba bean and wheat plants to salinity. In: HorstWJ (ed) Plant nutrition, 92. Springer, Netherlands, pp 422–423. doi:10.1007/0-306-47624-X_204.

Etesami, H. and Beattie GA (2018) Mining Halophytes for Plant Growth-Promoting Halotolerant Bacteria to Enhance the Salinity Tolerance of Non-halophytic Crops. Front. Microbiol. 9:148. doi: 10.3389/fmicb.2018.00148.

Etesami, H. (2018). Can interaction between silicon and plant growth promoting rhizobacteria benefit in alleviating abiotic and biotic stresses in crop plants? Agric. Ecosyst. Environ. 253, 98–112. doi: 10.1016/j.agee.2017.11.007

Etesami, H., and Beattie, G. A. (2017). "Plant-microbe interactions in adaptation of agricultural crops to abiotic stress conditions," in Probiotics and Plant Health, eds V. Kumar, M. Kumar, S. Sharma, and R. Prasad (Singapore: Springer), 163–200.

Fageria, N. K , Gheyi, H.R., Moreira, A. (2011) Nutrient bioavailability in salt affected soils, Journal of Plant Nutrition, 34:7, 945-962, DOI: 10.1080/01904167.2011.555578

FAO (1997). Production Yearbook, Food and Agricultural Organization, Rome, v. 48.

Glick B. R. (1995). The enhancement of plant growth by free living bacteria. Can. J. Microbiol., 41: 109–114.

Gouda, S., Kerry, R.G., Das, G., Paramithiotis, S., Shine H.S., Patra, J. K. (2018). Revitalization of plant growth promoting rhizobacteria for sustainable development in agriculture. Microbiological Research 206: 131–14.

Guerinot, M. (2010). Iron. In: Hell R, Mendel R-R (eds) Cell biology of metals and nutrients. Plant cell monographs, vol 17. Springer, Berlin, pp 75–94.

Gupta, G., Parihar, S.S., Ahirwar, N.K., Snehi, S.K., Singh, V. (2015). Plant Growth Promoting Rhizobacteria (PGPR): Current and Future Prospects for Development of Sustainable Agriculture. J Microb Biochem Technol 7:096-102. doi:10.4172/1948-5948.1000188

Gururani, M.A., Upadhyaya, C.P., Baskar, V., Venkatesh, J., Nookaraju, A., Park, S.W. (2012). Plant growth-promoting rhizobacteria enhance abiotic stress tolerance in Solanum tuberosum through inducing changes in the expression of ROS-scavenging enzymes and improved photosynthetic performance. J. Plant Growth Regul. 32: 245-258.

Habib, S.H., Kausar, H., Saud, H.M. (2016). Plant growth-promoting rhizobacteria enhance salinity stress tolerance in Okra through ROS-scavenging enzymes. Biomed Res. Int. 2016:6284547.

Hahm, M.S., Son J.S., Hwang, Y.J., Kwon, D.K., and Ghim, S.Y. (2017). Alleviation of Salt Stress in Pepper (*Capsicum annum* L.) Plants by Plant Growth-Promoting Rhizobacteria. J. Microbiol. Biotechnol. , 27(10), 1790–1797.

Han, H.S., Lee, K.D. (2005a). Physiological responses of soybean inoculation of *Bradyrhizobium japonicum* PGPR in saline soil conditions. Res J Agri Biol Sci 1(3):216–21.

Han, H.S., Lee, K.D. (2005b). Plant growth-promoting rhizobacteria: effect on antioxidant status, photosynthesis, mineral uptake and growth of lettuce under soil salinity. Res. J. Agric. Biol. Sci. 1: 210-215.

Han, Q.Q., Lu, X.P., Bai, J.P., Qiao, Y., Pare, P.W., Wang, S.M., Zhang, J.L., Wu, Y.N., *et al.*, (2014). Beneficial soil bacterium *Bacillus subtilis* (GB03) augments salt tolerance of white clover. Front. Plant Sci. 5, 525.

He M. and Dijkstra F. A. (2014). Drought effect on plant nitrogen and phosphorus: a metaanalysis. New Phytologist 204: 924–931 doi: 10.1111/nph.12952

Jha, Y., Subramanian, R.B., (2014). PGPR regulate caspase-like activity, programmed cell death, and antioxidant enzyme activity in paddy under salinity. Physiol. Mol. Biol. Plants 20 (2), 201–207.

Kang SM, Khan AL, Hamayun M, *et al.* (2012). Gibberellin-producing *Promicromonospora* sp. SE188 improves *Solanum lycopersicum* plant growth and influences endogenous plant hormones. J Microbiol. ;50:902–909.41.

Kang, S.-M., Khan, A. L., Waqas, M., You, Y.-H., Kim, J.-H., Kim, J.-G., *et al.* (2014). Plant growth-promoting rhizobacteria reduce adverse effects of salinity and osmotic stress by regulating phytohormones and antioxidants in *Cucumis sativus*. J. Plant Interact. 9, 673–682. doi: 10.1080/17429145.2014.894587

Kannan R, Damodaran T, Umamaheshwari S (2015). Sodicity tolerant polyembryonic mango root stockplants: a putative role of endophytic bacteria. Afr J Biotechnol 14(4):350–359.

Kapoor, R., and Kaur, M. (2016). Cytokinins production by fluorescent Pseudomonas isolated from rhizospheric soils of Malus and Pyrus. Afr. J. Microbiol. Res. 10, 1274–1279. doi: 10.5897/AJMR2016.8211

Karlidag, H, Yildirim, E, Turan, M, Pehluvan, M, Donmez, F. (2013). Plant growth promoting rhizobacteria mitigate deleterious effects of salt stress on strawberry plants (*Fragaria ananassa*). Hort. Sci. 48: 563-567.

Kavamura, V.N., Santos SN, Silva JL, *et al.* (2013). Screening of Brazilian cacti rhizobacetria for plant growth promotionunder drought. Microbiol Res. ;168:183–191.

Khan, M. S., A. Zaidi, M. Ahemad , M. Oves and P. A. Wani (2010) Plant growth promotion by phosphate solubilising fungi – current perspective, Archives of Agronomy and Soil Science, 56:1, 73-98, DOI: 10.1080/03650340902806469

Kim, S.Y., Lim, J.H., Park, M.R., Kim, Y.J., Park, T.I., Se, Y.W., Choi, K.G., Yun, S.J. (2005) Enhanced antioxidant enzymes are associated with reduced hydrogen peroxide in barley roots under saline stress. J Biochem Mol Biol 38(2):218–24. doi:10.5483/BMBRep.2005.38.2.218

Kloepper, J.W., Leong, J., Teintze, M. and Scroth, M. (1980). Enhanced plant growth by siderophores produced by plant growth-promoting rhizobacteria. Nature, 286: 885-886.

Kloepper J. W., Lifshitz R. and Zablotowicz R. M. (1989). Free-living bacterial inocula for enhancing crop productivity. Trends Biotechnol., 7: 39–43.

Kumar, A., Verma, J.P., (2017). Does plant—microbe interaction confer stress tolerance in plants: a review? Microbiol. Res. 207, 41–52.

Liu, F., Xing, S., Ma, H., Du, Z., and Ma, B. (2013). Cytokinin-producing, plant growth-promoting rhizobacteria that confer resistance to drought stress in *Platycladus orientalis* container seedlings. Appl. Microbiol. Biotechnol. 97, 9155–9164. doi: 10.1007/s00253-013-5193-2

Lucangeli, C., Bottini, R. (1996). Reversion of dwarfism in dwarf maize (*Zea mays* L.) and dwarf rice(*Oryza sativa* L.) mutants by endophytic *Azospirillum* spp. Biocell 20:223–228

Makela, A, Landsberg, J., Ek, A.R., Burk, T.E., Ter-Mikaelian, M., Agren, G.I. *et al.* (2000) Process-based models for forest ecosystem management: current state of the art and challenges for practical implementation. Tree Physiol 20(5–6):289–98. doi:10.1023/A:1004295714181

Malik, K.A.B., Rakhshanda, S., Mehnaz, G., Rasul, M.S., Mirza, S. (1997) Association of nitrogen-fixing plant-growth-promoting rhizobacteria (PGPR) with kallar grass and rice. Plant Soil 194 (1–2):37–44. doi:10.1023/A:1004742713538

Mandal, A.K., Sharma, R.C. and Singh, G (2009) Assessment of salt affected soils in India using GIS, Geocarto International, 24:6, 437-456, DOI:10.1080/10106040902781002.

Mapelli, F., Marasco, R., Rolli, E., *et al.* Potential for plant growth promotion of Rhizobacteria associated with Salicornia growing in Tunisian hypersaline soils. BioMed Res Int. (2013). http://dx.doi.org/10.1155/2013/248078.

Marulanda, A., Azcon, R., Chaumont, F., Ruiz-Lozano, J.M., Aroca, R. (2010) Regulation of plasma membrane aquaporins by inoculation with a *Bacillus megaterium* strain in maize (Zea mays L.) plants under unstressed and salt-stressed conditions. Planta 232(2):533–43. doi:10.1007/s00425-010-1196-8.

Mayak, S., Tirosh, T., Glick, B.R. (2004). Plant growth-promoting bacteria that confer resistance to water stress in tomatoes and peppers. Plant Sci. 166: 525-530.

Mittova, V., Tal, M., Volokita, M., Guy, M. (2002) Salt stress induces upregulationof an efficientchloroplast antioxidant system in the salttolerantwild tomato species *Lycopersicon pennellii* but not in thecultivated species. Physiol Plant 115(3):393–400. doi:10.1034/j.1399-3054.2002.1150309.x.

Mittova, V., Tal, M., Volokita, M., Guy, M. (2003) Up-regulation of the leafmitochondrial and peroxisomal antioxidative systems in response tosalt-induced oxidative stress in the wild salt-tolerant tomato species *Lycopersicon pennellii*. Plant Cell Environ 26(6):845–56. doi:10.1046/j.1365-3040.2003.01016.x

Nabti, E., Sahnoune, M., Ghoul, M., Fischer, D., Hofmann, A., Rothballer, M., Schmid, M. and Hartmann A. (2010). Restoration of growth of durum wheat (*Triticum durum* var. waha) under saline conditions due to inoculation with the rhizosphere bacterium *Azospirillum brasilense* NH and extracts of themarine alga *Ulva lactuca*. J Plant Growth Regul 29(1):6–22. doi:10.1007/s00344-009-9107-6

Nadeem, S.M., Hussain, I., Naveed, M., Ashgar, H.N., Zahir, Z.A., Arshad, M. (2006). Performance of plant growth promoting rhizobacteria containing ACC-deaminase activity for improving growth of maize under salt-stressed conditions. Pak J Agri Sci 43:114–121.

Nautiyal, C.S., Govindarajan, R., Lavania, M., Pushpangadan, P. (2008) Novel mechanism of modulating natural antioxidants in functional foods: Involvement of plant growth promoting rhizobacteria NRRL B-30488. J Agr Food Chem 56(12):4474–81. doi:10.1021/ jf073258i

Nayak, S.K., Nayak, S. and Patra, J.K., (2020). Rhizobacteria and its biofilm forsustainable agriculture: A concise review. In M.K.Yadav and B.P. Singh eds New and Future Developments in Microbial Biotechnology and Bioengineering, Elsevier publication, Netherlands, pp-165-175. Doi:10.1016/B978-0-444-64279-0.00013-X

Nehra, V. and Choudhary, M., 2015. A review on plant growth promoting rhizobacteria acting as bioinoculants and their biological approach towards the production of sustainable agriculture. Journal of Applied and Natural Science, *7*(1), pp.540-556.

Nehra, V., Saharan, B.S. and Choudhary, M., 2016. Evaluation of Brevibacillus brevis as a potential plant growth promoting rhizobacteria for cotton (*Gossypium hirsutum*) crop. SpringerPlus, 5(1), p.948.

Nunkaew, T., Kantachote, D., Nitoda, T., Kanzaki, H., and Ritchie, R. J. (2015). Characterization of exopolymeric substances from selected *Rhodopseudomonas palustris* strains and their ability to adsorb sodium ions. Carbohydr. Polym. 115,334–341. doi: 10.1016/ j.carbpol.2014.08.099

O'Brien, J. A., and Benková, E. (2013). Cytokinin cross-talking during biotic and abiotic stress responses. Front. Plant Sci. 4:451. doi: 10.3389/fpls.2013.00451.

Orhan F. (2016). Alleviation of salt stress by halotolerant and halophilic plant growth-promoting bacteria in wheat (*Triticum aestivum*). Brazilian Journal of Microbiology. 47(3):621-627. doi:10.1016/j.bjm.2016.04.001.

Ozturk, S. and B. Aslim: Modification of exopolysaccharide composition and production by three cyanobacterial isolates under salt stress. Environ. Sci. Pollut. Res., 17, 595-602 (2009).

Palaniyandi, S.A., Damodharan, K., Yang, S.H., Suh, J.W., 2014. Streptomyces sp. strain PGPA39 alleviates salt stress and promotes growth of 'Micro Tom' tomato plants. J. Appl. Microbiol. 117:766–773.

Pallai, R., Hynes, R. K., Verma, B., and Nelson, L. M. (2012). Phytohormone production and colonization of canola (*Brassica napus* L.) roots by *Pseudomonas fluorescens* 6-8 under gnotobiotic conditions. Can. J. Microbiol. 58, 170–178. doi: 10.1139/w11-120.

Paul D. and Lade H. (2014). Plant-growth-promoting rhizobacteria to improve crop growth in saline soils: a review. Agron Sustain Dev 34:737–752. DOI 10.1007/s13593-014-0233-6

Pii Y, Penn A, Terzano R, Crecchio C, Mimmo T, Cesco S. (2015). Plant-microorganism-soil interactions influence the Fe availability in the rhizosphere of cucumber plants. Plant Physiol Biochem.; 87:45-52. https://doi.org/10.1016/j.plaphy.2014.12.014

Qadir, M., Qureshi, R. H. & Ahmad, N. (2009). Nutrient availability in a calcareous saline sodic soil during vegetative bioremediation. Arid Soil Research and Rehabilitation, 11:4, 343-352, DOI: 10.1080/15324989709381487

Qurashi, A.W., Sabri, A.N. (2013). Osmolyte accumulation in moderately halophilic bacteria improves salt tolerance of chickpea. Pak J Bot.;45:1011–1016.

Qurashi, A.W., and Sabri, A.N. (2012). Bacterial exopolysaccharide and biofilm formation stimulate chickpea growth and soil aggregation under salt stress. Braz. J. Microbiol. 43, 1183–1191. doi: 10.1590/S1517-83822012000300046

Radzki, W., Gutierrez Mañero, F.J., Algar, E,, Lucas García, J.A., García-Villaraco, A., Ramos Solano, B. (2013). Bacterial siderophores efficiently provide iron to iron-starved tomato plants in hydroponics culture. Antonie Van Leeuwenhoek. Sep;104(3):321-30. doi: 10.1007/s10482-013-9954-9.

Rajput, L., Imran, A, Mubeen, F., Hafeez, F.Y. Salt-tolerant, P.G.P.R. strain cultivated in saline soil. Pak J Bot. (2013). *Planococcus rifietoensis* promotes the growth and yield of wheat (*Triticum aestivum*); Pak. J. Bot. 45:1955–1962. 35.

Rojas-Tapias, D., Moreno-Galvan, A., Pardo-Diaz, S., Obando, M., Rivera, D., Bonilla, R., (2012). Effect of inoculation with plant growth-promoting bacteria (PGPB) on amelioration of saline stress in maize (*Zea mays*). Appl. Soil Ecol. 61, 264–272.

Saha, M., Sarkar, S., Sarkar, B., Sharma, B.K., Bhattacharjee, S., and Tribedi, P. (2016) Microbial siderophores and their potential applications: a review. Environ Sci Pollut Res Int.; 23(5): 3984-99. Epub 2015 Mar 12.

Salama, Z.A., Shaaban, M.M., Abou El-Nour EA (1996). Effect of iron foliar application on increasing tolerance of maize seedlings to saline irrigation water. Egypt J Appl Sci 11(1): 169–175.

Saravanakumar, D. and Samiyappan, R. (2007). ACC deaminase from *Pseudomonas fluorescens* mediated saline resistance in groundnut (*Arachis hypogea*) plants. Journal of Applied Microbiology 102: 1283–1292.

Serraj R, and Sinclair TR (2002). Osmolyte accumulation: can it really help increase crop yield under drought conditions? Plant Cell Environ. 25(2):333–41. doi:10.1046/j.1365-3040.2002.00754.x.

Shrivastava, P., and Kumar, R. (2015). Soil salinity: a serious environmental issue and plant growth promoting bacteria as one of the tools for its alleviation. Saudi J. Biol. Sci. 22, 123–131. doi: 10.1016/j.sjbs.2014.12.001

Siddikee, M.A., Chauhan, P.S., Anandham, R., Han, G.H., Sa, T. (2010). Isolation, characterization, and use for plant growth promotion under salt stress, of ACC deaminase-producing halotolerant bacteria derived from coastal soil. J Microbiol Biotechnol. 20:1577–1584.

Siddiqui, I.A. and Shaukat, S.S. (2002). Resistance against damping-off fungus *Rhizoctonia solani* systematically induced by the plant-growth-promoting rhizobacteria *Pseudomonas aeruginosa* (1E-6S(+)) and *P. fluorescens* (CHAO). J. Phytopathol., 150: 500-506.

Singh, R.P., and Jha, P.N. (2016). The Multifarious PGPR *Serratia marcescens* CDP-13 Augments Induced Systemic Resistance and Enhanced Salinity Tolerance of Wheat (*Triticum aestivum* L.). PLoS ONE 11(6): e0155026. doi:10.1371/journal. pone.0155026

Singleton, P.W., Bohlool, B.B. (1984). Effect of salinity on nodule formation by soybean. Plant Physiol 74(1):72–6. doi:10.1104/pp. 74.1.72.

Tank, N. and Saraf, M. (2010). Salinity-resistant plant growth promoting rhizobacteria ameliorates sodium chloride stress on tomato plants, Journal of Plant Interactions, 5:1, 51-58, DOI: 10.1080/17429140903125848

Tewari, S., Arora, N.K., (2014). Multifunctional exopolysaccharides from *Pseudomonas aeruginosa* PF23 involved in plant growth stimulation, biocontrol and stress amelioration in sunflower under saline conditions. Curr. Microbiol. 69, 484–494.

Tripathy, B.C., Oelmüller, R. (2012). Reactive oxygen species generation and signaling in plants. Plant Signaling & Behavior.7(12):1621-1633. doi:10.4161/psb.22455.

Upadhyay, S. K. and Singh, D. P. (2015). Effect of salt-tolerant plant growth-promoting rhizobacteria on wheat plants and soil health in a saline environment Plant Biology 17 288–293

Upadhyay, S.K., Singh D.P., Saikia R. (2009). Genetic diversity of plant growth promoting rhizobacteria isolated from rhizospheric soil of wheat under saline conditions. Current Microbiology, 59, 489–496.

Upadhyay, S.K., Singh, J.S., Saxena, A.K., Singh, D.P. (2012). Impact of PGPR inoculation on growth and antioxidant status of wheat under saline conditions. Plant Biol 14(4):605–11. doi:10.1111/j.1438-8677.2011. 00533.x

Upadhyay, S.K., Singh, J.S., Singh DP (2011). Exopolysaccharide-producing plant growth-promoting rhizobacteria under salinity condition. Pedosphere 21(2):214–22. doi:10.1016/S1002-0160(11)60120-3

USDA, (1954). Agriculture handbook No. 60. Diagnosis and Improvement of Saline and Alkali Soils. 7-33.

Vivas, A., Marulanda, A., Ruiz-Lozano, J.M., Barea, J.M., Azcon R. (2003). Influence of a Bacillus sp. on physiological activities of two arbuscular mycorrhizal fungi and on plant responses to PEG-induced drought stress. Mycorrhiza, 13, 249–256.

Watanabe, M., Kawahara, K., Sasaki, K., and Noparatnaraporn, N. (2003). Biosorption of cadmium ions using a photosynthetic bacterium, *Rhodobacter sphaeroides* S and a marine photosynthetic bacterium, *Rhodovulum* sp. and their biosorption kinetics. J. Biosci. Bioeng. 95, 374–378. doi: 10.1016/S1389-1723(03)80070-1

Xu, Z.H., Saffigna, P.G., Farquhar, G.D., Simpson, J.A., Haines, R.J., Walker S *et al* (2000). Carbon isotope discrimination and oxygen isotope composition in clones of the F (1) hybrid between slash pine and Caribbean pine in relation to tree growth, water-use efficiency and foliar nutrient concentration. Tree Physiol 20(18):1209–17. doi:10. 1093/treephys/20.18.1209

Yaish, M.W. and Kumar, P.P. (2015). Salt tolerance research in date palm tree (*Phoenix dactylifera* L.), past, present, and future perspectives. Front. Plant Sci. 6:348. doi: 10.3389/fpls.2015.00348

Yang, J., Kloepper, J.W. and Ryu, C.-M. (2009). Rhizosphere bacteria help plants tolerate abiotic stress. Trends Plant Sci. 14, 1–4. doi: 10.1016/j.tplants.2008.10.004

Yao, L.X., Wu, Z.S., Zheng, Y.Y., Kaleem, I, Li C. (2010). Growth promotion and protection against salt stress by *Pseudomonas putida* Rs-198 on cotton. Eur J Soil Biol 46(1):49–54. doi:10.1016/j.ejsobi.2009.11.002

Yue, H., Mo W., Li C., Zheng Y., Li H. (2007). The salt stress relief and growth promotion effect of Rs-5 on cotton. Plant and Soil, 297, 139–145.

Zaidi, P.H., Rafique, S., Rai, P.K., Singh, N.N., Srinivasan, G. (2004). Tolerance to excess moisture in maize (*Zea mays* L.): susceptible crop stages and identification of tolerant genotypes. Field Crops Res. 90: 189-202.

Zuo, Y., Zhang F (2011). Soil and crop management strategies to prevent iron deficiency in crops. Plant Soil 339:83–95

5

Multiple Potential Traits of PGP *Pseudomonas* sp.

Padmavathi Tallapragada and Kavyashree B M

Department of Microbiology, School of Sciences, Jain (Deemed-to-Be University), 18/3, 9th Main, 3rd Block, Jayanagar, Bangalore- 560011, India

Abstract

In recent years, significant studies have been carried out on Plant growth promoting rizhobacteria (PGPR) to replace the usage of pesticides and fertilizers to improve the plant health. Plant growth promoting rizhobacteria (PGPR) is a group of rhizosphere bacteria, which are symbionts to plants and influence their growth by providing plant growth promoting substances. Different bacterial genera act as plant growth promoting rizhobacteria (PGPR): Pseudomonas sp., *Rhizobium* sp., *Azotobacter* sp., *Bacillus* sp. *Pseudomonads are gram negative rods, aerobic, fluorescent, motile and are able to produce siderophores which influence plant growth. There are multiple potential traits offered to the plants by plant growth promoting (PGP) Pseudomonas* sp. *are (1) Biocontrol agent against plant pathogens (2) Induced systemic resistance, which is a defense mechanisms in plants preceding to pathogenic attack (3) Influence of plant growth promoting (PGP) Pseudomonas* sp.*on agricultural crops (4) Siderophores, which are low molecular, high affinity iron chelating compounds which benefits the plants to acquire iron molecules (5) Nitrogen fixation (6) Phytoremediation of heavy metals (7) Solubilization of inorganic phosphates (8) Secretion of plant hormones (9) Bacteriocins(10) Antifungal factors (11) Antibiosis traits and (12) Endophytic PGP Pseudomonas* sp.

Keywords: Antifungal factors, Biocontrol, Plant growth promoting rhizobacteria (PGPR), PGP *Pseudomonas* sp., Phytoremediation, Plant hormones, Siderophores.

5.1. Introduction

The relationship between plants, soil and microorganisms have been improved during last decades by using PGPR as biofertilizers. PGPR are generally found in rhizosphere which is a zone of soil surrounding the roots of plants, containing organic compounds or root exudates secreted by the plants. Bacteria that are present in the rhizosphere and enhance plant growth by any mechanism are referred to as plant growth promoting rhizobacteria (PGPR). In both natural and man-made agroecosystems, interactions between plants and soil microorganisms have a profound effect on adaptation of plant to changing environment and plant growth. Selections of microbial isolates from naturally stressed environment or rhizosphere are considered as possible measures for improving crop health which can control different plant diseases and promote plant growth (Mayak *et al.*, 2004).

5.2. Plant Growth Promoting Rhizobacteria (PGPR)

Rhizobacteria is a group of bacteria with plant root zone habitat (rhizosphere) which has been researched and proven to improve soil fertility, increase plant resistance, and suppress plant pathogens. Rhizobacteria directly acts as a biological fertilizer and biological stimulant which produces essentials hormones to grow crops, such as IAA (Indoleacetic Acid), gibberellin, cytokinin, ethylene, dissolving minerals, and indirectly serves to prevent pathogenic microorganisms through the formation of siderophore and antibiotics. Furthermore, it has not been widely known to stimulate plant growth mechanisms (Alabouvette *et al.*, 1996). One of the Rhizobacteria which has already been investigated as PGPR and ISR is *Pseudomonas alcaligenes* isolate KtS1, TrN2, and TmA1. These isolates increase the growth and yield of tomatoes (*Solanum lycopersicum* L.) (Widnyana, 2011). Some problems investigated in this research are (1) how seed soaking with *P. alcaligenes* suspense the growth of swamp cabbage and (2) when the best time of seed soaking in promoting the growth of swamp cabbage is, with the purpose of obtaining information related to the benefits of seed soaking for swamp cabbage and best soaking time for swamp cabbage growth.

PGPR are naturally occurring soil bacteria that aggressively colonize plant roots and benefit plants by availing growth promotion (Glick, 1995). The indirect promotion of plant growth occurs when PGPR lessen or prevent the deleterious effect of plant pathogens on plants by production of inhibitory substances or by increasing the natural resistance of the host. PGPR involves in the process of nitrogen fixation, phosphorus solubilization, Hydrogen cyanide (HCN) production, production of phytohormones such as auxins, cytokinins and gibberellins, and lowering of ethylene concentration (Ali *et al.*, 2010). Among the PGPR,

Pseudomonas species stand out because of high level of genetic variability and competitiveness in soil. The most effective strains of *Pseudomonas* have been fluorescent *Pseudomonas* sp. which are characterized by their production of yellow green pigments, termed pyoverdines or pseudobactins, that fluorescence under UV irradiation and functions as siderophore. Apart from PGPR, AM fungi also influence plant growth by providing nutrition, antioxidants and certain enzyme (**Fig. 5.1**) (Nadeem *et al.*, 2014).

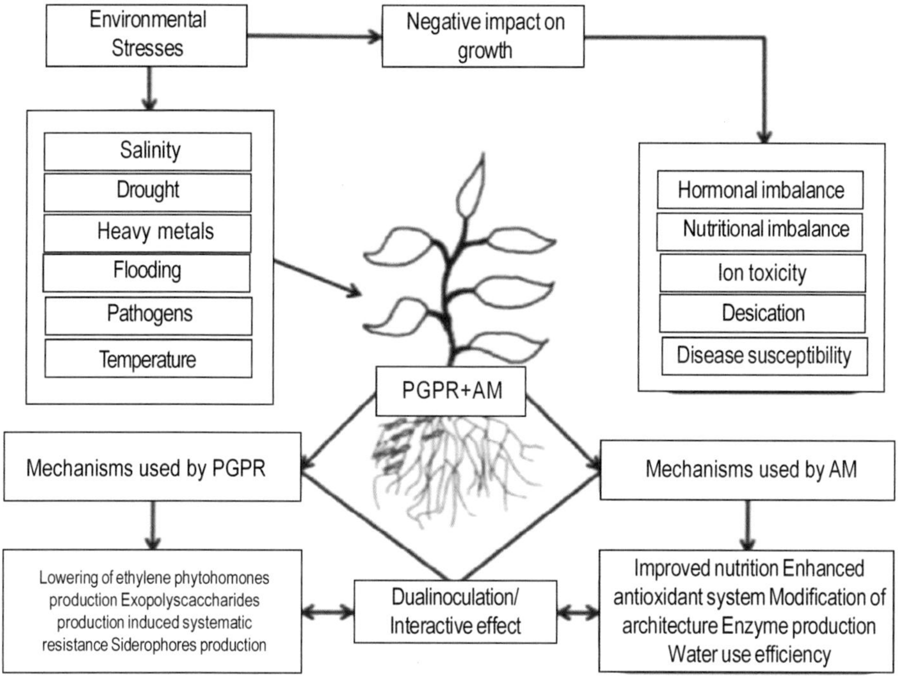

Fig. 5.1: Mechanism used by PGPR and mychorrhizae for enhancing plant growth under stress (Adopted from Nadeem et al., 2014)

5.2.1. Beneficial aspects of PGPR

PGPR residing in the soil environment can cause dramatic changes in plant growth by the production of growth regulators and/or improving plant nutrition by supplying and facilitating soil nutrient uptake (Zahir *et al.*, 2008). Rhizobacterial strains can also improve plant tolerance against salinity, drought, flooding, and heavy metal toxicity and, therefore, enable plants to survive under unfavorable environmental conditions (Zahir *et al.*, 2008; Mayak *et al.*, 2004; Sandhya *et al.*, 2009; Glick, 2010). These rhizobacteria can be used in different ways when plant growth promotion is required. The two major ways through which PGPR can facilitate plant growth and development include direct and indirect mechanisms (Glick, 1995).

Indirect growth promotion occurs when PGPR prevent or reduce some of the harmful effects of plant pathogens by one or more of the several different mechanisms (Glick and Bashan, 1997). Direct growth promotion takes place in different ways like providing beneficial compounds to the host plant synthesized by the bacterium and/or facilitating the uptake of nutrients from the soil environment (Kloepper *et al.*, 1987).

Despite these mechanisms, PGPR may also enhance plant growth and development by the virtue of their key enzymes (ACC-deaminase, chitinase) and also by the production of substances such as exopolysaccharides, rhizobitoxine, etc. that help plants to withstand stress conditions (Ashraf *et al.*, 2004; Glick *et al.*, 2007; Sandhya *et al.*, 2009). In general, PGPR induce plant growth by several ways. Some strains of PGPR uses more than one mechanism to withstand normal and stressful environment.

5.2.2. Harmful aspects of PGPR

The auxin production by the PGPR enhances plant growth (Patten and Glick, 2002), whereas at high level it inhibits root growth (Xie *et al.*, 1996). In certain cases, the application of PGPR and fungi triggers the pathogenic activity of the co-inoculated partner although the PGPR itself is non-pathogenic. PGP *Pseudomonas* species produce cyanide which is considered to have growth promotion as well as growth inhibition characteristics. On one hand, cyanide acts as a biocontrol agent against certain plant pathogens (Martínez-Viveros *et al.*, 2010) while on the other hand, it can also cause adverse effects on plant growth (Bakker and Schippers, 1987).

5.3. Biofertilizer

As the name suggest a biofertilizer is a preparation which uses certain efficient microorganisms that increases that soil fertility and improves plants growth by converting nutritionally important elements (nitrogen, phosphorus) from unavailable to available form through biological process such as nitrogen fixation and solubilization of rock phosphate (Rokhzadi *et al.*, 2008). There are different types of microorganisms which have been widely used as biofertilizers such as bacteria, fungi and algae.

5.3.1. History of Biofertilizer

Application of microorganism as inoculum is having an extensive past record that passed from generations of the agriculturalists. Basically, it began with culture of small-scale compost production that has evidently proved the ability of biofertilizer. This is recognized when the cultures accelerate the decomposition of organics residues and agricultural byproducts through various processes and gives healthy harvest of crops (Abdul Halim, 2009). In Malaysia, industrial scale microbial inoculants are started in the late 1940's and peaking up in 1970's taking guide by *Bradyrhizobium* inoculation on legumes. Government research institute, the Malaysian Rubber Board (MRB), Malaysia had been conducting research on *Rhizobium* inoculums for leguminous cover crops in the inter rows of young rubber trees in the large plantations. The commercial history of biofertilizer began with the launch of "Nitragin" by Nobbe and Hilther in 1895. This was followed by the discovery of *Azotobacter* and then Blue-green algae and a host of other microorganisms which are being used till date as biofertilizer (Kribacho, 2010). Biofertilizer are usually prepared as carrier-based inoculants containing effective microorganisms (Vessey, 2003). Microorganisms used as biofertilizer include: N_2 fixer e.g. *Rhizobium* sp., Cyanobacteria, and *Azotobacter chroococcum*, K solubilizers, e.g. *Bacillus mucilaginous*, P solubilizers e.g. *Bacillus megaterium*, *Aspergillus fumigatus*, PGPR, Vesicular Arbuscular Mycorrhiza (VAM) e.g. *Glomus mosseae* and S oxidizers e.g. *Thiobacillus* sp. (Itelima *et al*., 2018).

5.3.2. Significance of Biofertilizers

(1) Biofertilizers keep the soil environment rich in all kinds of macro and micronutrients via N fixation, P and K solubilization or mineralization, release of plant growth regulating substances, production of antibiotics and biodegradation of organic matter in the soil (Sinha *et al*., 2014).

(2) Beneficial microorganisms in biofertilizers accelerate and improve plant growth and protect plants from pests and diseases (El-Yazeid *et al*., 2007).

(3) These potential biological fertilizers would play the key role in productivity and sustainability of soil and also protects the environment as eco-friendly and cost effective inputs for the farmers (Khosro *et al*., 2012).

(4) The application of biofertilizer to the soil increases the biodiversity which constitutes all kinds of useful bacteria and fungi including the arbuscular mycorrhiza fungi (AMF) called PGPR and nitrogen fixers (Itelima *et al*., 2018).

5.4. PGP *Pseudomonas*

PGP *Pseudomonas* sp. are found in rhizosphere or agricultural soils and has multiple traits that make them well suited as PGPR. The most effective strains of *Pseudomonas* have been fluorescent *Pseudomonas* sp. There are studies exploit the potential of one group of bacteria that belong to Fluorescent pseudomonads (FLPs). FLPs help in the maintenance of soil health and are metabolically and functionally most diverse (Lata and Tilak, 2002; Lugtenberg and Dekkers, 1999). The presence of *Pseudomonas fluorescence* inoculant in the combination of microbial fertilizer plays an effective role in stimulating yield and growth traits of chickpea (Rokhzadi *et al*., 2008). Isolates of FLPs from roots, shoots, and rhizosphere soil of sugarcane provides significant increases in fresh and dry masses (Mehnaz *et al*., 2009). Field trials of a pseudomonad strain (GRP3) lead to a great increase in yield of legumes (Johri, 2001). Specific strains of the *Pseudomonas fluorescens/ putida* group have recently been used as seed inoculants on crop plants to promote growth and increase yields. These pseudomonads, termed PGPR, rapidly colonize plant roots of potato, sugar beet and radish, and cause statistically significant yield increases up to 144% in field tests (Kloepper *et al*., 2004; Suslow *et al*., 1979; Burr *et al*., 1978). The occurrence and activity of soil microorganisms are affected by a variety of environmental factors (e.g. soil type, nutrient abundance, pH, moisture content) as well as plant-related factors (species, age). So, while working on two winter wheat cultivars it was found that the genus *Pseudomonas* show higher counts, thus the population size of bacteria of the genus *Pseudomonas* depends on the development phase of wheat plants (Wachowska *et al*., 2006). Biosurfactant production by *Pseudomonas* sp. is an effective environmental trait which has a great potential for biotechnological and biomedical applications (Banat *et al*., 2010).

5.4.1. Isolation and Screening of PGP Pseudomonas

Soil samples were collected from rhizosphere. Physicochemical analyzed of samples based on soil texture, pH (7.5-8.8), and temperature (25 to 32°C) was examined. Microbial strains were isolated by the serial dilution method. One gram of dried soil was weighed and added to 9 ml of double distilled water (dd H_2O) in a sterile test tube and shaken well using vortex mixer; this stock solution was then diluted serially up to the dilution of $10^{-5}$and 0.1 mL of suitable diluted sample was spread plated on surface of selective King's B agar and incubated at 30°C for 2 days. After isolation of strains on selective King B medium the strains were identified and characterized by morphological, cultural, and biochemical tests using Bergey's manual as a reference (MacFaddin, 2000).

5.4.2. PGP Pseudomonas–plant interactions

This interaction can be considered to take place in four very broadly defined contact zones (**Fig. 5.2**).

5.4.2.1 Foliar surfaces colonized by epiphytic *Pseudomonas*

Foliar surfaces of a plant is the initial contact zone for pathogenic *Pseudomonas*. Surfaces of plants leaves are covered in a waxy cuticle which restricts water loss and contact between *Pseudomonas* and host cells. Bacteria live as saprotrophs on nutrients exuded from the plant, or organic matter deposited on surfaces, and are subject to high levels of fluctuating environmental stress such as temperature, dehydration and UV light. Bacteria can only enter plant tissues through natural openings such as wounds, stomata or hydathodes, but some *Pseudomonas* increase the incidence of damage to host tissues through ice nucleation (Wisniewski *et al.*, 1997; Beattie & Lindow, 1999; Lindow & Brandl, 2003).

5.4.2.2 Root surfaces colonized by rhizosphere Pseudomonas

Roots are designed for water uptake and present a large surface area that is not covered with a hydrophobic cutin layer. The lack of a cutin layer may offer greater potential for direct signaling between *Pseudomonas* and epidermal cells than on foliar surfaces. Roots release substantial quantities of root exudates, which are rich in sugars, dicarboxylic acids, amino acids and sloughed off root border cells, and which support a complex microflora and microfauna of saprotrophs, symbionts and predators (Gilroy & Jones, 2000; Hawes *et al.*, 2000). Roots also produce significant levels of secondary metabolites, many of which have anti-microbial activity (Flores *et al.*, 1999). In addition to direct interactions with plant cells, root colonizing *Pseudomonas* can affect plant physiology through interactions with other rhizosphere organisms, such as mycorrhizal fungi, soil- borne plant pathogens, and nitrogen-fixing and nitrogen-cycling bacteria (Lugtenberg *et al.*, 2001).

5.4.2.3 Intercellular spaces in leaves colonized by endophytic *Pseudomonas*

Endophytic *Pseudomonas* live on nutrients present in the apoplast of host cells, the acidic, non-living continuum provided by the continuous matrix of cell walls, or on nutrients released from dead cells during pathogenesis. Signal exchange between *Pseudomonas* and plant cells generally occurs across the barrier of the plant cell wall, rather than in the context of close contact between bacterial and host membranes as in many animal–bacteria interactions (for clear images of endophytic interactions see Bestwick *et al.*, 1997; Brown *et al.*, 2001).

5.4.2.4 Intercellular spaces in roots colonized by endophytic *Pseudomonas*

The properties of roots as habitats for endophytic *Pseudomonas* and plant roots are poorly understood, although *Pseudomonas fluorescens* and *Pseudomonas putida* are frequently isolated as endophytes from roots and tubers. *Pseudomonas* enter roots through wounds and natural openings, such as the point of emergence of lateral roots.

The interior of roots may have features in common with leaves, but they are also characterized by lack of photosynthetic tissue, less exposed surface area for gas exchange and synthesis of a wide range of anti-microbial secondary metabolites. Differences in the physiology of photosynthetic and non-photosynthetic tissue may have a substantial impact on the physiology of plant responses to *Pseudomonas*.

Many *Pseudomonas* sp. can live as epiphytes on the surface of leaves. Ice-nucleating strains of *Pseudomonas* promote frost damage, but epiphytic *Pseudomonas* can also act as biocontrol agents that suppress foliar pathogens by competition, exclusion and antibiosis (a, b) (**Fig. 5.2**). *Pseudomonas* invade leaves through wounds and natural openings to establish endophytic populations. Recognition of generic and host-specific elicitors produced by endophytic *Pseudomonas* primes and induces local defence responses and can elicit the hypersensitive response (HR) and systemic defence responses (c–f). Successful pathogens can evade or suppress recognition and cause disease symptoms at high bacterial densities (g). Damage caused by pathogens can also elicit systemic defences such as systemic acquired resistance (SAR) in roots and leaves (e).

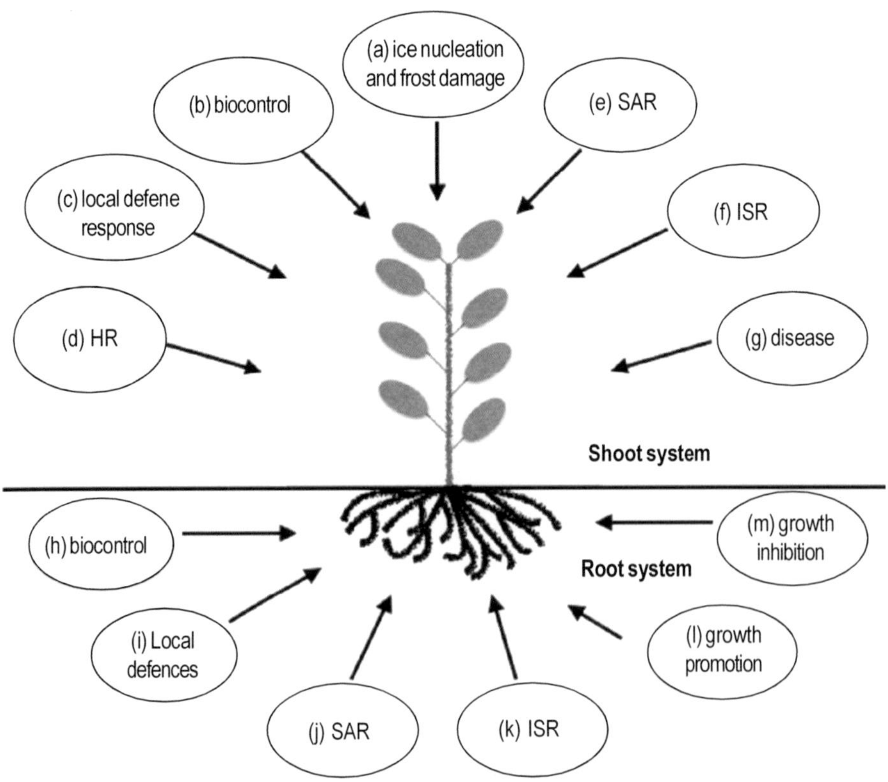

Fig. 5.2: PGP *Pseudomonas*–plant interactions

Many root colonizing *Pseudomonas* also have the capacity to suppress pathogens (h), but some also prime and elicit local and systemic defence responses such as induced systemic resistance (ISR) in roots and leaves (i–k).

Pseudomonas–plant interactions, including modulation and biosynthesis of plant hormones, can result in plant growth promotion or inhibition of plant growth (*l,m*), and is influenced by environmental and host factors, such as temperature, water availability, host genotype and plant health (Gail, 2004).

In general, mode of action of PGPP **(Fig. 5.3)** influence plant growth by nitrogen fixation, phosphate solubilisation, siderophore production, plant hormones production, nodulation and HCN. PGPP also manage diseases in plants by producing certain biocontrol agents which indirectly influence plant growth.

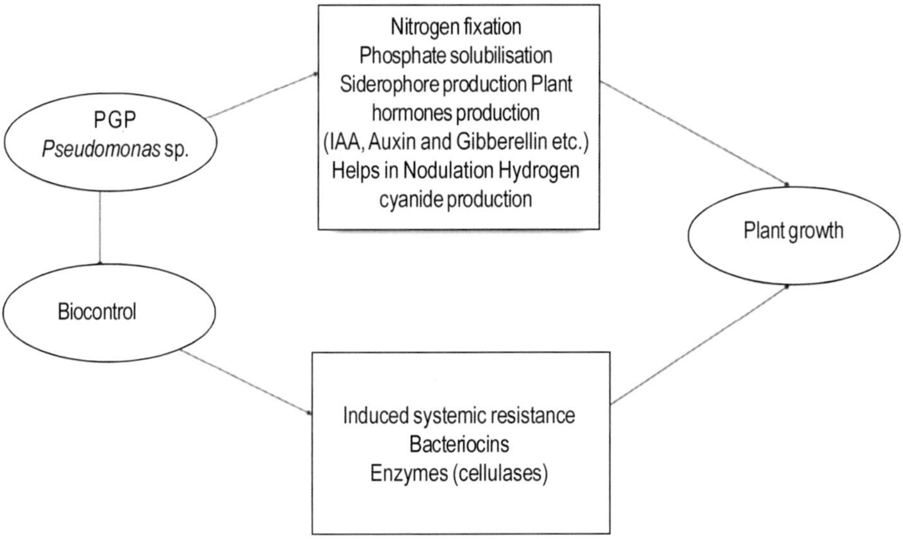

Fig. 5.3: Mode of action of PGPP

5.5. Multiple Potential Traits of PGP *Pseudomonas*

5.5.1 Biocontrol

Offensive PGPB colonization and defensive retention of rhizosphere niches are enabled by production of bacterial allelochemicals, including iron-chelating siderophores, antibiotics, biocidal volatiles, lytic enzymes, and detoxification enzymes (Bais *et al.*, 2004; Glick, 1995; Sturz and Christie, 2003).

5.5.2 Siderophores produced by PGP Pseudomonas

Iron is an essential growth element for all living organisms. The scarcity of bioavailable iron in soil habitats and on plant surfaces foments a furious competition (Loper and Henkels, 1997). Under iron-limiting conditions PGPB produce low-molecular-weight compounds called siderophores to competitively acquire ferric ion (Whipps, 2001).

Although various bacterial siderophores differ in their abilities to sequester iron, in general, they deprive pathogenic fungi of this essential element since the fungal siderophores have lower affinity (Loper and Henkels, 1999; O'Sullivan and O'Gara, 1992). Some PGPB strains go one step further and draw iron from heterologous siderophores produced by cohabiting microorganisms (Castignetti and Smarelli, 1986; Lodewyckx *et al.*, 2002; Loper and Henkels, 1999; Raaijmakers *et al.*, 1995; Wang *et al.*, 1993; Whipps, 2001).

Siderophore biosynthesis is generally tightly regulated by iron-sensitive Fur proteins, the global regulators GacS and GacA, the sigma factors RpoS, PvdS, and FpvI, quorum-sensing autoinducers such as *N*-acyl homoserine lactone, and site-specific recombinases (Cornelis and Matthijs, 2002; Ravel and Cornelis, 2003). However, some data demonstrate that none of these global regulators is involved in siderophore production. Neither GacS nor RpoS significantly affected the level of siderophores synthesized by *Enterobacter cloacae* CAL2 and UW4 (Saleh and Glick, 2001). RpoS is not involved in the regulation of siderophore production by *Pseudomonas putida* strain WCS358 (Kojic *et al.*, 1999). In addition, GrrA/GrrS, but not GacS/GacA, are involved in siderophore synthesis regulation in *Serratia plymuthica* strain IC1270, suggesting that gene evolution occurred in the siderophore-producing bacteria (Ovadis *et al.*, 2004). A myriad of environmental factors can also modulate siderophores synthesis, including pH, the level of iron and the form of iron ions, the presence of other trace elements, and an adequate supply of carbon, nitrogen, and phosphorus (Duffy and Defago, 1999) (**Table 5.1**).

Table 5.1: Siderophores produced by plant associated rhizosphere bacteria (Ranjani *et al.*, 2018)

Sl.No.	Siderophore	Bacteria
1	Pyoverdin	*Pseudomonas fluorescens*, *Pseudomonas putida*
2	Catechols	
	a) Agrobactin	*Agrobacterium tumefaciens*
	b) Enterobactin	Enterobacteriaceae family
	c) Azotochelin	*Azotobacter vinelandii*
3	Other types	
	Rhizobactin	*Rhizobium meliloti*
	Citric acid	*Bradyrhizobium japonicum*
	Azotobactin	*Azotobacter vinelandii*

5.5.3 Antibiosis

The biocontrol mechanism of PGPB has been studied in detail over the past two decades (Whipps, 2001). Compounds such as amphisin, 2,4-di-acetylphloroglucinol (DAPG), hydrogen cyanide, oomycin A, phenazine, pyoluteorin, pyrrolnitrin, tensin, tropolone, and cyclic lipopeptides produced by pseudomonads have been identified for antibiotic activity (Défago *et al*., 1993; de Souza *et al*., 2003; Nielsen *et al* 2002; Nielsen *et al* 2003; Raaijmakers *et al*., 2002).

The pathogen suppression was studied by biocontrol strains enabling antagonists to attain their ultimate objective. For example, biosynthesis of DAPG is stimulated and pyoluteorin is inhibited in *P. fluorescence* in the presence of glucose as a carbon source. Pyoluteorin becomes the more abundantly antimicrobial compound produced by this strain when glucose is repressed (Duffy *et al*., 1999). A degree of flexibilty is ensured for the antagonist when confronted with a different or a changeable environment. Antibiotic biosynthesis is influenced by biotic conditions (Duffy *et al*., 1997; Duffy *et al*., 2004; Haas *et al*., 2003; Notz *et al*., 2001; Pettersson *et al*., 2004). For example secondry metabolites produced by *P. fluorescence* CHA0 such as salicylates and pyoluteorin can affect DAPG production (Schnider- Keel *et al*., 2001). Antiobiotic production influenced by plant growth and development, since DAPG producers which are biologically more active do not produce exudate from young plant roots but it is induced by the exudates of older plant roots, which results in selective pressure against other rhizosphere microorganisms (Picard *et al*., 2000). Smith *et al*. demonstrated that significant role in the disease-suppressive interaction of plant with a microbial biocontrol agent was studied (Smith *et al*.,1999).

5.5.4 Induced systemic resistance

ISR is a phenomenon which is triggered by certain bacteria. It is phenotypically similar when compared to systemic acquired resistance (SAR). In response to primary infection by a pathogen, SAR is developed by plants in order to activate their defense mechanism, when a pathogen infects the tissue it induces hypersensitive reaction through which it becomes limited to local necrotic lesion (Van Loon *et al*.,1998). SAR and ISR are two different mechanisms, while ISR is infective against different types of pathogens. SAR is the mechanism where PGPB does not induce diseases symptoms on the host plant (Van Loon *et al*.,1998). The first observation of PGPB-elicited ISR was seen on carnation with reduced susceptibility to wilt caused by *Fusarium* sp. and reduced susceptibility to foliar disease caused by *Colletotrichum orbiculare* on cucumber (Van Peer *et al*., 1991; Wei *et al*., 1991).

Bacterial strains and host plants manisfest ISR (Kilic-Ekici *et al.*, 2004; Van Loon *et al.*,1998). PGPB-mediated ISR reactions have been reported by many researchers which involves free-living rhizobacterial strains and edophytic bacteria depicting ISR activity. Viswanathan *et al.*,1999 reported that *P. fluorescens* EP1 triggered ISR against *Colletotrichum falcatum* on sugarcane causing red rot sugarcane disease, *Burkholderia phytofirmans* PsJN against *Botrytis cinerea* on grapevine (Ait Barka *et al.*, 2000; Ait Barka *et al.*, 2002) and *Verticllium dahliae* on tomato (Sharma *et al.*, 1998), *P. denitrifcans* 1-15 and *P. putida* 5-48 against *Ceratocystis fagacearum* on oak (Brooks *et al.*, 1994), *P. fluorescens* 63-28 against *F. oxysporum* f. sp. *radicis lycopersici* on tomato (M'Piga *et al.*,1997) and *Pythium ultimum* and *F. oxysporum* f. sp. *pisi* on pea roots (Benhamou *et al.*, 1996), and *Bacillus pumilus* SE34 against *F. oxysporum* f. sp. *pisi* on pea roots (Benhamou *et al.*, 1996) and *F. oxysporum* f. sp. *vasinfectum* on cotton roots (Conn *et al.*, 1997).

5.5.5 Nitrogen fixer PGP Pseudomonas

Reduction of the atmospheric nitrogen (N2) in ammonium (NH4+) by nitrogenase is defined as nitrogen fixation. The process of nitrogen fixation introduces nitrogen into the biosphere. The natural fixing process is responsible for 65% and 25% of annual fixation and industrial fixation respectively. Eubacteria and archaeabacteria fixers nitrogen. These bacteria possess nitrogenase which is reponsible for the reduction of nitrogen gas to ammonia. In many bacterial species, characterisation and regulation of Nitrogen fixation genes were investigated (Elmerich, 1991; Dean & Jacobson, 1992; Merrick, 1992; Fischer, 1994; Lee *et al.*, 2000; and references therein). Young, 1992 reported that there were no nitrogen fixers among the strains of the genus *Pseudomonas* and most of these strains were putative nitrogen fixers. According to Chan *et al.*, 1994 *Pseudomonas* was reassigned to the genera in the α and β Proteobacteria.

It now seems that several strains unambiguously classified as true *Pseudomonas* sp. can be added to the list of nitrogen-fixers, on the basis of physiological properties, nitrogenase assays, phylogenetic studies and detection of nifH DNA by hybridization or PCR amplification can be as a criteria for nitrogen fixers and *Pseudomonas* sp. can classified in this catorgery (Chan *et al.*, 1994; Vermeiren *et al.* 1999). *P. stutzeri* CMT 9A were isolated from the roots of Sorghum (Krotzky & Werner 1987), whereas strain A15 was isolated from rice paddies in China (You & Zhou 1989; You *et al.*, 1991). Vermeiren *et al.*, 1999 reported that strain A15 was initially identified as an *Alcaligenes faecalis* and later identified as *P. stutzeri.*

5.5.6 Phytoremediation of heavy metals

The remediation of heavy metals is carried out by a biological technique called as bioremediation (Boopathy, 2000). For example, remediation of Cu and Zn is carried out by *Bacillus* spp. and *Pseudomonas aeruginosa* (Kumar *et al.*, 2011). The composting, land formation, bioreactors, bioaugmentation, bioventing, biofilters, biostimulation, intrinsic bioremediation, and pump and treat methods are few examples of bioremediation (Boopathy, 2000; Yang *et al.*, 2009).

In phytoremediation the insoluble metals compounds and minerals gets converted into bioavailable and soluble forms (Gadd, 2004). For example, in a case study, Solanum nigrum and *Pseudomonas* spp. LK9 was inoculated in shoots and roots of plants which increased Cd uptake ranging from 230 mg kg^{-1} to 292 mg kg^{-1} when compared with uninoculated plants (Sheng *et al.*, 2008). It was concluded the production of biosurfactants, siderophores and organic acids were responsible for the increased accumulation of metal (Chen et al., 2014). Fig. 5.4 shows a comparison between inoculated and uninoculated plants which concludes that plants inoculated with PGP bacteria accumulates heavy metals in greater amounts (Ullah *et al.*, 2015).

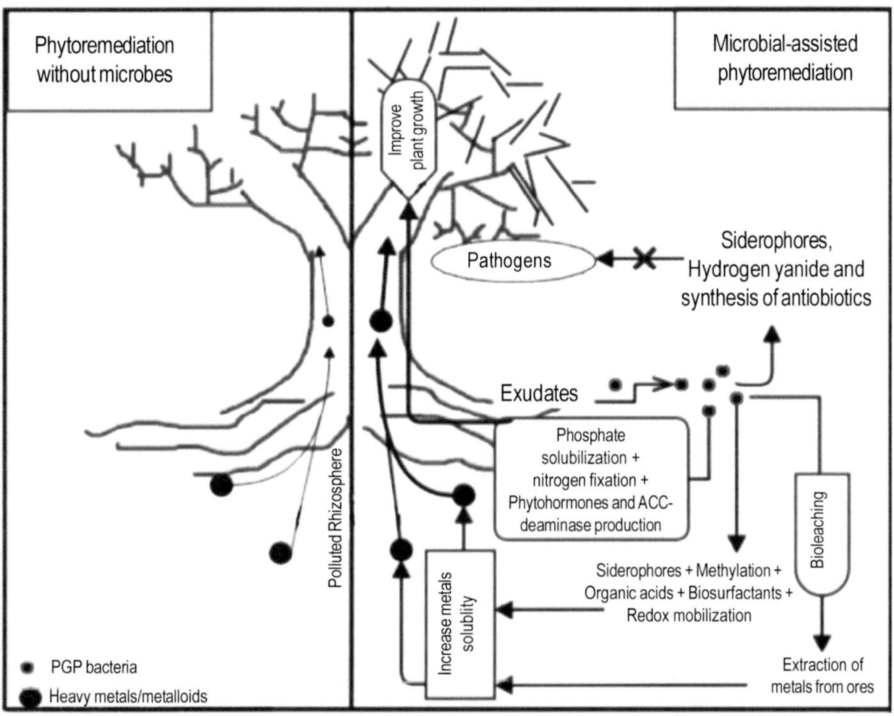

Fig. 5.4: Comparison between PGP bacteria inoculated and uninoculated plants (Ullah et al., 2015)

5.5.6.1 Mechanism of PGP bacterial enhanced phytoremediation

Recent studies shows that phytoremediation of heavy metals can either be carried out directly or indirectly with the help of plant growth promoting bacteria. The direct process involves solubilization, bioavailability and final accumulation of heavy metals by plants. Whereas indirect process helps in the accumulation of heavy metals enhance plant growth and protect plants from phytopathogens.

5.5.6.1.1 Direct promotion of phytoremediation

The heavy metals from solid matrices, such as soil, dumps, sediments and other industrial and municipal wastes can be removed by in direct processes which involves solubilization and bioavailability of heavy metals (Gadd, 2004; Yan-de *et al.*, 2007; Glick, 2010). The reactions and metabolic processes that occurs in biogeochemical cycling of nutrients, maintenance of soil structure, and detoxification of pollutants is due to soil microbes (Khan *et al.*, 2010). Plant growth promoting bacteria secretes siderophores, that has a capacity to bind to metals and thus improve their bioavailability in the rhizosphere (Yan-de *et al.*, 2007; Rajkumar *et al.*, 2010; Gadd, 2010) (**Fig. 5.5**).

5.5.6.1.2 Indirect promotion of phytoremediation

In indirect process, PGP bacteria increases the production of biomass and prevents plants from phytopathogens. It also helps the plants to remove pollutants (Luo *et al.*, 2012). The growth of the plant is retarted when nutrient uptake by plants is affected by heavy metals in the rhizosphere (Ouzounidou *et al.*, 2006). PGP bacteria have the capacity to enable essential nutrients to plants under nutrients deplete conditions. In the atmosphere about 78% Nis unavailable to growing plants. Under metal stressed conditions, diazotrophic bacteria can fix atmospheric nitrogen. Other examples are *Rhizobium leguminosarum bv trifolii* which can fix nitrogen in the presence of heavy metals (Nonnoi *et al.*, 2012). After nitrogen, phosphorus is the second most essential growth-limiting nutrient. The amount of phosphorus in soil is abundant but the plants cannot utilise it because it is in the insoluble form. Organic acids secreted by bacteria can solubilize phosphorus. *Rhizobium, Bacillus* and *Pseudomonas* are the best examples of phosphate-solubilizing bacteria (PSB) (Chen *et al.*, 2006; Rodriguez *et al.*, 2006). Fe which is usual form of occurrence of iron is forms insoluble hydroxides and oxyhydroxides that can not be utilized from plants and microorganisms. Siderophores are produced by rhizosphere bacteria that are capable of chelating Fe, which can be taken up as an iron nutrient by plant roots (Rajkumar *et al.*, 2010; Ma *et al.*, 2011). Moreover, the contribution of essential vitamins, mineral uptake, stomatal regulation, osmotic modification, and

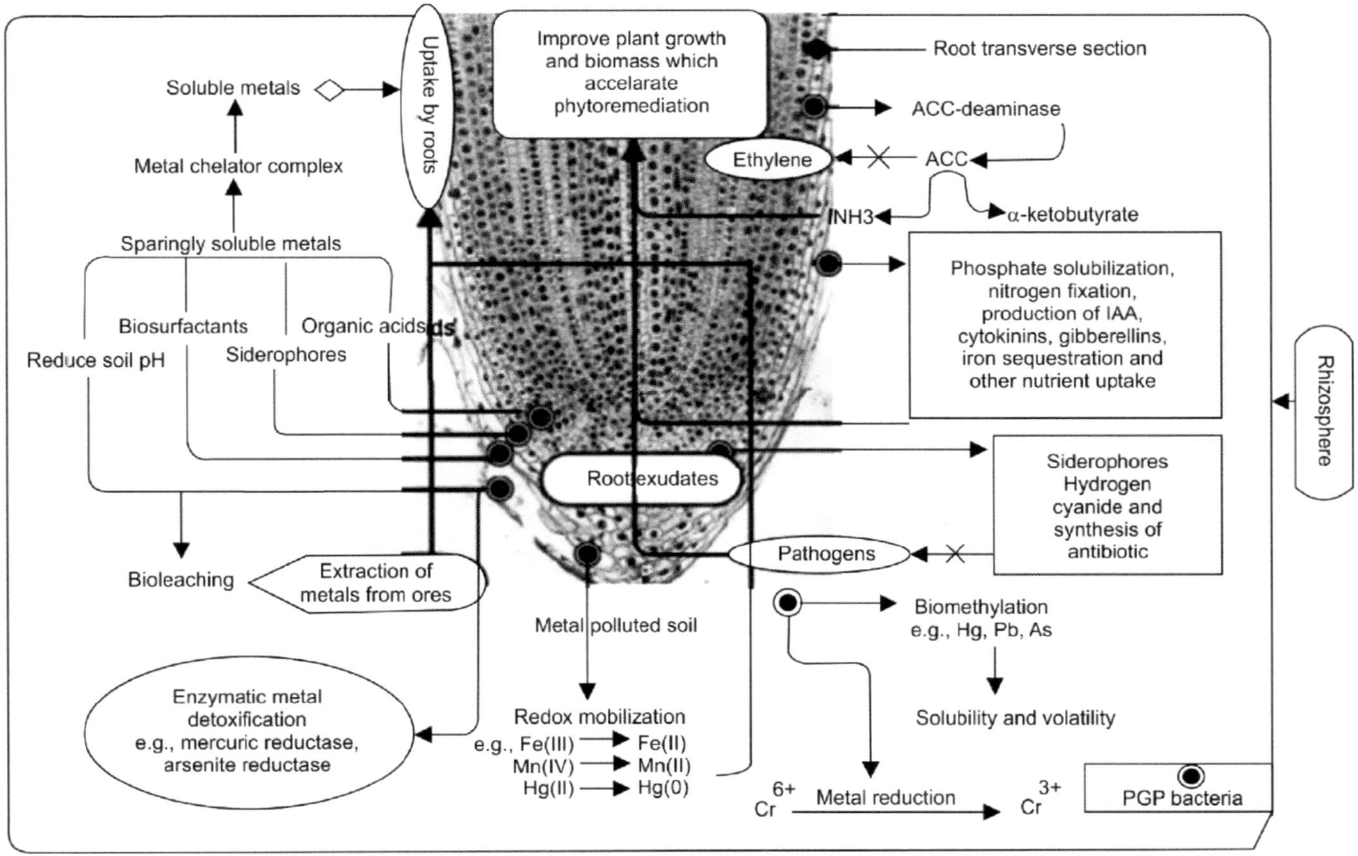

Fig. 5.5: Direct and indirect promotion of phytoremediation through PGP bacteria (Ullah et al., 2015)

adaptation of root morphology is influenced by PGP bacteria (Compant *et al.*, 2005 a,b). In general, these bacteria improve plant growth in metal/metalloid poluted soil, which in turn facilitates the phytoremediation process (Ullah *et al.*, 2015).

Moreover, PGP bacteria play a role in mineral uptake, the contribution of essential vitamins, stomatal regulation, osmotic modification, and adaptation of root morphology (Compant *et al.*, 2005a,b). In general, these bacteria improve plant growth in metal/metalloid polluted soil, which in turn facilitates the phytoremediation process (Ullah *et al.*, 2015).

5.5.7 Solubilization of inorganic phosphates

It was shown in several studies that *Pseudomonas fluorescens* was used as a soil bioinoculant that enhanced the plant growth by various mechanisms such as production of siderophore, antifungal compounds, plant hormones and solubilization of phosphate (Glick 1995; Nautiyal 1999). In few studies it was shown that the ability of *P. Solubilization* to solubilise phosphate and it's interaction with plant system at low temperatures were performed. A study has been performed to screen cold tolerant mutants of *P. fluorescens* to test their ability to solubilise phosphate and effects on plant growth under *in vitro* and *in situ* conditions by using the selected mutants.

5.5.7.1. Experimental application of PGP Pseudomonas sp.

Paper towel assay was set up to test the efficacy of mutants at low temperature. At a temperature of 28°C in trypticase soy broth (TSB) medium, *P. fluorescens* wild type strains, GRS1 and PRS9 and their respective mutants (CRPF2 and CRPF7) were subjected to seed bacterization of wheat var. Sonalika and mung bean var. PM-4. In the same way, sample of bacterized seeds with colony forming unit were counted on TSB medium with bacterial suspension adjusted to 1.90 × 108 CFU per seed. Pots were prepared containing containing sterile and unsterile field soil of pH 7.2. Bacterized seeds of mung bean were planted in pots at 10 cms diameter. These were grown containing in a greenhouse at a temperature range of 10–15°C. This was done to evaluate their ability to colonise the rhizoshpere. A second experiment was done to check how much these strains *in situ* could mobilise 'P' from two different sources of tri calcium phosphate (TCP) and KH. The 'P' sources were given as micronutrient once every week. The concentration of both TCP and KH Instead of soil ,the bacterized seeds were sown in pots containing quartz sand about 0.5–2 mm in diameter and theses were grown in a greenhouse. To maintain soil moisture content level, sterilized water was added daily to both the pots containing soil and sand. After three weeks, germination of seedlings from both the pots were recorded. Each treatment had four replications which was carried out by sowing 3 seeds per replicate in both pots. Analyzation of the data was done at 5% level of significance (Katiyar *et al.*, 2003).

5.5.7.1.1. Results

P. fluorescens strains GRS1, PRS9, ATCC13525 and their cold tolerant mutants grown in NBRIP broth (pH 7.0) for 2 days at 28°C were screened quantitatively for estimation of 'P' solubilization. 21-fold increase was exhibited by GRS1 mutant whereas CRPF7 and mutant of PRS9 showed a significant decrease (10.2-fold) in P solubilization activity. Paper towel assay was set up for studying phosphate solubilization and plant growth at 10°C to establish the relationship between mutants and wild types. Growth promotion of wheat and mung bean revealed the effect of mutants on these plants (Katiyar *et al.*, 2003).

5.5.8. Secretion of plant hormones

PGPR influence plant health in different ways. It has properties which is responsible for enhancing plant growth by producing certain plant hormones, siderophores, HCN and antibiotics. Among other hormones, Indole acetic acid (IAA) is an active auxins, which is obtained as an end product of L- tryptophan metabolism (Lambrecht *et al.* 2000; Persello-Cartieaux *et al.* 2003; Ryu *et al.* 2003). *Pseudomonas* can also cause harmful effects in plants indirectly, for example by producing enzymes that degrades the cell wall and release peptides and oligosaccharides, which affets the plant growth and signal transduction (Dumville & Fry 2000). In some interactions between plants and pathogens, synthesis of certain hormones can influence the development of galls in plants (Lindow & Brandl 2003; Persello-Cartieaux *et al.* 2003). The one of the main function of pathogen-induced galls is to take over the host's mechanism and direct the host to produce nutrients required by the pathogen to spread the infection. In PGPP, synthesis of plant hormones may function less disruptively to enhance the growth of plant roots and proliferation (Persello-Cartieaux *et al.* 2003). Bacteria that secretes hormones may also benefit from plants hormones such as IAA and cytokinin stimulation which provides methanol and saccharides as a nutrient source (Lindow & Brandl 2003).

5.5.8.1. Experimental application of PGP *Pseudomonas* sp.

Azotobacter and *Pseudomonas* sp. were isolated from rhizospheric soils of different crops such as wheat, berseem, mustard, cauliflower that was collected from Aligarh city, UP, India in October to December, 2002. The obtained isolates of *Azotobacter* and fluorescent *Pseudomonas* were screened on nitrogen free Jensen's medium and nutrient agar medium or King's medium respectively

(Holt *et al.*, 1994). These isolates were further screened for IAA production (Loper *et al.*, 1986). 200 ml of nutrient broth amended with 1 and 5 mg/ml of tryptophan was inoculated with 3 isolates of *Pseudomonas* sp. (Ps1, Ps4 and Ps7) and 3 strains of *Azotobacter* (Azs1, Azs6, Azs9) and incubated at 28 ± 2°C for 1 week on a shaker incubator. Centrifugation was carried out to separate bacterial cells from the supernatant at 10,000 rpm for 30 min. Using 1 N HCl, the supernatant was acidified to pH 2.5 to 3.0 and extracted twice by adding ethyl acetate with double the volume of the supernatant. Further extracted fraction was dried in a rotatory evaporator at 40°C. The obtained extract was dissolved in 300 ml of methanol and kept at –20°C. Purification was carried out using TLC (Ahmad *et al.*, 2005).

5.5.8.1.1. Results

21 rhizospheric isolates of *Azotobacter* and fluorescent *Pseudomonas* sp. were tentatively identified by biochemical tests and sugar fermentation. The range of IAA production without tryptophan by Azotobacter isolates was 2.68-10.80 mg/ml. There was an increase in the IAA production in the presence tryptophan (1, 2 and 5mg/ml) i.e. 1.47-11.88 mg/ml, 5.99-24.8 mg/ ml and 7.3-32.8 mg/ml respectively. Similarly 11 *Pseudomonas* isolates produced IAA in the range 5.34 to 22.4 mg/ml (Ahmad *et al.*, 2005). There was an increase in production IAA in the presence of 1, 2 and 5mg/ml of tryptophan i.e. 10.4-28.3 mg/ml, 20.8-37.5 mg/ml and 23.4-53.2 mg/ ml respectively. Therefore, isolates of fluorescent Pseudomonas produced greater amount of IAA when compared to Azotobacter isolates.

5.5.9. Bacteriocins produced by PGP Pseudomonas

Pseudomonas are commonly found in soil, fresh water, and marine ecosystems. Being an opportunistic pathogen, *Pseudomonas aeruginosa* causes diseases in humans. Most of the bacterial species belongs to this genus produce bacteriocins. The complete genome sequencing of *Pseudomonas* was attested in 2000 (Stover *et al.*, 2000).

A bacteriocin secreted from *P. aeruginosa* was first reported by François Jacob (Jacob, 1954). Jacob described that a strain of *P. aeruginosa* 10 was treated by a mutagen agent (i.e. UV irradiation, wavelength 253.7 nm) resulted in the production of an activated compound, which was obtained on bacterial cell lysis. This activated compound was found attached on the cell surface of susceptible bacteria and caused their death. Being structuraly similar to colicins, this compound was named as pyocin. Most of the *P. aeruginosa* species are capable of producing pyocins, but only few bacteria are involved in the production of pyocins in high quantity. The production of pyocins can be increased by treating the bacterial culture with mutagenic agents. Jacob (Jacob, 1954) used ultraviolet irradiation, while Kageyama (Kageyama, 1964) used mitomycin C to active growing cultures (1–2 μg/ml). The determination of pyocin and its kinetics activity are related in several publications (Takeya *et al* 1967; Higerd *et al.*, 1986; Kageyama, 1964; Ohkawa *et al.*, 1973).

5.5.9.1. Three types of pyocins

The first discovered pyocins was produced by *P. aeruginosa* R, called R-type pyocin which consists rod-like particle resembling a bacteriophage tail (Kageyama, 1964; Ishii *et al.*, 1965; Yui-Furihata, 1972). It was studied that all the R-type pyocins are resistant to enzymes such as nuclease and protease. These characteristics ease the purification of the pyocin and the production of specific antisera.

A second type pyocin was identified by (Takeya *et al* 1967) and named it as F. It consists flexuous rod-like structure. The F-type pyocin is also resistant to nuclease and protease. Futher, it was found that F-type pyocin was associated with R-type pyocins (Govan *et al.*,1974; Kuroda *et al.*,1979). When R-type pyocin from a bacterial strain is treated by an anti-R serum, it was then inoculated on a lawn of a susceptible strain, which results in wider zone of clearance (Ito *et al.*, 1970). The pyocin which was responsible for the above zone of inhibition was due to the third pyocin called as S-type pyocin. It a soluble pyocin which is only protease sensitive (Michel-Briand *et al.*, 2002).

The pyocin R appears to have an extended sheath (ES) consists of 34 annuli where each of them have six subunits (a). It also consists contracted sheath (CS), terminated by a base plate (BP) with tail fibres (TFi) and the core (C). Wheresas F-type pyocin appears to have flexuous rod like particle which consists of 23 annuli, a distal part (DP) and a fiber part (Fi) composed of three filaments with some globular structures **(Fig. 5.6)**.

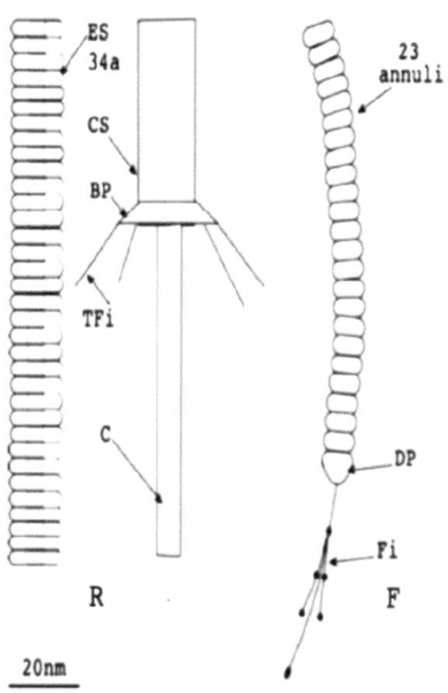

Fig. 5.6: Schematic representation of pyocins R and F.

5.6. Conclusion

To conclude, *Pseudomonas* contains multi-trait of PGP. The most intensively studied application for free living PGP *Pseudomonas* is agriculture. PGPP (Plant growth promoting *Pseudomonas*) influence plant growth by producing secondary metabolites such as siderophores, plant hormones (IAA, gibberellins) and certain biocontrol agents. PGPP also benefits agricultural crops with solubilization of phosphorous, induced systemic resistance and phytoremediation. In addition, PGPP produces biosurfactant that has a great potential for biotechnological applications. In future, the usage of PGPR replace chemical fertilizers which can cause side effects to agricultural crops. Further studies on PGPP nitrogen fixers would pave the way to find out more efficient strains that benefits the agricultural field. It is proposed that certain PGP *Pseudomonas* sp. can be used as an inoculant to obtain the desired PGP action in agricultural environment. Hence it is important to search for region specific microbial strains which can be used as a potential plant growth promoter to achieve desired product.

References

Abdul Halim NB (2009) Effects of using enhanced biofertilizer containing N-fixer bacteria on patchouli growth. Thesis. Faculty of Chemical and Natural Resources Engineering University Malaysia Pahang. p. 145

Ahmad F, Ahmad I, Khan MS, (2005) Indole acetic acid production by the indigenous isolates of Azotobacter and fluorescent Pseudomonas in the presence and absence of tryptophan. Turkish Journal of Biology, 29(1), 29-34.

AitBarka E, Belarbi A, Hachet C, Nowak J, Audran J C (2000) Enhancement of in vitro growth and resistance to gray mould of *Vitis vinifera* cocultured with plant growth-promoting rhizobacteria. FEMS Microbiol Lett 186:91-95

AitBarka E, Gognies S, Nowak, J. C. Audran, and A. Belarbi (2002) Inhibitory effect of endophyte bacteria on *Botrytis cinerea* and its influence to promote the grapevine growth. Biol Control 24(2): 135-142. Doi:10.1016/S1049-9644(02)00034-8

Alabouvette C, Lemanceau P, Steinberg C (1996) Biological control ofFusarium wilts: opportunities for developing a commercial product. Principles and practice of managing soilborne plant pathogens192-212

Ali B, Sabri AN, Hasnain S. (2010) Rhizobacterial potential to alter auxin content and growth of *Vigna radiate* (L.). World Journal of Microbiology and Biotechnology, 26(8): 1379-1384. Doi:10.1007/s11274-010-0310-1

Ashraf M, Berge SH, Mahmood OT (2004) Inoculating wheat seedling with exopolysaccharide-producing bacteria restricts sodium uptake and stimulates plant growth under salt stress. Biol Fertil Soils 40:157–62

Bais HP, Park SW, Weir TL, Callaway RM, and Vivanco JM (2004) How plants communicate using the underground information superhighway. Trends Plant Sci.9:26-32

Bakker AW, Schippers B (1987) Microbial cyanide production in the rhizosphere in relation to potato yield reduction and Pseudomonas spp. mediated plant growth-stimulation. Soil Biol Biochem19:451–7. Doi:10.1016/0038-0717(87)90037-X

Banat IM, Franzetti A, Gandolfi I, Bestetti G, Martinotti MG, Fracchia L, *et al* (2010) Microbial biosurfactants production, applications and future potential. Appl Microbiol Biotechnol 87(2):427–44. Doi:10.1007/s00253-010-2589-0

Beattie GA, Lindow SE (1999) Bacterial colonisation of leaves: a spectrum of strategies. Phytopathology 89:353–359. Doi: 10.1094/PHYTO.1999.89.5.353

Benhamou N, Belanger RR, Paulitz TC (1996) Induction of differential host responses by *Pseudomonas fluorescens* in Ri T-DNA-transformed pea roots after challenge with *Fusarium oxysporum* f. sp. *pisi* and *Pythium ultimum.* Phytopathology 86:114-178

Bestwick CS, Brown IR, Bennett MHR, Mansfield JW (1997) Localisation of hydrogen peroxide accumulation during the hypersensitive reaction of lettuce cells to *Pseudomonas syringae* pv. phaseolicola. Pl. Cell 9, 209–221

Boopathy R (2000) Factors limiting bioremediation technologies. Bioresour Technol 74:63–67

Brooks DS, Gonzalez CF, Apple DN, and Filer TH (1994) Evaluation of endophytic bacteria as potential biological control agents for oak wilt. Biol Control 4(4):373-381. Doi:10.1006/bcon.1994.1047

Brown IR, Mansfield JW, Taira S, Roine E, Romantschuk M (2001) Immunocytochemical localization of HrpA and HrpZ supports a role for the Hrp pilus in the transfer of effector proteins from *Pseudomonas syringae* pv. tomato across the host plant cell wall. Mol Pl–Microbe Inter- act 14, 394–404

Burr TJ, Schroth MN, Suslow T (1978) Increased potato yields by treatment of seed pieces with specific strains of *Pseudomonas fluorescens* and *P. putida.* Journal of Phytopathology 68:1377"1383

Castignetti D, Smarelli J (1986) Siderophores, the iron nutrition of plants, and nitrate reductase. FEBS Lett 209:147-151. Doi: 10.1016/0014-5793(86)81100-0

Chan YK, Barraquio WL, Knowles R (1994). N2-fixing pseudomonads and related soil bacteria. FEMS Microbiol Rev 13:95– 118

Chen Y, Rekha P, Arun A, Shen F, Lai WA, Young C (2006) Phosphate solubilizing bacteria from subtropical soil and their tricalcium phosphate solubilizing abilities. Appl Soil Ecol 34:33–41

Compant S, Duffy B, Nowak J, Clément C, Barka EA (2005a). Use of plant growth- promoting bacteria for biocontrol of plant diseases: principles mechanisms of action, and future prospects. Appl Environ Microbiol 71:4951–4959

Compant S, ReiterB, Sessitsch A, Nowak J, Clément C, Barka EA (2005b) Endophytic colonization of *Vitis vinifera* L. by plant growth-promoting bacterium Burkholderia sp. strain PsJN. Appl Environ Microbiol 71:1685–1693

Conn KL, Nowak J, and Lazarovits G (1997) A gnotobiotic bioassay for studying interactions between potato and plant growth-promoting rhizobacteria. Can J Microbiol 43:801-808

Cornelis P, Matthijs S (2002) Diversity of siderophore-mediated iron uptake systems in fluorescent pseudomonads: not only pyoverdines. Environ Microbiol 4:787-798

De Souza JT, de Boer M, de Waard P, van Beek TA, Raaijmakers JM (2003) Biochemical, genetic, and zoosporicidal properties of cyclic lipopeptide surfactants produced by *Pseudomonas fluorescens.* Appl Environ Microbiol 69(12): 7161-7172. DOI: 10.1128/aem.69.12.7161-7172.2003

Dean DR, Jacobson MR (1992) Biochemical genetics of nitrogenase. In Biological Nitrogen Fixation, pp. 763–834. Edited by G. Stacey, R. H. Burris & H. J. Evans. New York: Chapman & Hall

Défago G (1993) 2,4-Diacetylphloroglucinol, a promising compound in biocontrol. Plant Pathol 42:311-312. Doi:10.1111/j.1365-3059.1993.tb01506.x

Duffy B, Keel C, Défago G (2004) Potential role of pathogen signaling in multitrophic plant-microbe interactions involved in disease protection. Appl Environ Microbiol 70(3): 1836-1842 DOI: 10.1128/AEM.70.3.1836-1842.2004

Duffy BK, Défago G (1997) Zinc improves biocontrol of *Fusarium* crown and root rot of tomato by *Pseudomonas fluorescens* and represses the production of pathogen metabolites inhibitory to bacterial antibiotic biosynthesis. Phytopathology 87(12):1250-1257 DOI: 10.1094/PHYTO.1997.87.12.1250

Duffy BK, Défago G (1999) Environmental factors modulating antibiotic and siderophore biosynthesis by *Pseudomonas fluorescens* biocontrol strains. Appl Environ. Microbiol 65(6): 2429-2438

Dumville JC, Fry SC (2000) Uronic acid-containing oligosaccharins: their biosynthesis, degradation and signalling roles in non-diseased plant tissues. Pl Physiol Biochem 38:125 140

Elmerich C (1991) Genetics and regulation of Mo-nitrogenase. In Biology and Biochemistry of Nitrogen Fixation, pp. 103–141. Edited by M. J. Dilworth & A. R. Glenn. Amsterdam: Elsevier

El-Yazeid AA, Abou-Aly HA, Mady MA, Moussa SAM (2007) Enhancing growth, productivity and quality of squash plants using phosphate dissolving microorganisms (bio phosphor) combined with boron foliar spray. ResJ Agric Biol Sci 3(4): 274-286

Fischer HM (1994) Genetic regulation of nitrogen fixation in rhizobia. Microbiol Rev 58:352–386

Flores HE, Vivanco JM, Loyola-VargasVM (1999) 'Radicle' biochemistry: the biology of root-specific metab- olism. Trends Pl Sci 4:220–226

Gadd GM (2004) Microbial influence on metal mobility and application for bioremediation. Geoderma 122, 109–119

Gadd GM (2010) Metals: minerals and microbes: geomicrobiology and bioremediation. Microbiology 156: 609–643

Gail M. P. (2004) Plant Perceptions of Plant Growth-Promoting Pseudomonas. Philosophical Transactions: Biological Sciences 359(1446):907-918

Gilroy S, Jones DL (2000) Through form to function: root hair development and nutrient uptake. Trends Pl. Sci 5, 56–60

Glick B (1995) The enhancement of plant growth by free-living bacteria. Can. J. Microbiol 41:109-117

Glick BR (2010) Using soil bacteria to facilitate phytoremediation. Biotechnol. Adv. 28: 367–374.

Glick BR, Bashan Y (1997) Genetic manipulation of plant growth-promoting bacteria to enhance biocontrol of fungal phytopathogens. Biocontrol Adv 15:353–78

Glick BR, Cheng Z, Czarny J, Cheng Z, Duan J. (2007) Promotion of plant growth by ACC deaminase-producing soil bacteria. Eur J Plant Pathol;119:329–39. Doi:10.1007/s10658-007-9162-4

Govan JR (1974) Studies on the pyocins of Pseudomonas aeruginosa: morphology and mode of action of contractile pyocins. J Gen Microbiol 80(1): 1–15. DOI: 10.1099/00221287-80-1-1

Haas D, Keel C (2003) Regulation of antibiotic production in root-colonizing *Pseudomonas* spp. and relevance for biological control of plant disease. Annu Rev Phytopathol 41:117-153. DOI:10.1146/annurev.phyto.41.052002.095656

Hawes MC, Gunawardena U, MiyasakaS, ZhaoX (2000) The role of root border cells in plant defense. Trends Pl Sci 5: 128–133. DOI:10.1016/s1360-1385(00)01556-9

Holt JG, Krieg NR, Sneath PAP *et al* (1994) Bergy's Manual of Determinative Bateriology. 9th Ed, Williams and Wilkins Pub, Baltimore

Ishii SI, Nishi Y, Egami F (1965) The fine structure of a pyocin. J Mol Biol 13: 428–431

Itelima JU, Bang WJ, Onyimba IA, Sila MD, Egbere OJ (2018) Bio-fertilizers as key player in enhancing soil fertility and crop productivity: A Review. J Microbiol Biotechnol Rep. 2(1):22-28

Ito S, Kageyama M, Egami F (1970) Isolation and characterization of pyocins from several strains of Pseudomonas aeruginosa. J Gen Appl Microbiol 16:205–214

Johri BN (2001) Technology development and demonstration of a new bacterial inoculant (GRP3) for improved legume production. Uttar Pradesh Govternment, Project report

Kageyama M (1964) Studies of a pyocin. I. Physical and chemical properties. JBiochem (Tokyo) 55 49–53

Kageyama M, Ikeda K, Egami F (1964) Studies of a pyocin. III. Biological properties of the pyocin. J Biochem (Tokyo) 55:59–64

Katiyar V, Goel R (2003) Solubilization of inorganic phosphate and plant growth promotion by cold tolerant mutants of *Pseudomonas fluorescens*. *Microbiological research* 158(2):163-168. Doi:10.1078/0944-5013-00188

Khan S, Hesham AEL, QiaoM, Rehman S, He JZ (2010) Effects of Cd and Pb on soil microbial community structure and activities. Environ Sci Pollut Res 17:288–296.

Khosro M, Yousef S (2012) Bacterial bio-fertilizers for sustainable crop production: A review APRN Journal of Agricultural and Biological Science7(5):237-308

Kilic-Ekici O, Yuen GY (2004) Comparison of strains of *Lysobacter enzymogenes* and PGPR for induction of resistance against *Bipolaris sorokiniana* in tall fescue. Biol Control 30:446-455. Doi:10.1016/j.biocontrol.2004.01.014

Kloepper JW, Hume DJ, Scher FM, Singleton C, Tipping B, Lalibert EM, *et al* (1987) Plant growth-promoting rhizobacteria on canola (rapeseed). Phytopathology 71:42–6

Kloepper JW, Reddy MS, Rodriìguez-Kabana R, Kenney DS, Kokalis-Burelle N, Martinez-Ochoa N, Vavrina CS (2004) Application of Rhizobacteria in transplant production and yield enhancement. Acta Horticulturae 631: 217-229

Kojic M, Degrassi G, and Venturi V (1999) Cloning and characterization of the *rpoS* gene from the plant growth-promoting *Pseudomonas putida* WCS358: RpoS is not involved in siderophore and homoserine lactone production. Biochem Biophys Acta 1489(2-3): 413-420. DOI: 10.1016/s0167-4781(99)00210-9

Kribacho M (2010) Fertilizer ratios, krishak and bharati cooperative Ltd. Journal of Science 5 (8): 7 – 12.

Krotzky A, Werner D (1987) Nitrogen fixation in Pseudomonas stutzeri. Arch Microbiol 147:48–57. Doi:10.1007/BF00492904

Kumar A, Bisht BS, Joshi VD, Dhewa T (2011) Review on bioremediation of polluted environment: a management tool. Int J Environ Sci 1:1079–1093

Kuroda K, Kageyama M (1979) Biochemical properties of a new flexuous bacteriocin, pyocin F1, produced by Pseudomonas aeruginosa. J Biochem (Tokyo) 85:7–19

Lambrecht M, Okon Y, Vande Broek A, Vandeleyden J (2000) Indole-3-acetic acid: a reciprocal signalling molecule in bacteria–plant interactions. Trends Microbiol 8(7): 298–300. Doi: 10.1016/S0966-842X(00)01732-7

Lata, Saxena, A. K. and Tilak, K. V. B. R. (2002) Biofertilizers to augment soil fertility and crop production. In Soil Fertility and Crop Production (ed. Krishna, K. R.), Science Publishers, USA, pp. 279–312

Lee S, Reth A, MeletzusD, Sevilla M, Kennedy C (2000) Characterization of a major cluster of nif, fix and associated genes in a sugarcane endophyte, *Acetobacter diazotrophicus*. J Bacteriol 182(24):7088–7091. DOI: 10.1128/jb.182.24.7088-7091.2000

Lindow SE, Brandl MT (2003) Microbiology of the phyllosphere. Appl Environ Microbiol 69:1875–1883. DOI: 10.1128/AEM.69.4.1875-1883.2003

Lodewyckx C, Vangronsveld J, Porteous F, Moore ERB, Taghavi S, Mezgeay M, van der Lelie D (2002) Endophytic bacteria and their potential applications. Crit Rev Plant Sci 21:583-606. doi.org/10.1080/0735-260291044377

Loper JE, Henkels MD (1997) Availability of iron to *Pseudomonas fluorescens* in rhizosphere and bulk soil evaluated with an ice nucleation reporter gene. Appl Environ Microbiol 63:99-105

Loper JE, Henkels MD(1999) Utilization of heterologous siderophores enhances levels of iron available to *Pseudomonas putida* in the rhizosphere. Appl EnvironMicrobiol 65:5357-5363

Loper JE, Schroth MN (1986) Influence of bacterial source of indole-3- acetic acid of root elongation of sugar beet. Phytopathol 76: 386-389

Lugtenberg BJJ, Dekkers LC (1999) What makes Pseudomonas bacteria rhizosphere competent? Environ Microbiol 1(1): 9–13

Lugtenberg BJJ, DekkersL, BloembergGV (2001) Molecular determinants of rhizosphere colonization by *Pseudomonas*. A Rev Phytopathol 39:461–490. DOI:10.1146/annurev.phyto.39.1.461

Luo S, Xu T, Chen L, Chen J, Rao C, Xiao X, Wan Y, Zeng G, Long F, Liu C, Liu Y (2012) Endophyte-assisted promotion of biomass production and metal-uptake of energy crop sweet sorghum by plant-growth-promoting endophyte Bacillus sp. SLS18. Appl Microbiol Biotechnol.93:1745–1753. doi: 10.1007/s00253-011-3483-0. Doi: 10.1007/s00253-011-3483-0

M'Piga P, Belanger RR, Paulitz TC, Benhamou N (1997) Increased resistance to *Fusarium oxysporum* f. sp. *radicis-lycopersici* in tomato plants treated with the endophytic bacterium *Pseudomonas fluorescens* strain 63-28. Physiol Mol Plant Pathol 50(5): 301-320. DOI: 10.1006/pmpp.1997.0088

Ma Y, Prasad MNV, Rajkumar M, Freitas H (2011) Plant growth promoting rhizobacteria and endophytes accelerate phytoremediation of metalliferous soils. Biotechnol Adv29: 248–58

MacFaddin JF (2000) Biochemical Tests for Identi cation of Medical Bacteria. Williams and Wilkins, London, UK

Martínez-Viveros O, Jorquera MA, Crowley DE, Gajardo G, Mora ML (2010) Mechanisms and practical considerations involved in plant growth promotion by rhizobacteria. J Soil Sci Plant Nutr 10:293–319. Doi: 10.4067/S0718-95162010000100006

Mayak S, Tirosh T, Glick BR (2004) Plant growth-promoting bacteria that confer resistance to water stress in tomato and pepper. Plant Sci166:525–30

Mehnaz S, Weselowski B, Aftab F, Zahid S, Lazarovits G, Iqbal J (2009) Isolation, characterization, and effect of fluorescent pseudomonads on micropropagated sugarcane. Canadian Journal of Microbiology 55 (Suppl 8): 1007–1011

Michel-Briand Y, Baysse C (2002) The pyocins of Pseudomonas aeruginosa. *Biochimie*84(5-6):499-510. DOI:10.1016/s0300-9084(02)01422-0

Mishra M, Goel R (1999) Development of a cold resistant mutant of plant growth promoting Ps. fluorescens and its functional characterization. J Biotechnol 75:71–75. DOI: 10.1016/s0168-1656(99)00137-6

Nadeem SM ,Ahmad M, Zahir ZA, Javaid A,and Ashraf M (2014) The role of mycorrhizae and plant growth promoting rhizobacteria (PGPR) in improving crop productivity under stressful environments. Biotechnology Advances 32(2): 429–448. DOI:10.1016/j.biotechadv.2013.12.005

Nautiyal CS (1999) An efficient microbiological growth medium for screening phosphate solubilizing microorganisms. FEMS Microbiol Lett 17: 265–270. DOI: 10.1111/j.1574-6968.1999.tb13383.x

Nielsen TH, and Sørensen J (2003) Production of cyclic lipopeptides by *Pseudomonas fluorescens* strains in bulk soil and in the sugar beet rhizosphere. Appl Environ Microbiol 69(2):861-868. DOI: 10.1128/aem.69.2.861-868.2003

Nielsen TH, Sørensen D, Tobiasen C, Andersen JB, Christeophersen C, Givskov M, Sørensen J (2002) Antibiotic and biosurfactant properties of cyclic lipopeptides produced by fluorescent *Pseudomonas* spp. from the sugar beet rhizosphere. Appl Environ Microbiol 68(7):3416-3423. DOI: 10.1128/aem.68.7.3416-3423.2002

Nonnoi F, Chinnaswamy A, García de la Torre VS, Coba de la PeñaT, LucasMM, PueyoJJ, (2012) Metal tolerance of rhizobial strains isolated from nodules of herbaceous legumes Medicago spp. and Trifolium spp. growing in mercury contaminated soils Appl Soil Ecol 61:49–59

Notz R, Maurhofer M, Schnider-Keel U, Duffy B, Haas D, Défago G (2001) Biotic factors affecting expression of the 2,4-diacetylphloroglucinol biosynthesis gene *phlA* in *Pseudomonas fluorescens* biocontrol strain CHA0 in the rhizosphere. Phytopathology 91:873-881. DOI: 10.1094/PHYTO.2001.91.9.873

O'Sullivan DJ, O'Gara F (1992) Traits of fluorescent *Pseudomonas* spp. involved in suppression of plant root pathogens. Microbiol Rev 56(4):662-676

Ohkawa I, KageyamaM,Egami F (1973) Purification and properties of pyocin S2, J. Biochem. (Tokyo) 73:281–289

Ouzounidou G, Moustakas M, Symeonidis L, Karataglis S (2006) Response of wheat seedlings to Ni stress: effects of supplemental calcium. Arch Environ ContamToxicol 50: 346–352

Ovadis M, Liu X, Gavriel S, Ismailov Z, Chet I, CherninL(2004) The global regulator genes from biocontrol strain *Serratia plymuthica* IC1270: cloning, sequencing, and functional studies. J. Bacteriol 186(15):4986-4993. DOI: 10.1128/JB.186.15.4986-4993.2004

Pahari A., Pradhan A., Nayak S.K., Mishra B.B. (2017) Bacterial Siderophore as a Plant Growth Promoter. In: Patra J., Vishnuprasad C., Das G. (eds) Microbial Biotechnology. pp 163-180, Springer, Singapore. Doi:10.1007/978-981-10-6847-8_7

Patten CL, Glick BR (2002) Role of *Pseudomonas putida* indole acetic acid in development of the host plant root system. Appl Environ Microbiol 68:3795–801. DOI: 10.1128/aem.68.8.3795-3801.2002

Persello-Cartieaux F, Nussaume L, Robaglia C (2003) Tales from the underground: molecular plant–rhizobacteria interactions. Pl Cell Environ 26:189–199

Pettersson M, Bååth E (2004) Effects of the properties of the bacterial community on pH adaptation during recolonization of a humus soil. Soil Biol Biochem 36(9):1383-1388. Doi:10.1016/j.soilbio.2004.02.028

Picard C, Di Cello F, Ventura M, Fani R, Guckert A (2000) Frequency and biodiversity of 2,4-diacetylphloroglucinol-producing bacteria isolated from the maize rhizosphere at different stages of plant growth. Appl Environ Microbiol. 66(3):948-955. DOI: 10.1128/aem.66.3.948-955.2000

Raaijmakers JM, Vandersluis I, Koster M, Bakker PAHM, Weisbeek PJ, Schippers B (1995) Utilization of heterologous siderophores and rhizosphere competence of fluorescent *Pseudomonas* spp. Can J Microbiol 41:126-135

Raaijmakers JM, Vlami M, de Souza JT (2002) Antibiotic production by bacterial biocontrol agents. Antonie Leeuwenhoek 81:537-547. doi:10.1023/A:1020501420831

Rajkumar M, AeN, Prasad MNV, Freitas H (2010) Potential of siderophore- producing bacteria for improving heavy metal phytoextraction. Trends Biotechnol 28(3):142–149

RanjaniSN, NaikK, KushalaG (2018) Use of plant growth promoting microorganisms in plant propagation. Journal of Pharmacognosy and Phytochemistry7(3):478-481

Ravel J, Cornelis P (2003) Genomics of pyoverdine-mediated iron uptake in pseudomonads. Trends Microbiol 11(5):195-200. DOI: 10.1016/s0966-842x(03)00076-3

Rodriguez H, Fraga R, Gonzalez T, Bashan Y (2006) Genetics of phosphate solubilization and its potential applications for improving plant growth- promoting bacteria. Plant Soil 287:15–21

Rokhzadi A, Asgharzadeh A, Darvish F, Nour-Mohammadi G, Majidi E (2008) Influence of plant growth-promoting rhizobacteria on dry matter accumulation and yield of chickpea (*Cicer arietinum* L.) under field condition. Am-Euras J Agric Environ Sci 3(2): 253-257

Ryu CM, Farag MA, Hu CH, Reddy MS, Wei HX, Pare PW, Kloepper JW (2003) Bacterial volatiles promote growth in Arabidopsis. Proc Natl Acad Sci USA 100:4927–4932. DOI: 10.1073/pnas.0730845100

Saleh SS, and Glick BR (2001) Involvement of gacS and rpoS in enhancement of the plant growth promoting capabilities of *Enterobacter cloacae* CAL2 and UW4. Can J Microbiol 47:698-705

Sandhya V, Ali SKZ, Grover M, Reddy G, Venkateswarlu B (2009) Alleviation of drought stress effects in sunflower seedlings by the exopolysaccharides producing Pseudomonas putida strain GAP-P45. Biol Fertil Soils 46:17–26. Doi:10.1007/s00374-009-0401-z

Schnider-KeelU, Seematter A, Maurhofer M, Blumer C, Duffy B, Gigot-Bonnefoy C, Reimmann C, Notz R, DéfagoG, Haas D, C Keel (2000) Autoinduction of 2,4-diacetylphloroglucinol biosynthesis in the biocontrol agent *Pseudomonas fluorescens* CHA0 and repression by the bacterial metabolites salicylate and pyoluteorin. J Bacteriol 182(5):1215-1225. DOI: 10.1128/jb.182.5.1215-1225.2000

Sharma VK, Nowak J (1998) Enhancement of verticillium wilt resistance in tomato transplants by in vitro coculture of seedlings with a plant growth-promoting rhizobacterium (*Pseudomonas* sp. strain PsJN). Can J Microbiol 44:528-536

Sheng XF, Xia JJ, Jiang CY, He LY, Qian M (2008) Characterization of heavy metal-resistant endophytic bacteria from rape *Brassica napus* roots and their potential in promoting the growth and lead accumulation of rape. Environ Pollut 156:1164–1170.

Sinha RK, Valani D, Chauhan K, Agarwal S (2014). Embarking on a second green revolution for sustainable agriculture by vermiculture biotechnology using earthworms. International Journal of Agricultural Health Safety (1): 50 – 64

Smith KP, Handelsman J, Goodman RM (1999) Genetic basis in plants for interactions with disease-suppressive bacteria. Proc Natl Acad Sci USA 96(9): 4786-4790. DOI: 10.1073/pnas.96.9.4786

Stover CK, Pham XQ, Erwin AL, Mizoguchi SD, Warrener P, Hickey MJ, *et al* (2000) Complete genome sequence of *Pseudomonas aeruginosa* PA01, an opportunistic pathogen. Nature 406(6799): 959–964. DOI: 10.1038/35023079

Sturz AV, Christie BR (2003) Beneficial microbial allelopathies in the root zone: the management of soil quality and plant disease with rhizobacteria. Soil Tillage Res72(2):107-123

Suslow TV, Kloepper JW, Schroth MN, Burr TJ (1979) Beneficial bacteria enhance plant growth. CaliforniaAgriculture Online33 (Suppl 11): 15"17

Takeya K, Minamishima Y, Amako K, Ohnishi Y (1967) A small rod-shaped pyocin, Virology 31:166–168

Ullah A, Heng S, Munis MFH, Fahad S, Yang X(2015). Phytoremediation of heavy metals assisted by plant growth promoting (PGP) bacteria: a review. Environ Exp Bot 117:28-40

Van LoonL C, Bakker PAHM, Pieterse CMJ (1998) Systemic resistance induced by rhizosphere bacteria. AnnuRevPhytopathol 36:453-483. DOI: 10.1146/annurev.phyto.36.1.453

Van PeerR, Niemann GJ, Schippers B (1991) Induced resistance and phytoalexin accumulation in biological control of Fusarium wilt of carnation by *Pseudomonas* sp. strain WCS 417r. Phytopathology 81(7):728-734

Vermeiren H, Hai WL, Vanderleyden J (1998) Colonization and nifH expression on rice roots by Alcaligenes faecalis A15. In Nitrogen Fixation with Non-Legumes, pp. 167–177. Edited by K. A. Malik, M. S. Mirza & J. K. Ladha. Dordrecht: Kluwer.

Vermeiren, H., Willems, A., Schoofs, G., de Mot, R., Keijers, V., Hai, W. & Vanderleyden, J. (1999). The rice inoculant strain A15 is a nitrogen-fixing *Pseudomonas stutzeri* strain. Syst Appl Microbiol 22:215–224

Vessey JK (2003) Plant growth promoting Rhizobacteria as bio-fertilizers. Journal of Plant and Soil 25 (43): 511 – 586. doi:10.1023/A:1026037216893

Viswanathan R, Samiyappan R (1999). Induction of systemic resistance by plant growth-promoting rhizobacteria against red rot disease caused by *Colletotrichum falcatum* went in sugarcane. *Proceedings of the sugar technology association of India*, 61: 24-39.

Wachowska U, Okorski A, G³owacka K (2006) Population structure of microorganisms colonizing the soil environment of winter wheat. Plant, Soil and Environment52: 39–44.

Wang Y, Brown HN, Crowley DE, SzaniszloPJ(1993) Evidence for direct utilization of a siderophore, ferrioxamine B in axenically grown cucumber. Plant Cell Environ16:579-585.

Wei L, Kloepper JW, Tuzun S (1991) Induction of systemic resistance of cucumber to *Colletotrichum orbiculare* by select strains of plant growth-promoting rhizobacteria. Phytopathology 81:1508-1512

Whipps JM (2001) Microbial interactions and biocontrol in the rhizosphere. J Exp Bot 52:487-511

Widnyana, KI. (2011) Efforts to Obtain biocontrol agents Wilt Disease of Tomato *Fusarium oxysporum* f.sp *lycopersici* Through Exploration and Potential Test Isolate PGPR *Pseudomonas* spp. (in bahasa) Jurnal Bumi Lestari Lingkungan Hidup Vol.11/ No.2/ Agustus/ 2011 ISSN. 1411-9668 hal. 265 – 276.

Wisniewski, M., Lindow, S. E. & Ashworth, E. N. 1997 Observations of ice nucleation and propagation in plants using infrared video thermography. Pl Physiol113:327–334

Xie H, Pasternak JJ, Glick BR (1996) Isolation and characterization of mutants of the plant growth-promoting rhizobacterium *Pseudomonas putida* GR12-2 that overproduce indole acetic acid. Curr Microbiol ;32(2): 67–71.

Yan-de J, Zhen-li H, Xiao-e Y (2007) Role of soil rhizobacteria in phytoremediation of heavy metal contaminated soils. J Zhejiang Univ Sci B8:192–207

Yang SZ, Jin HJ, Wei Z, He RX, Ji YJ, Li XM, Yu SP (2009) Bioremediation of oil spills in cold environments: a review. Pedosphere 19:371–381

You CB, Song HX, Wang JP, Lin M, Hai WL (1991)Association of Alcaligenes faecalis with wetland rice. Plant Soil 137:81–85

Young JPW (1992) Phylogenetic classification of nitrogen-fixing organisms. In Biological Nitrogen Fixation, pp. 43–86. Edited by G. Stacey, R. H. Burris & H. J. Evans. New York: Chapman & Hall.

Yui-Furihata C (1972) Structure of pyocin R. II. Subunits of sheath, J Biochem (Tokyo) 72:1–10

Zahir ZA, Munir A, Asghar HN, Shahroona B, Arshad M (2008) Effectiveness of rhizobacteria containing ACC-deaminase for growth promotion of peas (*Pisum sativum*) under drought conditions. J Microbiol Biotechnol, 18:958–63.

6

Use of Biofertilizers for Sustainable Agriculture

Arup Sen[1], Kaushik Batabyal[1], Sanchita Mondal[2]
Arindam Sarkar[1] and Dibyendu Sarkar[1]

[1]*Department of Agricultural Chemistry and Soil Science*
Bidhan Chandra Krishi Viswavidyalaya, Mohanpur, Nadia
West Bengal-741 252
[2]*Department of Agronomy, Bidhan Chandra Krishi Viswavidyalaya*
Mohanpur, Nadia, West Bengal-741 252

Abstract

Sustainable agriculture is vital in today's world as it offers the potential to meet our agricultural needs, something that conventional agriculture fails to do. Current soil management strategies are mainly dependent on inorganic chemical fertilizers, which caused a serious threat to human health and environment. To overcome the ecological problems resulting from the loss of plant nutrients and to increase crop yields in the absence of resources for obtaining costly fertilizers, microscopic organisms that allow more efficient nutrient use or increase nutrient availability can provide sustainable solutions for the present and future agricultural practices. Microbial populations are instrumental to fundamental processes that drive stability and productivity of agro-ecosystems. Several investigations addressed at improving understanding of the diversity, dynamics and importance of soil microbial communities and their beneficial and cooperative roles in agricultural productivity. Thus the technique, which uses microscopic organism as biofertilizer, is environment friendly and ensures safe and healthy agricultural products. In this chapter we aim to provide a brief overview of potential use of biofertilizer for attaining sustainability of agro-ecosystem.

Keywords: Sustainable agriculture, chemical fertilizer, microscopic organism, biofertilizer

6.1. Introduction

Conventional agriculture plays a significant role in meeting the food demands of growing human population, which has also led to an increasing dependence on chemical fertilizers and pesticides (Santos *et al.*, 2012). These increases, however, have come with tremendous environmental costs. High and indiscriminate fertilizer consumption coupled with intensive farming have created environmental problems such as deterioration of soil quality, surface water, and groundwater quality, as well as air pollution, reduced biodiversity, and impaired ecosystem functions (Schultz *et al.*, 1995; Vance, 2001). Efficient and sustainable practices are needed to allow cost-efficient and adequate agricultural production for nutrition of the growing populations (Gentili and Jumpponen, 2005). Globally, there is urgent need for sustainable agricultural practices. Accordingly, scientists and researchers are arguing in favour of organic fertilizers and biofertilizer as the best solution to avoid soil pollution and many other threats to environment and life caused by overuse of chemical fertilizers.

The rhizospheric soils contain diverse type of efficient microbes with beneficial effects on crop productivity. Such microorganisms may comprise of mixed population of naturally occurring microbes that can be applied as inoculants to increase soil microbial diversity. Investigations have shown that the inoculation of efficient microbial community to the soil ecosystem improves soil health, growth, yield and quality of crops. The biofertilizer production technology includes isolation of suitable microbial strains, selection of beneficial organism, preparation of mother and seed culture, inoculants production, carrier preparation and their mixing, followed by curing, packaging, storage and dispatch. So biofertilizer is essentially a substance containing living organism *viz.*, bacteria, algae, and fungi either in isolation or in combination for application to seeds, plant surfaces, or to soil near the rhizosphere and this may help for increasing the supply or availability of primary nutrients to the host plant. Biofertilizer have shown great potential for nitrogen fixation, solubilizing phosphorus, and stimulating plant growth through the synthesis of growth promoting substances. Biofertilizer is sustainable or environmental friendly, low cost, renewable source of plant nutrient. The microorganisms in biofertilizer help restore the soil's natural nutrient cycle (N and P mainly) and build soil organic matter. In this chapter we aim to provide a brief overview of potential use of various biological agents with crop-yield-improving properties.

6.2. Efficient soil microbes used as biofertilizers

Such microorganisms may comprise of mixed populations of naturally occurring microbes that can be applied as inoculants to increase soil microbial diversity. Inoculation of efficient microbial community to the soil ecosystem improves soil quality, soil health, growth, yield and quality of crops. These microbial populations may consist of selected species of microorganisms including N_2-fixing microorganism, P mobilizing microorganism, plant growth promoting rhizobacteria, etc.

6.2.1. N_2 fixing microorganisms as biofertilizer

Although abundant and ubiquitous in the air, N is the most limiting nutrient to plant growth because the atmospheric N is not available for plant uptake. Some bacteria are capable of N_2 fixation from the atmospheric N pool. Some of these are free-living N_2 fixing bacteria, some have adapted to form symbiotic associations with plants and others live in close association in the plant root zone (rhizosphere) without forming symbiotic association with plant. In the following section, we discuss the different types of microorganisms as potential biofertilizers which are capable of allowing plant access to the atmospheric N pool.

6.2.1.1. Symbiotic N_2 fixing microorganism

Rhizobium

As per Fred *et al.* (1932) the genus *Rhizobium* was established in 1889 by Frank based on its ability to form nodules on the roots of legumes. The genus *Rhizobium* along with the genera *Agrobacterium* and *Chromobacterium* comprise the family Rhizobiaceae (Breed *et al.*, 1957). *Rhizobium* is an aerobic and heterotrophic bacteria found in the nodules of legumes which fixes the atmospheric nitrogen (N_2) symbiotically. It colonizes on the root of specific legumes. It's a symbiotic association in which bacteria derive carbohydrates from the root tissues of the host plant which in return obtain nitrogenous compound synthesized by the bacteria for their growth and development. *Rhizobium* bacteria enters in to the root hairs and develop nodules varies from whitish to brown, green and pink. Pink colour nodules are the sign of presence of leghemoglobin, which efficiently fixes the atmospheric nitrogen (N_2) and called as effective nodules. There is considerable specificity between *Rhizobium* species and the host plants. The classification of these organisms is based on plant infection into seven cross inoculation groups to form nodules as per Fred *et al.* (1932) and are listed in **Table 6.1**.

Table 6.1: Cross-inoculation groups of *Rhizobium*

Rhizobium species	Cross-inoculation group	Legumes
R. trifolii	Clover group	*Trifolium*
R. meliloti	Alfalfa group	*Trigonella, Melilotus, Medicago*
R. phaseoli	Bean group	*Phaseolus*
R. lupine	Lupine group	*Lupinus, Ornithopus*
R. leguminosarum	Pea group	*Pisum, Vicia, Lens*
R. japonicum	Soybean group	*Glycine*
Rhizobium sp. (miscellany)	Cowpea (miscellany) group	*Vigna, Arachis*

The N_2 fixing capability of rhizobia varies greatly (up to 450 kg N/ha) among host plant species and bacterial strains (Stamford *et al.*, 1997; Unkovich and Pate, 2000). Therefore, selection of the best strain must take rhizobia-host compatibility into account for production of biofertilizers. They must have a high N_2 fixation rate and be able to compete with the indigenous rhizobia to maximize infection of the target crops (Stephens and Rask, 2000). From a practical perspective, the inoculums must be easily produced and have a high survival rate in field condition once inoculated on target seeds (Date, 2000).

Rhizobium inoculants were reported to significantly increase the grain yields of different pulse crops in different locations and soil types. An experiment conducted in Tamil Nadu showed the benefits of *Rhizobium* inoculation in pigeon pea and chick pea (**Table 6.2**). A comparison between *Rhizobium* inoculation and 25 kg N/ha as chemical fertilizer without *Rhizobium* inoculation showed that in pulses *Rhizobium* inoculation is more beneficial than chemical fertilizer. The status of soil N is also improved by growing pulses inoculated with appropriate culture.

Table 6.2: Increase in yield and soil nitrogen by cultivation of *Rhizobium* inoculated pigeon pea and chickpea (Source: Gaur, 2006).

Treatment	Grain yield		Soil N after cultivation (kg/ha)	Gain in N (kg/ha)
	kg/ha	% increase over uninoculated		
Pigeon pea (soil N status was 122 kg N/ha before cultivation)				
Uninoculated	941	-	128	6
Rhizobium inoculated	1043	10.8	152	30
Fertilizer N at 25 kg/ha	963	2.3	147	25
Chickpea (soil nitrogen status 188 kg N/ha) before cultivation				
Uninoculated	532	-	219	31
Rhizobium inoculated	581	9.2	238	50
Fertilizer N at 25 kg/ha	604	13.5	216	28

Cyanobacteria

Due to evolutionary antiquity of cyanobacteria, they are widely adapted to survive against various extreme environmental conditions such as drought, salinity, low to high temperatures, etc. An aquatic cyanobacterium, *Trichodesmium* contributes approximately 36% of global N_2 fixation (Gallon, 2001). Cyanobacterial N_2 fixation has been found essential in rice cultivation system. Azollae in association with blue green algae can fix atmospheric N, amounting to 100-150 kg/ha/year in rice field. Until the end of the 1970s, *Azolla-Anabaena* symbiosis was the major N source for the 6.5×10^6 ha of rice cultivation in China. In many parts of Asia, the cyanobacteria (mainly *Nostoc* and *Anabaena*) still have importance for rice-field fertility (Kundu and Ladha, 1995).

Besides the contribution to nitrogen fertilization, growth promoting substances liberated by these algae play important role in sustaining the crop yield. Production of auxin-like substances and vitamins by *Cylindrospermum musicola* increased the root growth and yield of rice (Venkataraman and Neelakantan, 1967). A number of other growth promoting substances such as amino acids, sugars, polysaccharides, vitamins, growth hormones (IAA, 3- methyl Indole) towards improving and sustaining crop growth have been documented in the literature (Kaushik, 1998).

Dry land soils in desert and semi-arid regions suffer from major constraints like poor physical properties, low organic matter and fertility and higher water deficiency. Cyanobacterial application to such organically poor soils played a significant role in improving the status of carbon, nitrogen and other nutrients (Nisha *et al.*, 2007). Diazotrophic cyanobacteria which are photoautotrophic and N_2-fixing improve crop production by acting as natural biofertilizers through increase in both C and N status of soils.

Cyanobacteria can be used to rehabilitate and reclaim the saline soils as they form a thick stratum on the soil surface during the favourable months of rainy and winter seasons (Pandey *et al.*, 2005a;2005b). The metabolite synthesized by cyanobacteria when incorporated in the soil, may help to conserve organic C, organic N, and organic P as well as moisture, and convert Na^+-clay complexes to Ca^{2+}-clay complexes and enhance soil properties (Vaishampayan *et al.*, 2001). Organic matter and N added by cyanobacteria may bind the soil particles, and thus improve soil permeability, aeration and fertility. Application of mixed cyanobacterial inoculum to saline soils showed a significant decrease in pH with increase in total soil N, P and organic-C (**Table 6.3**).

Table 6.3: Changes in physico-chemical characteristics of saline soil due to application of cyanobacterial inoculum (*Nostoc calcicola*) (*Source*: Pandey *et al*., 2005a; 2005b).

Soil properties	Uninoculated cyanobacterial growth condition	Inoculated cyanobacterial growth condition
pH	10.40	5.0–8.80
Organic-C (%)	0.12	0.59–0.66
Total-N (%)	0.02	0.17–0.18
Total-P (%)	0.03	0.03–0.3
C/N ratio	6.00	3.28–4.00
Na^+ (mg/kg)	0.78	0.60

There are several metal pollutants viz., Cu, Zn, Ni, Co, Pb, Cr, Cd, which are commonly encountered in the soil systems and the application of cyanobacteria is helpful in reducing the heavy metal pollution in soil (Kaushik *et al*., 1999). Among the photoautotrophs, cyanobacteria are relatively more tolerant to heavy metals (Fiore and Trevors, 1994). Some efficient cyanobacteria reported for the remediation of heavy metals polluted soils has been presented in **Table 6.4**.

Table 6.4: Cyanobacteria reported for the remediation of heavy metals from soil systems.

Cyanobacteria	Toxicants	References
Microcystis aeruginosa f. flosaquae strain C3–40	Cd, Cu, Pb, Mn and Zn	Parker *et al*. (2000)
Synechococcus sp.	Cu, Pb, Ni and Cd	Yee *et al*. (2004)
Limnothrix planctonica, *Synechococcus leopoldiensis* and *Phormidium limnetica*	Hg	Lefebvre *et al*. (2007)
Nostoc calcicola and *Chroococcus* sp.	Cr	Anjana *et al*. (2007)
Lyngbya and *Gloeocapsa*.	Cr	Kiran *et al*. (2008)

Cyanobacterial biofertilizers mobilize nutritionally important elements such as P from a non-usable to a usable form through biological processes (Hegde *et al*., 1999). Thus, there are tremendous role of cyanobacteria as a biofertilizer for supporting sustainable agricultural practices in various environments (Kannaiyan, 2002).

6.2.1.2. Non-symbiotic N_2 fixing Microorganisms

Azospirillum

Azospirillum is an associative micro-aerophilic N_2 fixer which colonizes the root mass and fixes N in loose association with plants. *Azospirillum* colonizes a great variety of annual and perennial plants, many of which have never been reported to be colonized by N_2-fixing bacteria. Accordingly, *Azospirillum* possesses a great potential as a general root colonizer, whose use is not limited by host specificity (Bashan and Holguin, 1997). This fixes the N in low oxygen tension. These bacteria induce the plant root to secrete mucilage which creates the low oxygen environment for binding the atmospheric N. They are efficient for N-fixation in low land rice. Indeed, several studies indicate that *Azospirillum* can increase the growth of various crops. These include sunflower, carrot, oak, sugar beet, tomato, eggplant, pepper, and cotton in addition to wheat and rice (Bashan and Holguin, 1997). In two decades of field experiments, general consensus is that in 60-70% of the cases *Azospirillum* application resulted significant increase in crop yield (Okon and Labandera-Gonzalez, 1994). The yield increases can be substantial to the tune of up to 30%, but generally range from 5-30%. Rai and Gaur (1982) tested one strain of *Azospirillum lipoferum* having high N fixing capacity as inoculant to supplement the nitrogen need of a wheat crop in field condition with different doses of N with and without the inoculant (**Table 6.5**). The results showed that the inoculant could contribute about 40 kg N/ha to the crop.

Table 6.5: Effect of *Azospirillum lipoferum* on the yield and nitrogen uptake of wheat crop

Treatments	Grain		Straw	
	Yield (q/ha)	N uptake (kg/ha)	Yield (q/ha)	N uptake (kg/ha)
Control	12.6	24.8	17.8	8.2
Inoculation	20.7	42.7	21.1	14.9
40 kg N	23.7	51.2	32.4	18.7
40 kg N + inoculation	31.1	65.5	38.2	21.9
80 kg N	29.6	61.0	36.9	18.9
80 kg N + inoculation	41.5	85.4	48.9	26.5
C.D. at 5%	7.3	-	6.8	-

Though there are enough evidences of N_2 fixation by *Azospirillum* in soil, some scientists believe that the yield increases by *Azospirillum* are possibly a result of the production of growth-promoting substances rather than N_2 fixation (Okon, 1985). Tien *et al.* (1979) found increased yield of pearl millet due to *Azospirillum* derived IAA, GA and cytokinin like growth promoting substances. Indole acetic acid and indole lactic acids were formed from tryptophane by *A. brasilense*. Its growth promoting properties are fairly well documented, and its commercial production as well as field application are also simple. Inoculums can be produced and applied as in peat formulation, production of which is inexpensive. The peat formulation can also be directly utilized in field research and agricultural applications. However, it requires further research to allow selection of a reliable and effective means for inoculums production and field application (Vande Broek *et al.*, 2000).

Azotobacter

Azotobacter is free living heterotrophic aerobic bacteria and fixes atmospheric N in the rhizosphere. It is susceptible to water logged conditions. *Azotobacter* species are gram-negative bacteria found in neutral and alkaline soils and sensitive to acidic reactions (pH less than 6.0). Its inoculation is useful to increase germination rate, root and shoot length, grain yield, improve post-harvest seed quality and nitrogen nutrition in cereals and non-leguminous crop plants. Meshram and Shende (1982) showed increased N uptake by maize due to *Azotobacter* inoculation (**Table 6.6**).

Table 6.6: Total uptake of nitrogen by maize crop (kg/ha) as influenced by *Azotobacter* inoculation, FYM and levels of nitrogen

Treatment			Average of 2 years (1979 and 1980)
N (kg/ha)	FYM (1000 kg/ha)	Azotobacter (packets/ha)	
0	0	0	39.27
0	10	0	41.20
0	0	3	53.27
0	10	3	61.06
40	0	0	68.21
80	0	0	75.72
120	0	0	74.73
40	10	0	69.31
80	10	0	77.25
120	10	0	78.34
40	0	3	72.40
80	0	3	76.03
120	0	3	73.55
40	10	3	82.27
80	10	3	89.12
120	10	3	79.98

The cultures of *A. chroococcum* synthesizes considerable quantities of biologically active growth promoting substances and most of them belong to B group vitamins viz., nicotinic and panthothenic acids, biotin and heteroauxins and gibberellins (Gaur, 2006). It also produces cytokinin like substances. So, it is possible that the detected growth response was also due to the bacterial synthesis of secondary growth-promoting compounds, such as plant growth hormones (Polyanskaya *et al.*, 2002). Besides, *Azotobacter* also produces antibiotic and the value of *Azotobacter* for the recovery of plant diseases is already established (Meshram, 1984).

6.2.2. Phosphorus mobilizing microorganism as biofertilizer

6.2.2.1. Phosphorus solubilizing microorganism

Total P content in soil is usually high, but most of this soil P pool is not in forms available for plant uptake. Bacteria that can mobilize P from unavailable soil pools and increase P availability to plants are of great importance (**Table 6.7**).

Table 6.7: List of some phosphorus solubilizers or phytase producers

Genus	Reference
P solubilizers	
Aspergillus, Penicillium, Trichoderma	Barthakur, (1978)
Bradyrhizobium, Rhizobium	Antoun *et al.*, (1998)
Enterobacter	Kim *et al.*, (1997b)
Gordonia	Hoberg *et al.*, (2005)
Pantoea	Deubel *et al.*, (2000)
Pseudomonas	Deubel *et al.*, (2000); Hoberg *et al.*, (2005)
Rahella	Kim *et al.*, (1997a)
Phytase producers	
Aspergillus, Emmericella, Penicillum	Yadav and Tarafdar, (2003)
Peniophora	George *et al.*, (2007)
Pseudomonas	Richardson and Hadobas, (1997)
Telephora, Suillus (ectomycorrhizal fungi)	Colpaert *et al.*, (1997)

Most predominant phosphorus solubilizing bacteria (PSB) belong to the genera *Bacillus* and *Pseudomonas* (Richardson, 2001). In several pot and field experiments, inoculation with P solubilizing microorganisms resulted in increased plant growth and P uptake (Kumar and Narula, 1999). Also PSB application to Sandy loam soil has been found to increase the grain yield and P_2O_5 uptake of soybean (**Table 6.8**).

Table 6.8. Effect of phosphate solubilizing microorganisms on grain yield and uptake of phosphorus by soybean crop (Source: Gaur, 2006).

Treatment	Grain(q/ha)	Total P_2O_5 uptake (kg/ha)
Control	16.50	7.458
Aspergillus awamori	17.24	10.530
Pseudomonas striata	17.86	11.578
Bacillus polymyxa	17.30	10.210
Rock Phosphate (200 kg P_2O_5/ha)	17.38	9.706
RP+*A. awamori*	18.14	11.568
RP+*P. striata*	18.92	13.170
RP+*B. polymyxa*	18.92	12.772
Superphosphate (80 kg P_2O_5/ha)	17.40	10.908
C.D. at 5% level	1.38	0.698

Besides, phosphate solubilizing microorganisms also produce growth promoting substances and biocontrol agents. Naumova *et al.* (1962) reported that *Bacillus megaterium* var. *phosphacticum* and *B. fluorescens* could accumulate some biologically active compounds such as auxins, gibberellin, 'vitamins, etc. in the medium, which can stimulate plant growth and inhibit the growth of fungi like *Fusarium* and *Alternaria*. Barea *et al.* (1976) found that out of 50 PSB, 20 synthesized all the three types of plant hormones, 43 produced IAA, 29 formed gibberellins and 45 cultures produced cytokinin-like substances. Thus, PSB may play significant role in increasing use of cheaper P sources (e.g., rock phosphate instead of superphosphate) that reduces much dependency on chemical P fertilizers.

6.2.2.2. Phosphorus absorber

Mycorrhizas are symbiotic, generally mutualistic organisms which establish balanced associations between soil fungi and most vascular plants where both partners exchange nutrients and energy (Brundrett, 2002). The mycorrhizal fungi colonize the root cortex of most plant species and, once biotrophically established into the root tissues, they develop an extra radical mycelium which overgrows the soil surrounding plant roots. This hyphal net is a structure specialized for the acquisition of mineral nutrients from the soil, particularly those whose ionic forms have poor mobility or are present in low concentration in the soil solution, as is the case with P (Barea, 1991). It provides the plant with an adaptive strategy for P acquisition in soils with low P availability.

The widespread and ubiquitous mycorrhizal type is characterized by the tree like symbiotic structures, termed as “arbuscules”, that the fungus develops within the root cortical cells, and where most of the nutrient exchange between the fungus and the plant is thought to occur. Arbuscular mycorrhiza (AM) inoculation enhances plant growth by improving mineral nutrition. These AM associations are especially important for P acquisition by plants growing in both agricultural and natural systems (Barea, 1991). The AM hyphae can grow beyond the root in P depletion zone and deliver the intercepted P to the plant and possibly that is the reason why AM increase P accumulation and plant growth in soils with low P availability. The increased plant growth with AM inoculation is also due to greater uptake of other secondary- and micro-nutrients like zinc, molybdenum, copper, potassium, sulphur, iron, manganese. The AM symbiosis not only influences nutrient cycling in soil-plant systems but also improves plant health through increased protection against environmental stresses, whether they be biotic (e.g., pathogen attack) or abiotic (e.g., drought, salinity, heavy metals, organic pollutants), and enhancing soil structure through the formation of the aggregates necessary for good soil tilth (Rillig and Mummey, 2006). Thus, a commercial inoculum of AM fungi could be a good substitute for chemical fertilizer.

6.2.3. Plant growth promoting rhizobacteria (PGPR) as biofertilizer

Various bacteria together can promote plant growth. Collectively such bacteria are called plant-growth-promoting rhizobacteria (PGPR). The PGPR exert a direct effect on plant growth by production of phytohormones, solubilization of inorganic phosphates, increased iron nutrition through iron-chelating siderophores and the volatile compounds that affect the plant signalling pathways. A group of biofertilizers comparising beneficial rhizobacteria, identified as PGPR, are strains from genera of *Pseudomonas*, *Azospirillum*, *Azotobacter*, *Bacillus*, *Burkholderia*, *Enterobacter*, *Rhizobium*, *Erwinia* and *Flavobacterium* (Rodriguez and Fraga, 1999). A list of some PGPR regulating various growth parameters of crop plants has been given in **Table 6.9**.

Table 6.9. Plant growth promoting rhizobacteria regulating various growth parameters/yields of crop/fruit plants.

PGPR	Crop parameters	References
Pseudomonas putida G 12–2	Early developments of canola seedlings	Glick *et al.* (1997)
P. fluorescens strain	Growth of pearl millet	Niranjan *et al.* (2003)
P. putida strain	Growth stimulation of tomato plant	Gravel *et al.* (2007)
P. alcaligenes PsA15, *Bacillus polymyxa* BcP26, and *Mycobacterium phlei* MbP18,	Enhance uptake of N, P and K by maize crop in nutrient deficient calcisol soil	Egamberdiyeva (2007)
P. putida strains R-168 and DSM-291; *P. fluorescens* strains R-98 and DSM-50090;	Improves seed germination, seedling growth and yield of maize	Nezarat and Gholami (2009)
P. fluorescens strain R-93, *P. fluorescens* DSM 50090, *P.putida* DSM291, *P. fluorescens* strains, CHA0 and Pf1	Increase growth, leaf nutrient contents and yield of banana cv. Virupakshi (*Musa* spp. AAB) plants	Kavino *et al.* (2010)

The soil-borne *Pseudomonas* have received paramount attention because of their catabolic versatility, efficient root-colonising ability and capacity to produce a wide range of enzymes and metabolites that help the plant withstand varied biotic and abiotic stresses. Kohler *et al.* (2006) demonstrated the beneficial effect of PGPR *Pseudomonas mendocina* strains on stabilization of soil aggregate. The three PGPR isolates *Pseudomonas alcaligenes* PsA15, *Bacillus polymyxa* BcP26 and *Medicago sativa* MbP18 were able to tolerate high temperatures and salt concentrations and thus confer on them potential competitive advantage to survive in arid and saline soils such as calcisol (Egamberdiyeva, 2007). Relatively recently, it was discovered that many PGPB contain the enzyme 1-aminocyclopropane-1-carboxylic acid (ACC) deaminase that cleave the ethylene precursor ACC to α-ketobutyrate and ammonia and thereby lower the ethylene levels in developing or stressed plants (Saleem *et al.*, 2007). The plants treated with bacteria containing ACC-deaminase may have relatively extensive root growth due to lowered ethylene levels thus leading to resistance against various stresses (Safronova *et al.*, 2006).

The PGPR may also protect plants against pathogens either by direct antagonistic interactions with the pathogen or through induction of host resistance or by both. PGPR that indirectly enhance plant growth via suppression of phytopathogens do so by a variety of mechanisms. These include, the ability to produce siderophores to chelate iron, making it unavailable to pathogens, to synthesize antifungal metabolites such as antibiotics, fungal cell wall lysing enzymes, or hydrogen cyanide that suppress the growth of fungal pathogens, to successfully compete with pathogens for nutrients or specific niches on the root and to induce systemic resistance (Persello Cartieaux *et al.*, 2003).

6.2.4. Multiple inoculations and interactions among biofertilizers

Many scientists refer to the practice of inoculation and introduction of more than one fungus and/or bacterium into the target crops as multiple inoculations. For example, some bacterial genera e.g., *Paenibacillus* are able to stimulate mycorrhizal colonization and are often isolated from rhizosphere of mycorrhizal plants (Poole *et al.*, 2001). These bacteria may also provide good crop yield with plant-growth-promoting effect. Ruiz-Lozano and Bonfante (2001) hypothesized that the bacterial association with the AM fungi may positively influence the host plant nutrient uptake and nutrient transport from the AM fungus to the plant. Clearly, there is some evidence that bacteria may influence mycorrhizal fungi and fungal colonization of plant. Co-inoculation with AM fungi and *Frankia* stimulated N_2 fixation and AM development in *Hippophaë tibetana* (Tian *et al.*, 2002). In alfalfa *(Medicago sativa)*, N and P acquisition was stimulated by inoculation with *Rhizobium,* AM fungi, and phosphate solubilizing bacteria (Toro *et al.*, 1998). Kohler *et al.* (2009) investigated the positive influence of inoculation with a PGPR, *Pseudomonas mendocina*, in combination with an AM fungus, *Glomus intraradices* or *Glomus mosseae* on growth and nutrient uptake and other physiological activities of *Lactuca sativa* affected by salt stress (**Table 6.10**). Thus, it can be stated that fungi and bacteria have various facilitative interactions, which may be of use in biofertilizer applications once the compatible combinations of fungi and bacteria are identified.

Similar interactions also occur among bacteria. For example, we briefly list a few experimental systems that present evidence for facilitative interactions among bacteria. *Pseudomonas fluorescens* increased nodulation and

Table 6.10: Effect of inoculation with *G. intraradices*, *G. mosseae* and *P. mendocina* on foliar nutrient contents of *L. sativa* seedlings grown at three levels of salinity (n = 5) (Source: Kohler et al., 2009).

	N (g/kg)	P (mg/kg)	K (mg/kg)	Na (mg/kg)	Fe (mg/kg)
Without NaCl					
Control	16.8a	1987ab	48,750d	7339b	341c
Fertilized	16.7a	2535c	50,929de	6807ab	283bc
P. mendocina	18.6ab	2177b	50,536de	6039a	226bc
G. intraradices	17.4a	2733d	49,250de	6743ab	253bc
G. mosseae	19.5ab	2257bc	52,929ef	8682b	371c
P. mendocina + *G. intraradices*	21.0b	3222e	53,250f	6396a	252bc
P. mendocina + *G. mosseae*	20.0b	2420cd	57,321g	8043b	279bc
2 g NaCl /kg soil					
Control	23.1bc	1886a	45,286bc	34,629d	321bc
Fertilized	27.4de	2196b	45,286bc	33,307d	234bc
P. mendocina	22.9bc	2038ab	42,929bc	27,507c	403c
G. intraradices	23.9bc	2325bc	43,571bc	25,121c	321bc
G. mosseae	22.2bc	2393bc	44,786bc	26,621c	272bc
P. mendocina + *G. intraradices*	23.8bc	2166b	44,321bc	27,807c	190ab
P. mendocina + *G. mosseae*	24.8cd	2294bc	44,714bc	27,479c	212b
4 g NaCl/ kg soil					
Control	24.1cd	2046ab	36,750a	48,393f	373c
Fertilized	31.5e	2463cd	42,214ab	37,929de	157a
P. mendocina	26.3cd	2171b	42,393bc	41,500e	231bc
G. intraradices	28.6de	2020ab	39,643a	38,500e	201b
G. mosseae	26.2cd	2287bc	43,679bc	40,429e	248bc
P. mendocina + *G. intraradices*	28.7de	2204bc	44,643bc	36,643de	135a
P. mendocina + *G. mosseae*	29.0e	2602d	46,714c	37,643de	169ab

Values in columns followed by the same letter are not significantly different (Tukey, $p < 0.05$).

nitrogenase activity of *Bradyrhizobium japonicum* in a soybean culture system (Chebotar *et al.*, 2001). In a study on the co-inoculation of the N_2-fixing *Phyllobacterium* sp. and the P-solubilizing *Bacillus licheniformis* in mangrove, Rojas *et al.* (2001) suggested that the interaction effects among the different rhizosphere bacteria should be considered when evaluating the growth-promoting effects of these bacteria.

The above examples strongly suggest the possibility of a complex web of interactions between root-associated fungi, rhizosphere bacteria and host plants. The numerous benefits provided by various bacteria and fungi to plant growth and crop yield may open new avenues for developing different biologically active fertilizers. Combinations of growth and yield-promoting bacteria and fungi as well as combinations of organisms facilitating establishment of plant-derived benefit from inoculation deserve further study.

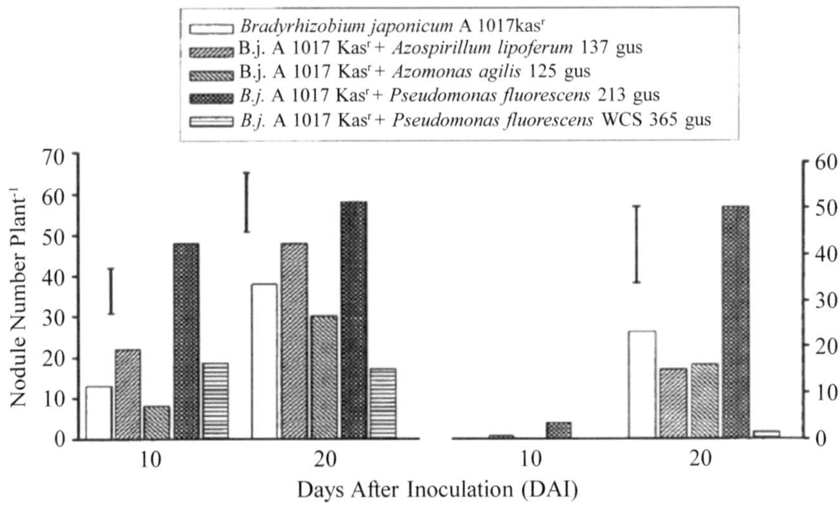

Fig. 6.1: Nodulation and acetylene reduction activity (*ARA*) of soybean at 10 and 20 DAI as influenced by coinoculation with rhizobacteria and *Bradyrhizobium japonicum* A1017kas^r (Source: Chebotar et al., 2001). Data are means of three replicates. *Bars* represent least significant differences at the 5% level.

6.3. Conclusion

Use of biofertilizer offers unmatched prospects for sustaining agriculture production system. These are environment friendly and cheaper than the chemical fertilizers and help to maintain soil health in long run through the multifunctional effects. They help mitigate different biotic and abiotic stresses, enhance growth and nodulation and reduce disease incidence to crop plants, besides lowering the load of heavy metals and other toxins in soil. Biofertilizers make soil more healthy, active and live through modifying beneficial microbial population, influencing nutrient cycling and improving soil structures through different secretions. Thus, biofertilizer can be a good substitute for chemical fertilizer if it is used following standard application protocol.

References

Anjana K, Kaushik A, Kiran B *et al* (2007) Biosorption of Cr(VI) by immobilized biomass of two indigenous strains of cyanobacteria isolated from metal contaminated soil. J Hazard Mater 148:383–386.

Antoun H, Beauchamp CJ, Goussard N *et al* (1998) Potential of *Rhizobium* and *Bradyrhizobium* species as plant growth promoting rhizobacteria on non-legumes: effect on radishes (*Raphanus sativus* L.). Plant Soil 204:57–67. Doi:10.1023/A:1004326910584

Barea JM (1991) Vesicular-arbuscular mycorrhizae as modifiers of soil fertility. In: Stewart BA (ed) Advances in soil science. Springer, New York, p 1–40.

Barea JM, Navaro E, Montoya E (1976) Production of plant growth regulators by rhizosphere phosphate-solubilizing bacteria. J Appl Bacteriol 40:129–134.

Barthakur HP (1978) Solubilization of relatively insoluble phosphate by some fungi isolated from the rhizosphere of rice. Indian J Agric Sci 48:762–766.

Bashan Y, Holguin G (1997) Azospirillum-plant relationships: Environmental and physiological advances (1990-1996). Can J Microbiol 43:103-121.

Breed RS, Murray EGD, Smith NR (1957) Bergey's manual of deteminative bacteriology, 7th edn. Williams & Wilkins, Baltimore.

Brundrett MC (2002) Coevolution of roots and mycorrhizas of land plants. New Phytol 154:275–304.

Chebotar VK, Asis CA, Akao S (2001) Production of growth promoting substances and high colonization ability of rhizobacteria enhance the nitrogen fixation of soybean when inoculated with *Bradyrhizobium japonicum*. Biol Fertil Soils 34:427–432. Doi:10.1007/s00374-001-0426-4

Colpaert JV, Van Laere A, Van Tichelen KK *et al* (1997) The use of inositol hexaphosphate as a phosphorus source by mycorrhizal and non-mycorrhizal Scots pine (*Pinus sylvestris*). Funct Ecol 11:407–415.

Date RA (2000) Inoculated legumes in cropping systems of the tropics. Field Crops Res 65:123-136.

Deubel A, Gransee A, Merbach W (2000) Transformation of organic rhizosdeposits by rhizosphere bacteria and its influence on the availability of tertiary calcium phosphate. J Plant Nutr Soil Sci 163:387–392.

Egamberdiyeva D (2007) The effect of plant growth promoting bacteria on growth and nutrient uptake of maize in two different soils. Appl Soil Ecol 36:184–189.

Fiore MF, Trevors JT (1994) Cell composition and metal tolerance in cyanobacteria. Biometals 7:83–103.

Fred EB, Baldwin IL, McCoy E (1932) Root nodule bacteria and leguminous plants. Wisconsin University Studies in Science No. 5. The University of Wisconsin Press. Madison.

Gallon JR (2001) N2 fixation in phototrophs: Adaptation to a specialized way of life. Plant Soil 230:39-48. Doi:10.1023/A:1004640219659

Gaur AC (2006) Biofertilizers in sustainable Agriculture. Directorate of information and publications of agriculture, Indian Council of Agricultural Rsearch, New Delhi.

Gentili F, Jumpponen A (2005) Potential and possible uses of bacterial and fungal biofertilizers, Handbook of Microbial Biofertilizers, Food Products. The Haworth Press, Inc, New York • London • Oxford p 1-28.

George TS, Simpson RJ, Gregory PJ *et al* (2007) Differential interaction of *Aspergillus niger* and *Peniophora lycii* phytases with soil particles affects the hydrolysis of inositol phosphates. Soil Biol Biochem 39:793–803. Doi:10.1016/j.soilbio.2006.09.029

Glick BR, Chagping L, Sibdas G (1997) Early development of canola seedlings in the presence of plant growth-promoting bacterium *Pseudomonas putida* GR12-2. Soil Biol Biochem 29:1233–1239. Doi:10.1016/S0038-0717(97)00026-6

Gravel V, Antoun H, Tweddell RJ (2007) Growth stimulation and fruit yield improvement of greenhouse tomato plants by inoculation with *Psedomonas putida* or *Trichoderma atroviride*: possible role of Indole Acitic Acid (IAA). Soil Biol Biochem 39(8):1968–1977. Doi:10.1016/j.soilbio.2007.02.015

Hegde DM, Dwivedi BS, Babu SNS (1999) Biofertilizers for cereal production in India-a review. Indian J Agric Sci 69:73–83.

Hoberg E, Marschner P, Lieberei R (2005) Organic acid exudation and pH changes by *Gordonia* sp. and *Pseudomonas fluorescens* grown with P adsorbed to goethite. Microbiol Res 160:177–187.

Kannaiyan S (2002) Biofertilizers for sustainable crop production. In: Kannaiyan S (ed) Biotechnology of biofertilizers. Kluwer Academic Publishers, AA Dordrecht, The Netherlands with Narosa Publishing House New Delhi India, p 9–49.

Kaushik BD (ed) (1998) Soil plant microbe interactions in relation to nutrient management. Venus publishers, New Delhi p 55-63.

Kaushik S, Sahu BK, Lawania RK *et al* (1999) Occurrence of heavy metals in lentic water of Gwalior region. Pollut Res 18:137–140.

Kavino M, Harish S, Kumar N *et al* (2010) Effect of chitinolytic PGPR on growth, yield and physiological attributes of banana (*Musa* spp.) under field conditions. Appl Soil Ecol 45:71–77.

Kim KY, Jordan D, Krishan HB (1997a) *Rahella aquatilis*, a bacterium isolated from soybean rhizosphere, can solubilize hydroxyapatite. FEMS Microbiol Lett 153:273–277.

Kim KY, McDonald GA, Jordan D (1997b) Solubilization of hydroxyapatite by *Enterobacter agglomerans* and cloned *Escherichia coli* in culture medium. Biol Fert Soils 24:347–352.

Kiran B, Kaushik A, Kaushik CP (2008) Metal-salt co-tolerance and metal removal by indigenous cyanobacterial strains. Process Biochem 43:598–604.

Kohler J, Caravaca F, Carrasco L *et al* (2006) Contribution of *Pseudomonas mendocina* and *Glomus intraradices* to aggregates stabilization and promotion of biological properties in rhizosphere soil of lettuce plants under field conditions. Soil Use Manage 22:298–304.

Kohler J, Hernandez JA, Caravaca F *et al* (2009) Induction of antioxidant enzymes is involved in the greater effectiveness of a PGPR versus AM fungi with respect to increasing the tolerance of lettuce to severe salt stress. Environ Exp Bot 65:245–252.

Kumar V, Narula N (1999) Solubilization of inorganic phosphates and growth emergence of wheat as affected by *Azotobacter chroococcum* mutants. Biol Fert Soils 28:301–305.

Kundu DK, Ladha JK (1995) Efficient management of soil and biologically fixed N2 in intensively-cultivated rice fields. Soil Biol Biochem 27:431-439.

Lefebvre DD, Kelly D, Budd K (2007) Biotransformation of Hg (II) by Cyanobacteria. Appl Environ Microbiol 73:243–249.

Meshram SU (1984) Suppressive effect of Azotobacter chroococcum on *Rhizoctonia solani* infestation of potatoes. Neth J Plant Pathol 90:127-32.

Meshram SU, Shende ST (1982) Total nitrogen uptake by maize with *Azotobacter* inoculation. Plant Soil 69:275-280. Doi:10.1007/BF02374522

Naumova AN, Mishustin EN, Marienko VM (1962) Bulletin of Academy Sciences, USSR 5:709-17.

Nezarat S, Gholami A (2009) Screening plant growth promoting rhizobacteria for improving seed germination, seedling growth and yield of maize. Pak J Biol Sci 12:26–32.

Niranjan SR, Deepak SA, Basavaraju P *et al* (2003) Comparative performance of formulations of plant plant promoting rhizobacteria in growth promotion and suppression of downy mildew in perlmillet. Crop Prot 22:579–588.

Nisha R, Kaushik A, Kaushik CP (2007) Effect of indigenous cyanobacterial application on structural stability and productivity of an organically poor semi-arid soil. Geoderma 138(1-2): 49–56. Doi:10.1016/j.geoderma.2006.10.007

Okon Y (1985) Azospirillum as a potential inoculant for agriculture. Trends Biotechnol 3(9):223-228. Doi:10.1016/0167-7799(85)90012-5

Okon Y, Labandera-Gonzalez CA (1994) Agronomic applications of *Azospirillum*: An evaluation of 20 years worldwide field inoculation. Soil Biol Biochem 26:1591-1601.

Pandey P, Kang SC, Maheshwari DK (2005b) Isolation of endophytic plant growth promoting Burkholderia sp. MSSP from root nodules of Mimosa pudica. Curr Sci 89:170–180.

Pandey P, Kapil D, Shukla PN *et al* (2005a) Cyanobacteria in alkaline soil and the effect of cyanobacteria inoculation with pyrite amendments on their reclamation. Biol Fertil Soils 41:451–457.

Parker DL, Mihalick JE, Plude JL *et al* (2000) Sorption of metals by extracellular polymers from the cyanobacterium *Microcystis aeruginosa* f. *flos aquae* strain C3-40. J Appl Phycol 12:219–224. Doi:10.1023/A:1008195312218

Persello Cartieaux F, Nussaume L, Robaglia C (2003) Tales from the underground: molecular plant–rhizobacteria interactions. Plant Cell Environ 26:189–199.

Polyanskaya LM, Vedina OT, Lysak LV *et al* (2002) The growth-promoting effects of *Beijerinckia mobilis* and *Clostridium* sp. cultures on some agricultural crops. Microbiol 71:109-115.

Poole EJ, Bending GD, Whipps JM *et al* (2001) Bacteria associated with *Pinus sylvestris-Lactarius rufus* ectomycorrhizas and their effects on mycorrhiza formation *in vitro*. New Phytol 151:743-751. Doi:10.1046/j.0028-646x.2001.00219.x

Rai SN, Gaur AC (1982) Nitrogen fixation by *Azospirillum* spp. and effect of *Azospirillum lipoferum* on the yield and N-uptake of wheat crop. Plant Soil 69:233-283.

Richardson AE (2001) Prospects for using soil microorganisms to improve the acquisition of phosphorus by plants. Aust J Plant Physiol 28:897-906.

Richardson AE, Hadobas PA (1997) Soil isolates of *Pseudomonas* spp. that utilize inositol phosphates. Can J Microbiol 43: 509–516.

Rillig MC, Mummey DL (2006) Mycorrhizas and soil structure. New Phytol 171:41–53.

Rodriguez H, Fraga R (1999) Phosphate solubilizing bacteria and their role in plant growth promotion. Biotech Adv 17:319–339. DOI: 10.1016/s0734-9750(99)00014-2

Rojas A, Holguin G, Glick BR *et al* (2001) Synergism between *Phyllobacterium* sp. (N2-fixer) and *Bacillus licheniformis* (P-solubilizer), both from a semiarid mangrove rhizosphere. FEMS Microbiol Ecol 35:181-187.

Ruiz-Lozano JM, Bonfante P (2001) Intracellular *Burkholderia* strain has no negative effect on the symbiotic efficiency of the arbuscular mycorrhizal fungus *Gigaspora margarita.* Plant Growth Regul 34:347-352.

Safronova VI, Stepanok VV, Engqvist GL *et al* (2006) Root-associated bacteria containing 1-minocyclopropane-1-carboxylate deaminase improve growth and nutrient uptake by pea genotypes cultivated in cadmium supplemented soil. Biol Fertil Soils 42:267–272.

Santos VB, Araujo SF, Leite LF *et al* (2012) Soil microbial biomass and organic matter fractions during transition from conventional to organic farming systems. Geoderma 170:227–231. Doi: 10.1016/j.geoderma.2011.11.007

Schultz RC, Colletti JP, Faltonson RR (1995) Agroforestry opportunities for the United States of America. Agrofor Syst 31:117-142. Doi:10.1007/BF00711720

Stamford NP, Ortega AD, Temprano F *et al* (1997) Effects of phosphorus fertilization and inoculation of *Bradyrhizobium* and mycorrhizal fungi on growth of Mimosa caesalpiniaefolia in an acid soil. Soil Biol Biochem 29:959-964.

Stephens JHG, Rask HM (2000) Inoculant production and formulation. Field Crops Res 65:249-258.

Tian C, He X, Zhong Y *et al* (2002) Effects of VA mycorrhizae and *Frankia* dual inoculation on growth and nitrogen fixation of *Hippophaë tibetana*. For Ecol Manage 170:307-312. Doi:10.1016/S0378-1127(01)00781-2

Tien T, Gaskin M, Hubbel D (1979) Plant growth substances produced by *Azospirillum brasilense* and their effect on the growth of pearl millet (*Pennisetum americanum* L.). Appl Environ Microbiol 37:1016–1024.

Toro M, Azcón R, Barea JM (1998) The use of isotopic dilution techniques to evaluate the interactive effects of *Rhizobium* genotype, mycorrhizal fungi, phosphate-solubilizing rhizobacteria and rock phosphate on nitrogen and phosphorus acquisition by *Medicago sativa*. New Phytol 138:265-273.

Unkovich MJ, Pate JS (2000) An appraisal of recent field measurements of symbiotic N2 fixation by annual legumes. Field Crops Res 65:211-228.

Vaishampayan A, Sinha RP, Hader DP *et al* (2001) Cyanobacterial biofertilizers in rice agriculture. Bot Rev 67:453–516. Doi:10.1007/BF02857893

Vance CP (2001) Symbiotic nitrogen fixation and phosphorus acquisition. Plant nutrition in a world of declining renewable sources. Plant Physiol 127:390-397.

Vande Broek A, Dobbelaere S, Vanderleyden J *et al* (2000) *Azospirillum*-plant root interactions: Signaling and metabolic interactions. In: Triplett EW (ed) Prokaryotic nitrogen fixation: a model system for analysis of a biological process. Wymondham, UK: Horizon Scientific Press, p 761-777.

Venkataraman GS, Neelakantan S (1967) Effect of cellular constituents of the nitrogen fixing blue-green algae. *Cylindrospermum nusciola* on the root growth of rice seedlings. J General Appl Microbiol 13:53–61.

Yadav RS, Tarafdar JC (2003) Phytase and phosphatase producing fungi in arid and semi-arid soils and their efficiency in hydrolyzing different organic P compounds. Soil Biol Biochem 35(6): 1–7. Doi:10.1016/S0038-0717(03)00089-0

Yee N, Benning LG, Phoenix VR *et al* (2004) Characterization of metal cyanobacteria sorption reactions: a combined macroscopic and infrared spectroscopic investigation. Environ Sci Technol 38:775–782.

7

Biological Nitrogen Fixation Mechanism and Applications

Avishek Pahari[1], Suraja Kumar Nayak[2], Avishek Banik[3] Priti Binita Lakra[4] and Bibhuti Bhusan Mishra[5]

[1]Centre for Wildlife Health, CVSc. & AH, OUAT, Bhubaneswar, Odisha
[2]Department of Biotechnology, College of Engineering and Technology Biju Patnaik, University of Technology, Bhubaneswar, Odisha
[3]Department of Life Sciences, Presidency University, Kolkata, West Bengal
[4]AICRP on Seed Technology Research, NSP (crops), OUAT, Bhubaneswar Odisha, India
[5]Department of Microbiology, CBSH, OUAT, Odisha

Abstract

Nitrogen (N_2) is an essential nutrient for every organisms. It is the building block of many biological molecules. Atmosphere contains 79% of N_2 gas which cannot be used by plants directly as it is chemically inert. Many diazotrophs like bacteria and cyanobacteria have the capability to convert atmospheric dinitrogen into a usable form by means of Biological nitrogen fixation (BNF). N_2 fixation is generally three types i.e. Asymbiotic, Associative and Symbiotic. Asymbiotic N_2-fixation also known as free living N_2-fixation is under taken by some bacteria and cyanobacteria. In associative N_2-fixation, some bacteria fix nitrogen in association with root of some grasses and cereal plants. In symbiotic nitrogen fixation, Rhizobium form root nodule in legumes and fix atmospheric N_2 and convert it into the ammonia (NH_3) by the help of Nitrogenase enzyme. This oxygen sensitive enzyme consist of two components i.e. di-nitrogenase (MoFe protein) and di-nitrogenase reductase (Fe protein) which are highly conserve in sequence and structure. The N_2 fixing genes e.g. nod gene (nodulation gene) and the nif gene (nitrogen-fixing gene) are located in the plasmid which is also called mega plasmid. Symbiotic association between legume and rhizobium have agronomic importantance and reduce the use of N-fertilizers or even avoided. The long term use of the nitrogenous biofertilizer in the agricultural field is a promising approach *to develop and fulfil N_2 demand of the growing crop without causing any environmental hazard.*

Keywords: Diazotrophs, Biological nitrogen fixation (BNF), *Rhizobium*, nitrogenase

7.1 Introduction

Agriculture in developed and developing nations have undergone large change with application of fertilizer and various agrochemicals like pesticides, weedicides, herbicides etc. with a view to increase productivity. Population explosion coupled with industrialization and urbanization has demanded enhanced productivity of agricultural produces with a decreasing availability of cultivable land. Green revolution I, led by the Noble Laurate Norman Borlaug witnessed large increase in productivity with use of high yielding variety and agrochemicals. Green revolution in India allowed to overcome poor agriculture productivity, elevating from the backdrop of food crisis to food surplus country. But over a span of more than 40 years not only India but also the developed countries are harvesting the impact of Green revolution I. According to Mulvaney *et al* (2009), long-term use the synthetic fertilizers drastically reduced the recycling of organics or other wastes which led to a decline the organic carbon levels in soils, damage soil Physico-chemicalproperties, reduced the soil biodiversity leading to crop yield and increase pathogen attack on the crops. Today scientists are emphasizing on Green revolution II, where organic farming is prioritized and soil is amended with organic manures, biofertilizers to increase productivity. Emphasis is being given to reduce application of fertilizer and/or agrochemicals in agriculture.

The rhizosphere is the narrow region of soil which is directly influenced by root secretions and associated with soil microorganisms. The bacteria that colonize in the rhizosphere region as well as plant roots and enhance the plant growth by different mechanisms known as plant growth promoting rhizobacteria (PGPR) (Pahari *et al*, 2017). Several biotic factors like genetic traits of host plant, the colonizing organism and abiotic factors *viz.* soil humidity, soil pH, temperature etc. enhance the root colonization. PGPR can be classified into two different classes' i.e. extracellular plant growth-promoting rhizobacteria (ePGPRs) and intracellular plant growthpromoting rhizobacteria (iPGPRs) (Viveros *et al*, 2010). The ePGPRs may exist in the rhizosphere, on the rhizoplane, or in the spaces between the cells of root cortex, and iPGPRs are located generally inside the specialized structures of root cells like nodules. Bacterial genera such as *Agrobacterium*, *Arthrobacter*, *Azotobacter*, *Azospirillum*, *Bacillus*, *Burkholderia*, *Caulobacter*, *Erwinia*, *Pseudomonas* and *Serratia* belongs to ePGPR (Ahemad and Kibret, 2014). The iPGPR belongs to the family of

Rhizobiaceae and includes species of *Rhizobium*, *Allorhizobium*, *Mesorhizobium* and endophytes, and *Frankia* all of which can symbiotically fix atmospheric nitrogen in association with the higher plants (Bhattacharyya and Jha, 2012; Pradhan *et al*, 2017). PGPR can enhance the plant growth and productivity by different mechanisms like by biological nitrogen-fixation, IAA production, phosphate solubilization, siderophore, production, HCN production, cytokinins and gibberellic acid production (Glick, 1995; Pahari and Mishra, 2017). It can increase the germination percentage, total biomass of plants, early flowering, seedling vigor, root and shoot growth, seed weight, fodder and fruit yields; can serve as a potential source of bio-fertilizers (Boddey and Dobereiner, 2000; Ramamoorthy *et al*, 2001).

7.2 Biological nitrogen fixation (BNF): Key to sustainable agriculture

Nitrogen (N_2) is a component in many building blocks like amino acids, proteins, nucleic acids etc. essential for maintenance of life. Moreover it is also present in chlorophylls, alkaloids and cytochromes. Air constitute approximately 79% of nitrogen. Plant cannot directly use the atmospheric di-nitrogen gas (N=N) and can only take up the nitrogen in the form of ammonium and nitrates (Döbereiner, 1997). Hence plant growth and productivity is affected due to nitrogen deficiency. For which, the use of N-based synthetic fertilizer is increased globally for the better production (Dobermann, 2007; Westhoff, 2009). According to Jensen *et al* (2012), at one time plant can take up very small percentage of applied supplements and majority of the supplemented nitrogen fertilizer i.e. 30% to 50% is wasted and runs off into waterbodies causing environmental pollution through supporting of algal blooms formation and eutrophication. Moreover, excessive application of N fertilizers responsible for the maximum N_2O emissions because nitrogen is broken down by soil bacteria by denitrification process and liberate N_2O, which reacts with oxygen and converted to Nitric oxide (NO). This nitric oxide which in turn reacts with ozone (O_3) and also responsible for Global warming.

Biological nitrogen fixation (BNF), a microbiological process in the biosphere converts atmospheric dinitrogen into a plant-usable form. Some bacteria and cyanobacteria have the microbial enzyme 'nitrogenase' which convert the N_2 to ammonia (NH_3). Subsequently, ammonia is used in the synthesis of essential elements. Nitrogen input through BNF can help maintain soil N reserves as well as substitute for N fertilizer to attain large crop yields (Peoples *et al*, 2009). Thus, agricultural sustainability will require maximum use of BNF for a long period of time because it is a major source of nitrogen for plants. Moreover, BNF offers an economical and ecologically sound means of reducing external nitrogen input, improving the quality & quantity of internal resources (Figueiredo *et al*, 2013). According to Freitas (2007), BNF can be symbiotic when there is a mutualistic association between plant species and the members of rhizobia or it can be done by free-living fixing microorganisms like *Azotobacter* and *Beijerinckia* asymbiotically. Rhizobia are distributed in different taxonomic groups i.e. α- and β-rhizobia according to their morphological, physiological, genetic, and phylogenetic characteristics (Lindström *et al*, 2006). Similarly in *Azolla,* a heterosporous pteridophyte water fern, contains endosymbiont *Anabaena azollae*, a nitrogen fixing cyanobacterium (Nostocaceae family). It is present in the cavities within the leaves of the ubiquitous water fern (*Azolla filiculoides*) and fix atmospheric nitrogen (N_2) into ammonia (NH_3) heterocysts, the site of nitrogen-fixation. *Anabaena azollae* can make nitrogen available to autotrophic plants and ultimately to all members of the ecosystem. Although *Azolla* can absorb nitrates from the water, it can also absorb ammonia secreted by *Anabaena* within the leaf cavities (Watanabe, 1982)

7.3 Types of Nitrogen fixation

Biological nitrogen fixation is carried out by different group microorganisms like bacteria, cyanobacteria and actinomycetes. According to their mode of nitrogen fixation, it is differentiate into three main categories, namely, Asymbiotic (Free living), Associative and Symbiotic nitrogen fixation.

7.3.1 Asymbiotic N2-Fixation

Many heterotrophic bacteria like *Azotobacter*, *Bacillus*, *Clostridium*, *Klebsiella* and cyanobacteria like *Anabaena* and *Nostoc* living freely in the soil and fix significant levels of nitrogen (Wagner, 2011). According to Vadakattu and Paterson (2006), this free-living organisms behave as anaerobes or micro-aerophiles while fixing nitrogen because can be inhibited by Oxygen. Due to scarcity of suitable carbon and energy sources for these organisms, their contribution to global nitrogen fixation rates is generally considered minor. However, a recent study in Australia of an intensive wheat rotation farming system demonstrated that free-living microorganisms contributed 20 kg/ha per year to the long-term nitrogen needs of this cropping system (30-50% of the total needs). This group of microorganisms must find their own source of energy either by oxidizing organic molecules released by other organisms or from decomposition. Some of the microbes have chemolithotrophic capabilities and can thereby utilize inorganic compounds as a source of energy.

The free-living cyanobacteria are considered to be fairly important nitrogen fixers and they grow mainly in the agricultural crop fields. Under suitable conditions *viz.* water, oxygen and nutrients, they may fix ten times much nitrogen as of the other free-living bacteria (Rai *et al*, 2000). Some workers claim that cyanobacteria are mainly responsible for maintaining the fertility and productivity of rice fields. They have heterocyst for the nitrogen fixation and protect oxygen sensitive nitrogen fixing enzyme i.e. nitrogenase. Some unicellular and non-heterocystous cyanobacteria have some specialized cells by which they can fix nitrogen by reducing the oxygen level. Typically, they fix nitrogen in dark and photosynthesize in light (Kumari and Sinha, 2011).

7.3.2 Associative Nitrogen Fixation

There is an intermediate biological system referred to as associative N_2 fixation, and they are usually found in association with the roots of grasses and cereal plants such as rice, wheat, corn, oats, and barley. In this type of association, no nodules are formed and bacteria fix appreciable amounts of nitrogen within the rhizosphere of the host plants. For example, Species of *Azospirillum* are able to form close associations with several members of the *Poaceae* (grasses). Other examples of this group are *Bacillus, Klebsiella, Enterobacter, Pseudomonas azotogensis* etc. The efficiency of nitrogen fixation by microbes is determined by several factors like soil temperature (*Azospirillum* sp. favours more temperate and/or tropical environments), the ability of the host plant to provide a rhizosphere environment low in oxygen pressure, the availability of host photosynthates for the bacteria, the competitiveness of the bacteria, and the efficiency of nitrogenase (Vlassak and Reynders, 1979).

7.3.3 Symbiotic Nitrogen Fixation

Leguminous plant provide an option for reducing heavy reliance on nitrogen fertilizers and legumes have the ability to form a symbiotic relationship with soil bacteria called "Rhizobia" (Jensen *et al*, 2012). There are 13000 different known species of legumes like soya beans, alfalfa, peas, vetch beans, clover lupines and peanuts etc. can fix atmospheric N_2 in association with the specific rhizobia. The rhizobia invade the root & multiply within its cortex and form "nodules" where they fix atmospheric nitrogen to ammonia. In return, the plant supplies all the necessary nutrients like carbohydrates and energy source for nitrogen fixation. The process enormously reduces the agriculture's dependence on nitrogen fertilizers. It has been estimated that biological nitrogen fixation produces roughly 200 million tonnes of nitrogen annually (Graham, 2003; Peoples *et al*, 2009). In nonleguminous plants like *Casuarina, Datisca* and *Alnus*, the actinorhizal microorganism i.e. *Frankia* form nodule and fix N_2. Some of these actinorhizal plants are especially attractive for cultivation, because they thrive under arid conditions.

The rhizobia have long been classified on the basis of the legumes they infect. The legumes such as *Pisum, Cajanus cajan, Cicer arietinum, Glycine soja* etc. form symbiotic relationship with *Bradyrhizobium, Allorhizobium, Azorhizobium, Sinorhizobium, Photorhizobium, Mesorhizobium* and *Rhizobium.* The plants were divided into 'cross-inoculation groups', and a specific rhizobium species would inoculate plants within a cross-inoculation group but there are also exceptions such as sweet clover and alfalfa are infected by the same strain. An organism nodulates a leguminous plant does not ensure that the association will be highly effective in N_2 fixation. Differences in the effectiveness of the bacteria are called as 'strain variation' (Kumari and Sinha, 2011). *Rhizobium* is a free living, Gram-negative, aerobic, rod shaped bacteria which forms nodules in the leguminous plant. It is a fast growing bacterium whereas *Bradyrhizobium* is a slow growing strain which possesses sub-polar flagella. The nodule formation by *Rhizobium* follows a very specific pattern. In the first step, the nodule is formed which provide the correct environment for housing of *Rhizobium*; second, the division of symbiotic tissue (*i.e.*, nodule numbers) by internal and external factors, and the third conversion of atmospheric nitrogen into ammonia by the invading bacteria using the nitrogenase enzyme complex and its associated biochemical machinery. The steps of nodule formation are detailed below (**Fig. 7.1**).

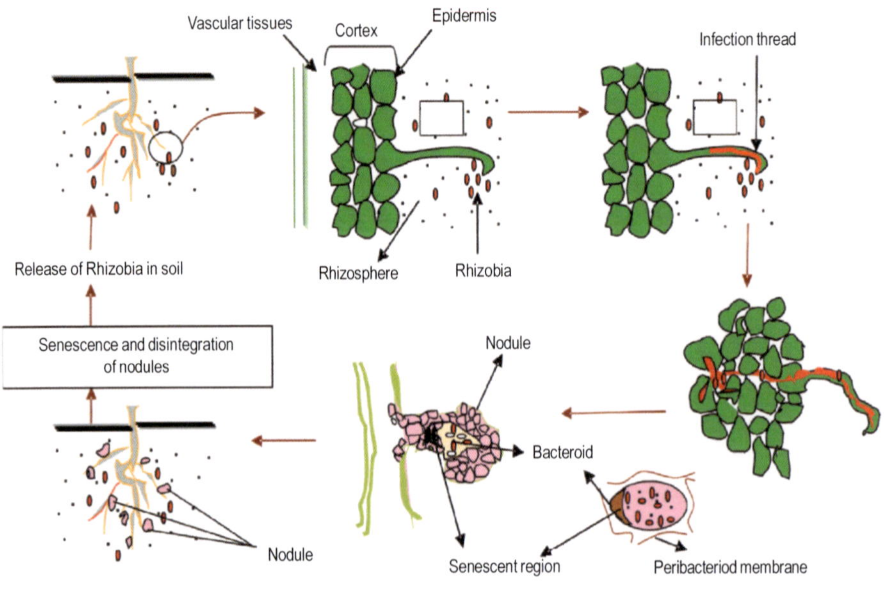

Fig. 7.1: Schematic view of N_2-fixing bacterial association with leguminous plant and development of root nodules by *Rhizobium* sp. (Adapted from Kumari and Sinha 2011)

7.3.3.1 Reorganization and attachment

Infection by *Rhizobium* involves chemotaxis of the organisms towards the roots in response to the growth stimulating substances like organic acids, thiamine, amino acids, sugars, biotin and flavonoids etc. secreted by the root of the symbiont which enhances the growth of rhizobia over other microbes. This process requires a coordinated exchange of signals between the two symbiotic partners (van Rhyn and Vanderleyden, 1995; Cheng and Walker, 1998). The interaction between polysaccharide (callose) present on the surface of rhizobia and the lectin secreted by the root hairs helps in the recognition of the correct host plant by the specific *Rhizobium*. Moreover the root exudates also contain, some other compounds (specific flavonoids) i.e. lipochitin oligosaccharides which is responsible for the activation of *nod* genes in the bacteria. In response to flavonoids from legume roots, the rhizobia produce lipooligosaccharides i.e. Nod factors that function to induce the production of root nodules. (Mahdhi *et al*, 2008). Several biochemical and genetic mechanisms are present by which regulatory and structural nodulation genes control host specificity. The regulatory *nodD* genes encode DNA-binding proteins that activate the transcription of the other *nod* operons and determine the first level of host specificity (Fisher and Long, 1992). NodD proteins are activated by plant signals and according to the plant host, their nature and abundance in root exudates may vary (Horvath *et al*, 1987). Moreover structural *nod* genes include both species-specific gene and genes that are common to all *Rhizobium* species. The *nodABC* genes encode enzymes involved in the synthesis of the lipo-oligosaccharide core. Briefly NodC is an N-acetylglucosaminyl transferase that determines the synthesis of the chitin oligomer backbone (Spaink *et al*, 1994), NodB is a chito oligosaccharide deacetylase (John *et al*, 1993) and NodA is required for the N-acylation of the amino sugar backbone. Hence, nodABC genes are functionally equivalent between Rhizobium species (Kondorosi *et al*, 1989).

7.3.3.2 Curling of Root Hairs

After the attachment of *Rhizobium* on the root surfaces, the root-hair curls due to the release of release *nod* factors. Some of the Scientists believe that curling is also affected by Indole acetic acid (IAA), a plant hormone. After root-hair deformation, the *Rhizobium* penetrate and enter into the root-hair and inner cortical cell division leading to nodule formation. It also induce the plant to develop a cellulosic tube, called as 'Infection thread' (Sprent and Faria, 1988). The *Rhizobium* cells spread within the infection thread and move into the underlying root cells and finally released into cytoplasm of the host cell. After that the branching of infection thread occurred due to the continuous production of *nod* factors which induce controlled production of a nodule (Kumari and Sinha, 2011).

7.3.3.3 Bacteroid formation and Development of mature nodule

When the bacteria are released from the infection thread into the host cell cytoplasm, they get transformed into dormant, swollen, irregular-shaped called as 'bacteroids'. After that the bacteroids are surrounded by the plant-derived membrane, called as peribacteroid membrane (PBM) and the entire structure is called 'Symbiosome', which is the actual site of nitrogen fixation. *Rhizobium* released mucopolysaccharide which reacts with the component of root hair for the production of polygalacturonase. This compound stimulate the inner cortical tissue to divide rapidly forming the core of the nodule. However application of Nitrogen fertilizers in the agricultural field inhibit the nodule formation as well as reduce nitrogen fixation. Generally nodules are produced in two forms i.e. indeterminate cylindrical nodules and determinate spherical nodules. Both the nodules are originated from different types of cortical cells. Determinate nodules are characteristic of soybean and common bean. Legumes such as alfalfa and garden pea present indeterminate nodules (Liu *et al*, 2018). The core of the nodule is tetraploid in nature which is a peculiar characteristic and *Rhizobium* occupy the central position in the nodule after released from the infection tube.

The root nodules are distinctly red or pink but in young stage, it is generally white or gray and unable to fix nitrogen. As nodules grow in size, they gradually turn pink or reddish in color, indicating initiation of nitrogen fixation. The pink or red color is caused by Leghaemoglobin that regulates oxygen flow to the bacteria and protect oxygen liable nitrogenase enzyme. Inside the nodule, oxygen level is controlled by leghaemoglobin. Current evidence suggests that the 'haeme' component of the proptein is synthesized by *Rhizobium* while the 'globin' comes from the legume. When the nodule dies, bacteria are released into soil. Leghaemoglobin is chemically similar to hemoglobin found in blood, however it has ten times higher affinity for oxygen than human haemoglobin. It supplies O_2 to the respiring symbiotic bacterial cells. It enhances the transport of oxygen at low partial pressure and also provides the protection to nitrogenase against oxygen and stimulates ATP production needed for N_2-fixation.

7.4 Nitrogenase enzyme complex

Nitrogen is reduced to ammonia during biological N_2-fixation by the 'nitrogenase' enzyme complex. This reaction is dependent on reduced ferredoxin and requires hydrogen and energy from ATP, resulting in the formation of molecular hydrogen. Nitrogenase enzyme is mainly present within the bacteroids and it very oxygen sensitive. The enzyme contains two components **(Fig. 7.2)** i.e. dinitrogenase [molybdenum-iron (MoFe) protein] and dinitrogenase reductase (Fe protein) (Cheng, 2008). In some microorganisms, vanadium is present instead of Mo ion. The overall stoichiometry of dinitrogen reduction by nitrogenase (EC 1.18.2.1) is: N_2 + 8H+ +8e-+16MgATP → 2NH3 + H2 + 16MgADP + 16Pi. Nitrogenase turnover requires an electron (e^-) donor in addition to adenosine triphosphate (ATP). Electrons are generated *in vivo* either oxidatively or photosynthetically, depending on the organism. For the reduction of one molecule of N_2 to two molecule of NH_3, nitrogenase enzyme require 16 molecules of ATP and one molecule of H_2 is produced as a by-product. The activity and synthesis of nitrogenase is regulated by post-translation modification of the Fe protein (Reich *et al* 1986; Smith *et al*, 1987). The two proteins i.e., Mo-Fe and Fe cannot function independently. In dinitrogenase (MoFe), the number of Fe-centers varies whereas in dinitrogenase reductase (Fe protein), there is four Fe-centers. In Mo-Fe, two Mo present and Fe-protein donates electron to Mo-Fe protein which is conserved in all nitrogenase systems, but the electron donor to Fe-protein is not conserved (Shah *et al*, 1983). According to Bishop *et al* (1985), different microorganisms such as *Azotobacter vinelandii* and *A. chroococcum* contain different kind of nitrogenase enzyme. This occurs due to deletion of normal gene (*nif YKDH*) for nitrogenase.

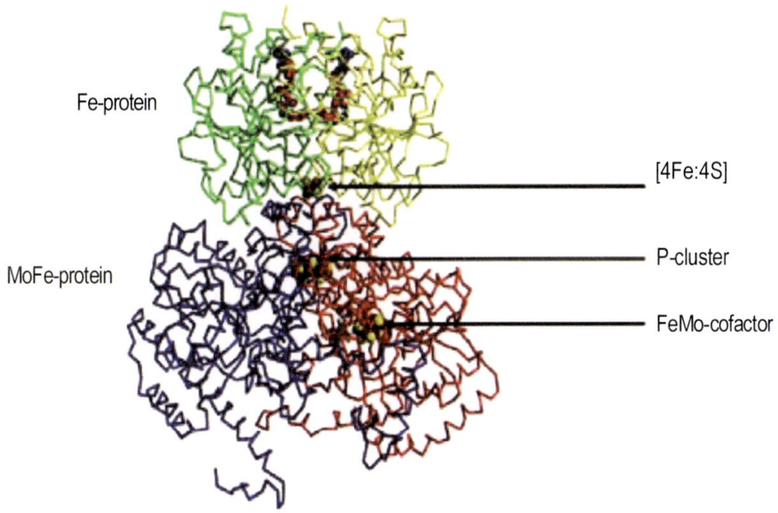

Fig. 7.2: Complex of the nitrogenase proteins (Adapted from Cheng 2008)

7.5 Assimilation and transportation of biologically fix ammonia

Plants absorb biologically fix nitrogen in the form of ammonium ($NH4^+$) and nitrate ($NO3^-$) which is transported from the root to the shoot via the xylem elements for amino acid synthesis. Apart from its cellular toxicity, elevated level of $NH4^+$ represses the nitrogenase complex in nitrogen fixing bacteria (NFB) (Britto *et al*, 2002). Hence, the superfluous $NH4^+$ must be excreted into the surrounding medium. For endophytic NFB, the $NH4^+$ is excreted in to the plant cytoplasm and assimilated by plants for cellular macromolecule synthesis (Poole *et al*, 2018). These nitrogenous compounds are transported in the form of glutamine, asparagine, 4-methylene glutamine, citrulline, allantoic acid etc. in different plant species (Tegeder and Masclaux Daubresse, 2018). Assimilation of ammonia is catalyzed by combined activity of glutamine synthetase (GS) and GS-glutamine oxoglutrate aminotransferase (GOGAT) which remain repressed in bacteroides but induced in host cells (Esteves-Ferreira *et al*, 2018). Two different kinds of mutations enhance ammonia excretion by diazotrophic bacteria: disruption of the *nifA/nifL*-dependent control of *nif* gene expression or partial inhibition of GS or GOGAT for ammonia assimilation (Colnaghi *et al*, 1997). However, modification of GS–GOGAT pathway of ammonia assimilation is difficult probably because of bacteria depend exclusively on this pathway for N-assimilation and therefore, glutamine transport deficient constitutive/ conditional mutant strains isolation is difficult (Colnaghi *et al*, 1997). Thus, understanding the regulation of N_2 fixation and assimilation of ammonia by bacteria are prime interests for development of versatile bio-fertilizer (Ortiz-Marquez *et al*, 2014, Banik *et al*, 2019). The most well-established system studied in root nodules of legume, where bacteroids fix atmospheric N_2 in a low-oxygen environment, in return they receive carbon in form of malate. Inside the infected roots, the bacteroid is surrounded by a peribacteroid membrane (PBM), which is mainly originate from the plants, is very permeable to NH_3 and $NH4^+$. The PBM plays a very decisive role to establish the symbiotic relationship between the host plant and bacteroid as they possess H^+-ATPase to pump H^+ to create a pH gradient across the PBM (Slatni *et al*, 2012). On the other hand, generally endophytic N_2 fixers harbor at intercellular spaces of different pant tissues and diffuse excess ammonia (Banik *et al*, 2016a, 2016b) that secreted ammonia is transported by different ammonia transporters inside the cell and assimilated to form different cellular macromolecule.

7.6 Nitrogen fixing bacteria: Biofertilizer formulation and technology

For crop growth and development, generally farmers apply approximately 100 kg of nitrogen fertilizer per hectare. However, synthetic fertilizers are not cost effective and create soil and environmental pollution (Deaker *et al*, 2004). In contrast, biofertilizers are the living formulations of microbes, exhibit numerous advantages such as nutrient recycling and enhance soil fertility & crop productivity (Brahmaprakash and Sahu, 2012). Hence, it can be an ideal supplement or substitute to chemical fertilizers. In the present scenario, development of carrier based formulation of one or more beneficial bacteria is an effective way to improve growth and productivity of (Bashan, 1998; Pahari *et al*, 2017). Peat is the most frequently used carrier for rhizobial inoculant due to its' higher water-holding capacity and large surface area, which support rhizobial growth and survival in large numbers (Smith, 1992). Beside peat based formulation, rhizobia need stirring and aeration to reach a high population density in liquid culture. A basal nutrient medium with mannitol as the carbon source and including yeast extract is the standard formulation for culturing rhizobia (Vincent, 1970). Different additives are added to liquid inoculant formulations for protecting the cells on seeds at high temperatures and during desiccation. Different biopolymers are used for inoculant production due to their ability to limit heat transfer, good rheological properties, and high water activities (Mugnier and Jung, 1985). According to Roughley (1968), *Rhizobium meliloti*, *Rhizobium trifolii*, and a cowpea *Rhizobium* reach similar ceiling populations in peat after 3 weeks of incubation, when liquid cultures of different ages are used. Carrier based formulation of *Rhizobia* can be applied on them legume seeds (Deaker *et al*, 2004). In Brazil, carrier based formulation of the bio-inoculants currently saves approximately 11 billion dollars per year when considering only soybean crops (Hungria, 2012). Singh and Rao (1979), observed that seed inoculation with *Azospirillum brasilense* significantly increase the nodulation and grain yield of soybeans in unsterilized soil with different levels of urea ranging from 0 to 80 kg N/ha. In recent years, biopolymers (proteins, nucleic acids, or polysaccharides) produced by living organisms have been used as bacterial carriers for microbial inoculants (Silva *et al*, 2009; Borschiver *et al*, 2008).These formulations encapsulate living cells, protect microorganisms against many environmental stresses, and gradually release the cells into soil in large quantities where the polymers are degraded by soil microorganisms.

According to Peng *et al* (2002), rhizobial inoculation in rice significantly increase the plant biomass, nitrogen content and grain yield. Similarly, Yanni *et al* (1997) isolated 11 strains of *Rhizobium leguminosarum* from rice roots in Egypt where rice has been grown in rotation with clover for generations and it increases the total N content, grain yield, grain N content and harvest index of rice (99% confidence). Moreover, certain species of rhizobia can improve rice growth and yield through the mechanism that improve single leaf net photosynthetic rate. However, some of the *Rhizobium* isolates inhibited rice seedling growth and development (Perrine *et al*, 2005). Similarly, *Azolla* is used as a practicable biofertilizer throughout the world in conventional farming because of faster vegetative growth rate during spring, fixing more nitrogen. Moreover, it increase the efficiency of nitrogen use and reduce water pollution, by changing strategies of mineral use and by integrating organic fertilizers in the rice production system (Biswas *et al*, 2005).

Diazotrophs with plant growth promoting traits can improve soil nitrogen status and also with solubilization of fixed forms of phosphates in soil, production of phytohormones like gibberellins and indole acetic acid (IAA) and production of siderophores can enhance crop growth and productivity. So, combined application of nitrogen fixing bacteria and other PGPR can be beneficial for agricultural yield. Dubey (1998), reported that co-inoculation of *Rhizobium* sp. with phosphate solubilizing bacteria (PSB) or arbuscular mycorrhizal fungi (AMF) has been found to be significantly better than rhizobial inoculation alone. Similarly combine use of *Bradyrhizobium*, AMF and PSB also increase the soybean yield from 0.95 Mg/ha to 1.21 Mg/ha in control. So application of PGPR in the agricultural field is the most effective for those cultivars which have a higher yield potential.

7.7 Conclusion

Nitrogen is an essential element in agriculture. Deficiency of nitrogen have many adverse effect on growth and productivity of the crop. Use of chemical fertilizer since Green Revolution-I has adverse effect on the soil health and soil microflora. On account of that application of diazotrophs as biofertilizer and microbes with PGPR traits can be very effective not only in reducing application of chemical fertilizer but also increase soil health with regards to microbial diversity of the soil.

References

Ahemad M and Kibret M. (2014) Mechanisms and applications of plant growth promoting rhizobacteria: current perspective. J King Saud Univ Sci. 26:1–20.

Banik A, Dash GK, Swain P, Kumar U, Mukhopadhyay SK, Dangar TK. (2019) Application of rice (*Oryza sativa* L.) root endophytic diazotrophic *Azotobacter sp.* strain Avi2 (MCC 3432) can increase rice yield under green house and field condition. Microbiological research 219:56-65.

Banik A, Mukhopadhaya SK, Dangar TK. (2016a) Characterization of N2-fixing plant growth promoting endophytic and epiphytic bacterial community of Indian cultivated and wild rice (*Oryza* spp.) genotypes. Planta. 243: 799-812.

Banik A, Mukhopadhaya SK, Sahana A, Das D, Dangar TK. (2016b) Fluorescence resonance energy transfer (FRET)-based technique for tracking of endophytic bacteria in rice roots. Biology and Fertility of Soils. 52: 277-82.

Bashan Y. (1998). Inoculants of plant growth-promoting bacteria for use in agriculture. Biotechnol. Adv. 16:729–770.

Bhattacharyya PN and Jha DK. (2012) Plant growth-promoting Rhizobacteria PGPR: emergence in agriculture. J World Microbiol Biotechnol. 28:1327–1350.

Bishop PE, Rizzo TM and Bott K. (1985). Molecular cloning of *nif* DNA from *Azotobacter vinelandii*. J. Bacteriol. 162**:** 21-28.

Biswas M, Parveen S, Shimozawa H and Nakagoshi N. (2005) Effects of Azolla species on weed emergence in a rice paddy ecosystem," Weed Biology and Management. 5(4):176–183.

Boddey RM and Dobereiner J. (2000) Nitrogen fixation associated with grasses and cereals. Recent Progress and Propectives for the Future Fert. Res. 42:241-250.

Borschiver S, Almeida LFM and Roitman T (2008) Monitoramento Tecnológico e Mercadológico de Biopolímeros. Polím: Ciênc Tecnol 18:256–261.

Brahmaprakash GP and Sahu PK. (2012) Biofertilizers for sustainability. J Indian Inst Sci 92:37–69.

Britto DT, Kronzucker HJ. (2002) NH4+ toxicity in higher plants: a critical review. Journal of Plant Physiology. 159: 567-84.

Cheng HP and Walker GC. (1998) Succinoglycan is required for initiation and elongation of infection threads during nodulation of alfalfa by *Rhizobium meliloti*. Journal of Bacteriology. 180:5183-5191.

Cheng Qi. (2008) Perspectives in Biological Nitrogen Fixation Research. Journal of Integrative Plant Biology.50 (7): 784–796.

Colnaghi R, Green A, He L, Rudnick P, Kennedy C. (1997) Strategies for increased ammonium production in free-living or plant associated nitrogen fixing bacteria. Plant and Soil. 194:145-54.

Deaker R, Roughley RJ and Kennedy IR. (2004) Legume seed inoculation technology – a review. Soil Biol Biochem. 36:1275–1288.

Döbereiner J. (1997) Importância da fi xação biológica de nitrogênio para a agricultura sustentável. Biotecnol Ciênc Desenvolv 1:2–3 (Encarte Especial).

Dobermann A. (2007) Nutrient use efficiency-measurement and management in a time of new challenges. In: Proceedings of the IFA International Workshop on Fertilizer Best Management Practices. Fertilizer Best Management Practices, Brussels, Belgium. International Fertilizer Industry Association. pp 1–28.

Dubey SK. (1998). Response of soybean to biofertilizer with and without nitrogen, phosphorus and potassium on swell shrink soil. Indian J Agron. 43:546-549.

Esteves-Ferreira AA, Inaba M, Fort A, Araújo WL, Sulpice R. (2018) Nitrogen metabolism in cyanobacteria: metabolic and molecular control, growth consequences and biotechnological applications. Critical reviews in microbiology 44: 541-60.

Figueiredo MVB, Mergulhão ACES, Sobral JK, Junior MAL and Araújo ASF. (2013) Biological Nitrogen Fixation: Importance. N.K. In: Arora (ed.), Plant Microbe Symbiosis: Fundamentals and Advances, DOI Springer, India. Pp 267-289. 10.1007/978-81-322-1287-4_10

Fisher RF and Long SR. (1992) *Rhizobium-plant* signal exchange. Nature (London) 357:655–660.

Freitas SS. (2007) Rizobactérias promotoras de crescimento de plantas. In: Silveira APD, Freitas SS (eds) Microbiota do solo e qualidade ambiental. Instituto Agronômico de Campinas, Campinas. pp 1–20

Glick BRX. (1995) The enhancement of plant growth by free living bacteria. Can J. Microbiol. 41:109–117.

Graham PH and Vance CP. (2003) Legumes: Importance and constrains to greater use. Plant Physiol.131:872–877.

Horvath B, Bachem CWB, Schell J and Kondorosi A. (1987) Host-specific regulation of nodulation genes in Rhizobium is mediated by a plant-signal, interacting with the nodD gene product EMBO J. 6:841–848.

Hungria M. (2012) Fixação biológica do N2 na agricultura de baixo carbono. In: Anais do FertBio 2012. UFAL, Maceió.

Jensen ES, Peoples MB, Boddey RM, Gresshoff PM, Henrik HN, Alves BJR and Morrison MJ. (2012) Legumes for mitigation of climate change and the provision of feedstock for biofuels and bio-refineries. A review. Agron Sustain Dev. 32:329–364.

John M, Röhrig H, Schmidt J, Wieneke U and Schell J. (1993) *Rhizobium* NodB protein involved in nodulation signal synthesis is a chitooligosaccharide deacetylase. Proc Natl Acad Sci USA. 90:625–629.

Kondorosi É, Gyuris J, Schmidt J, John M, Duda E, Hoffman B, Schell J and Kondorosi Á. (1989) Positive and negative control of *nod* gene expression in *Rhizobium meliloti* is required for optimal nodulation. EMBO J.8:1331–1340.

Kumari S and Sinha RP. (2011) Symbiotic and Asymbiotic N2 Fixation. In: P Sinha RP, Sharma NK and Rai AK (eds) Advances in Life Sciences, I.K. International Publishing House Pvt. Ltd., New Delhi. pp 134-148.

Lindström K, Kokko-Gonzales P, Terefework Z and Räsänen LA. (2006) Differentiation of nitrogen fixing legume root nodule bacteria (Rhizobia). In: Cooper JE, Rao JR (eds) Molecular approaches to soil, rhizosphere and plant microorganism analysis. CABI, Wallingford/ Cambridge, pp 236–258.

Liu A, Contador CA, Fan K and Lam HM. (2018) Interaction and Regulation of Carbon, Nitrogen, and Phosphorus Metabolisms in Root Nodules of Legumes. Frontiers in Plant Science. 9:1-18.

Mahdhi M, de Lajudie P and Mars M. (2008) Phylogenetic and symbiotic characterization of rhizobial bacteria nodulating *Argyrolobium uniflorum* in Tunisian arid soils. Can. J. Microbiol. 54: 209-217.

Mugnier J and Jung G. (1985) Survival of bacteria and fungi in relation to water activity and the solvent properties of water in biopolymer. Appl Environ Microbiol. 50:108–114.

Mulvaney RL, Khan SA and Ellsworth TR. (2009) Synthetic Nitrogen Fertilizers Deplete Soil Nitrogen: A Global Dilemma for Sustainable Cereal Production. J Environ Quality. 38:2295-2314.

Ortiz-Marquez JC, Do Nascimento M, Curatti L. (2014) Metabolic engineering of ammonium release for nitrogen-fixing multispecies microbial cell-factories. Metabolic engineering 23:154-64.

Pahari A and Mishra BB. (2017) Characterization of Siderophore producing Rhizobacteria and its Effect on Growth Performance of Different Vegetables. Int.J.Curr.Microbiol.App.Sci. 6(5): 1398-1405. https://doi.org/10.20546/ijcmas.2017.605.152

Pahari A and Mishra BB. (2017). Antibiosis of Siderophore Producing Bacterial Isolates against Phythopathogens and Their Effect on Growth of Okra. Int.J.Curr.Microbiol.App.Sci. 6(8): 1925-1929. https://doi.org/10.20546/ijcmas.2017.608.227

Pahari A, Pradhan A, Maity S and Mishra BB. (2017) Carrier Based Formulation of Plant Growth Promoting Bacillus species and Their Effect on Different Crop Plants. Int.J.Curr.Microbiol.App.Sci. 6(5):379-385. http://dx.doi.org/10.20546/ijcmas.2017.605.043

Pahari A, Pradhan A, Nayak S and Mishra BB. (2017) Bacterial Siderophore as a Plant Growth Promoter. In: Patra JK, Vishnuprasad CN and Das G (eds) Microb Biotechnol, Vol 1. Springer-Nature, Singapore. pp 163–180.

Peng SB, Biswas JC, Ladha JK, Yaneshwar PG and Chen Y. (2002) Influence of rhizobial inoculation on photosynthesis and grain yield of rice. Agron J. 94:925–929. 10.2134/agronj2002.0925

Peoples MB, Brockwell J, Herridge DF, Rochester IJ, Alves BJR, Urquiaga S, Boddey RM, Dakora FD, Bhattarai S, Maskey SL *et al.* (2009) The contributions of nitrogen-fixing crop legumes to the productivity of agricultural systems. Symbiosis. 48:1–17.

Perrine-Walker FM, Hocart CH, Hynes MF and Rolfe BG. (2005) Plasmidassociated genes in the model micro-symbiont Sinorhizobium meliloti 1021 affect the growth and development of young rice seedlings. Environ Microbiol. 7:1826–1838.

Poole P, Ramachandran V, Terpolilli J. (2018) Rhizobia: from saprophytes to endosymbionts. Nature Reviews Microbiology 16: 291.

Pradhan A, Pahari A, Mohapatra S and Mishra BB. (2017) Phosphate-Solubilizing Microorganisms in Sustainable Agriculture: Genetic Mechanism and Application. In: T. K. Adhya *et al.* (eds.), Advances in Soil Microbiology: Recent Trends and Future Prospects, Microorganisms for Sustainability, Springer Nature Singapore Pte Ltd. Pp 81-97. https://doi.org/10.1007/978-981-10-7380-9_5.

Rai AN, So¨derba¨ck E and Bergman B. (2000). Cyanobacterium–plant symbioses. New Phytologist. 147: 449–481.

Ramamoorthy V, Viswanathan R, Raguchander T, Prakasan V and Samiyappan R. (2011) Introduction of systemic resistance by Plant growth promoting rhizobacteria in crop plants against pests and diseases. 20: 1-11.

Reich S, Almon H and Böger P. (1986). Short-term effect of ammonia on nitrogenase activity of *Anabaena variabilis* (ATCC29413). FEMS Microbiol. Lett. 34: 53-56.

Roughley RJ. (1968) Some factors influencing the growth and survival of root nodule bacteria in peat culture. J. Bacteriol. 31:259-265.

Shah VK, Stacey G and Brill WJ. (1983) Electron transport to nitrogenase. Purification and characterization of pyruvate: flavodoxin oxidoreductase. The *nifJ* gene product. J. Biol. Chem. 258: 12064-12068.

Silva MF, Oliveira PJ, Xavier GR, Rumjanek NG and Reis VM. (2009) Inoculantes formulados compolímeros e bactérias endofíticas para a cultura da cana-de-açúcar. Pesqui Agropecu Bras 44:1437–1443.

Singh CS and Subba Rao NS. (1979) Plant Soil 53: 387. https://doi.org/10.1007/BF02277872

Slatni T, Dell'Orto M, Salah IB, Vigani G, Smaoui A, Gouia H, Zocchi G, Abdelly C. Immunolocalization of H+-ATPase and IRT1 enzymes in N2-fixing common bean nodules subjected to iron deficiency. Journal of Plant Physiology. 2012 Feb 15;169(3):242-8.

Smith RL, Van Baalen C and Tabita FR. (1987) Alteration of the Fe protein of nitrogenase by oxygen in the cyanobacterium *Anabaena* sp. strain CA. J. Bacteriol. 169: 2537-2542.

Smith RS. (1992) Legume inoculant formulation and application. Can J Microbiol. 38:485–492.

Spaink HP, Wijfjes AHM, van der Drift KMGM, Haverkamp J, Thomas-Oates JE, Lugtenberg BJJ. (1994) Structural identification of metabolites produced by the NodB and NodC proteins of *Rhizobium leguminosarum*. Mol Microbiol. 3:821–831.

Sprent JI and Faria SM. (1988) Mechanisms of infection of plants by nitrogen fixing organisms. Plant and Soil 110: 157–165.

Tegeder M, Masclaux Daubresse C (2018) Source and sink mechanisms of nitrogen transport and use. New Phytologist 217: 35-53.

Vadakattu G and Paterson J. (2006) Free-living bacteria lift soil nitrogen supply. Farming Ahead. 169:40.

Van Rhyn P and Vanderleyden, J. (1995) The Rhizobium-plant symbiosis. Microbiological Reviews. 59:124-142.

Vincent JM. (1970) A manual for the practical study of root-nodule bacteria. International Biological Programme handbook no. 15. Blackwell Scientific Publications, Ox ford, England.

Viveros OM, Jorquera MA, Crowley DE, Gajardo G and Mora ML. (2010) Mechanisms and practical considerations involved in plant growth promotion by rhizobacteria. J Soil Sci Plant Nutr.10:293–319.

Vlassak K. and Reynders R. (1979) Agronomic aspects of biological dinitrogen fixation by Azoxpirillum spp. In: Vose PB and Ruschel AP (eds) Associative N2 Fixation, Vol 1. Boca Raton, FL: CRC Press. pp 93-102.

Wagner SC. (2011) Biological Nitrogen Fixation. Nature Education Knowledge. 3(10):15.

Watanabe I. (1982) Azolla–Anabaena symbiosis–its physiology and use in tropical agriculture. In: Dommergues YR and Diem HS (eds) Microbiology of Tropical Soils and Plant Productivity ., Martinus Nijhoff, The Ague, The Netherlands. pp 169–185.

Westhoff P. (2009). The economics of biological nitrogen fixation in the global economy. In: Emerich DW, Krishnan HB (eds) Nitrogen fixation in crop production. Agronomy Monograph No. 52. Madison, WI: American Society of Agronomy, pp 309–328.

Yanni YG, Rizk R, Corich V, Squartini A, Ninke K, Philip-Hollingsworth S, Orgambide G, De Bruijn F, Stoltzfus J and Buckley D. (1997) Natural endophytic association between Rhizobium leguminosarum bv.trifolii and rice roots and assessment of its potential to promote rice growth. Plant and Soil. 194(1–2):99–114.

8

Potassium Solubilizing Microorganisms (KSM) A Very Promising Biofertilizers

Padmavathi Tallapragada and Titus Matthew

Department of Microbiology, School of Sciences, Jain University (Declared as Deemed-to-be university u/s 3 of the UGC Act, 1956) 18/3 9th Main, 3rd block, Jayanagar, Bangalore- 560011, India

Abstract

Potassium is a very essential nutrient element that is required for proper health, growth, functioning and productivity of plants. However, up to 98% of this very important mineral is held up in insoluble, unavailable forms in the soil and with only approximately 2% soluble form available for plant use. Biofertilizers add nutrients to the soil for plant use by some mechanisms like the breakdown of phosphorus, potassium, fixing nitrogen, and also synthesizing some substances which in turn enhance plant growth. These processes take place naturally. Microorganisms that solubilize potassium have demonstrated an increased potassium uptake by plants when added as bio-inoculants consequently resulting in increased yield by these plants. This article reviewed the methods of isolation of potassium solubilizing bacteria and fungi and their use as biofertilizers. Data are presented showing summary of researches conducted on potassium solubilizing microorganisms and their findings. Figures on production efforts of biofertilizers in India per year and by states are also added.

Keywords: Potassium, Microorganisms, Biofertilizers

8.1 Introduction

Population of the world continues to rise and with it an increasing challenge of a balance in food needs of this ever increasing human population so as to cater for its nutrition. Today the emergence of genetic engineering has opened up easier and faster ways of food propagation to meet this challenge. However, as good and fascinating as the technology is, it also faces some ethical challenges and the limited availability to the majority of the farmers who are the feeders of the population. This in essence still leaves the farmer with the only option of cultivating his crops on the soil. Land (soil) remains a limited resource which cannot be expanded. Over the time, farmers have continued to cultivate this limited resource thereby depleting and exhausting it of the mineral. This has resulted in increasing demand for chemical fertilizers for cultivation of crops in developing countries. Such fertilizers are injurious to the soil, the environment, thereby putting the health and lives of plants, animals, microbes and man in danger. This makes it so important to experiment safer practices to provide needed nutrition for plants with minimal or no adverse effects. Methods that are friendly to the ecosystem thereby sustaining agriculture.

Research over the years has proven that some microorganisms when grown along with plants as their hosts complement their hosts in growth. Several species of bacteria in soil are found to do so well in the soil around plants. These bacteria in many ways enhance the growth of plants. Such bacteria are referred to as plant growth promoting rhizobacteria (PGPR). These PGPR are used today as biofertilizers. Potassium solubilizing microorganisms (KSM) are part of such plant growth stimulating/enhancing microorganisms that are being explored and exploited as alternative to chemical fertilizers. Potassium (K) exists in several forms in the soil, including mineral potassium, non-exchangeable potassium, exchangeable potassium and dissolved or solution potassium (K^+ ions). Plants can only directly take up the solution potassium (Shanware *et al.*, 2014). The present chapter emphasises on characterization, mechanism, mass production and use of KSB as biofertilizer with few examples.

8.2 Biofertilizer

The advent of biofertilizers has come as a potential that can be very good supplements for enhancing plant growth without adversely affecting the environment. Biofertilizer is a formulation consisting of active microbes that inhabits the rhizosphere or internal parts of plants. This happens when it is added to the soil, seeds or plant body thereby making the required nutrients available to the plant which in turn promotes growth. It is different from other forms of plant nutrient suppliers like chemical fertilizer, organic fertilizer and manure.

Strictly speaking biofertilizer must contain living organisms which helps to make nutrients available to the plant in a soluble form. It has come to be a more environmentally adequate supplement enamours potential for supply of nutrients to plants and reducing the use chemical fertilizers (Rana *et al.*, 2013).

How biofertilizers work

- Atmospheric nitrogen in the root nodules of leguminous plants is fixed by them for use by the plant.
- They are able to transform the rock minerals to soluble form called as biotransformation.
- They search for and collect minerals from soil strata.
- They produce hormones and antimetabolites which promote root growth.
- They are the primary cause of organic matter degradation and add to the soil mineral content.
- The yield of plant is increased by 10-25% with addition of biofertilizer because of increase in the mineral content of the soil, without affecting the soil (SAMCRC, 2010).

Advantages of biofertilizer

- A form of nutrient source that is renewable.
- Strengthen the health of soil.
- Augments chemical fertilizers.
- Supplant chemical fertilizers by 25-30%.
- The yield of is increased by 10-40%.
- Helps in the decay of plant residues and hence balances Carbon: Nitrogen ratio of soil.
- The water holding capacity, structure and texture of the soil is improved
- Does affect plant growth and soil fertility negatively.
- They secret growth hormones that invigorate plant growth.
- Fungistatic and other substances that are like antibiotics are secreted by them.
- Make nutrients soluble and increase their bioavailability.
- Eco-friendly and economically feasible (https://www.slideshare.net/gunjan_rjt/biofertilizer-18234085).

8.3 Potassium in thc soil

For healthy plant growth, Potassium (K) is a very essential macronutrient. 90% out of 100% of soil potassium is found as insoluble part of rocks and silicate minerals. This makes the presence of soluble potassium which is essential for plant use very low. But it remains among seven other elements which are essentially required for plant growth and reproduction and often referred as "the regulator" for its involvement with >60 different plant enzyme systems and several metabolic processes. Drought and disease resistance are also increased due to this element (Cakmak, 2005; Billore *et al.*, 2009). Potassium is the third important plant nutrient next only to Nitrogen (N) and Phosphorus (P) (Rehm and Schmitt, 2002). The exchangeable potassium on the surface of the soil particles and that dissolved in soil water system which often amounts to < 100 ppm out of 20000 ppm and comprise only 0.1-2% of the total potassium in the soil is the only available potassium for plants (George & Michael, 2002). Others are bound in insoluble minerals like mica and feldspar. This is further compounded by the imbalance in fertilizer application where the ratio of potassium to other minerals like phosphorus and nitrogen is very small.

In Indian soil, the soluble potassium form is present in approximately 2% and the other insoluble form is present in the range of 98% in form of minerals such as vermiculite, mica, muscovite, feldspar, and biotite (Goldstein, 1994). This presents an apparent need to search for alternative sources of potassium for plant uptake and use as well as maintaining its availability in the soil for a sustained use. Reports show that soil microbes play important role in the natural potassium cycle; presence of potassium solubilizing microorganisms in the soil could therefore play vital alternative role in making potassium available for use by plants (Rogers *et al.*, 1998).

8.4 Potassium solubilizing microorganisms (KSM)

Potassium solubilizing microorganisms (KSM) are microbes found in the rhizosphere that solubilizes the insoluble (unavailable) forms of potassium (K) to available (soluble) form to be used by plants for nutrition, development and productivity. Several bacteria and fungi have been isolated and implicated as potassium solubilizers with effective interactions between soil and plant systems.

8.4.1 Potassium solubilizing bacteria (KSB)

Several rhizosphere bacteria solubilizes potassium. *Bacillus circulatans*, *Burkholderia* sp., *Bacillus mucilaginosis*, *Pseudomonas* sp., *Bacillus edaphicus*, *Acidothiobacillus ferrooxidans*, and *Paenibacillus* sp. (Parmar and Sindhu, 2013) are few established bacteria solely employed for the potassium solubilisation.

8.4.2 Isolation of KSB

Method of serial dilution plate in modified Aleksandrov medium which contained (5.0 g Glucose, 0.5 g $MgSO_4.7H_2O$, 0.1g $CaCO_3$, 0.006 g $FeCl_3$, 2.0 g Ca_3PO_4, 3.0g insoluble mica powder as potassium source and 20.0g agar) in 1 litre of deionized water was used to isolate rhizobacteria from rhizosphere soil (Hu *et al.*, 2006). From several locations (five samples each from six sites) of plant rhizosphere soil samples were collected in random order. The five samples so collected were mixed together to give a composite sample. 0.1ml of up to 10^{-5} dilution of the composite soil was plated on Aleksandrov medium. Using biological oxygen demand (BOD) incubator, these plates were incubated at 28±2 °C in for 3-4 days. Colonies showing morphological difference were selected (Sivaramaiah *et al.*, 2007; Sahu and Sindhu, 2011). The isolates were maintained by periodic transfer on Aleksandrov agar medium slants and stored at 4 °C for further use.

8.4.3 Rhizobacterial isolates screened for potassium

Rhizobacterial isolates were screened on modified Aleksandrov medium plates by spot test method (Sindhu *et al.*, 1999). Modified Aleksandrov medium plates were prepared, mica powder which is an insoluble form of potassium was added to plate A while K_2HPO_4 (soluble form of potassium) was added to plate B and to each of the plates was added a loopful 10 μL of 10^6 CFU mL^{-1} of 48hour old isolates. Plates were incubated at 28±2 °C for 3 days. Potassium solubilization by the isolates was discerned by halo zone formation.

8.4.3.1 Quantitative value of potassium release

To 25ml Aleksandrov broth in 50 ml flask modified by either sugar; galactose, glucose, arabinose and xylose was inoculated a loopful of 48 hours old isolate. These were incubated at 28±2°C for 10 days. Suspension of the isolate growth was centrifuged at 7,000 rounds for 10 minutes. This separated the supernatant from the cell growth and insoluble potassium. 1ml of the supernatant was pipetted into a 50 ml volumetric flask, distilled water was added to the flask and the volume made to 50 ml. The content was mixed and fed to atomic absorption spectrometer for determination of potassium content (Manib *et al.*, 1986). Several concentrations of 10 ppm (0.5, 1.0 and 1.5 ppm) KCl solution were used to determine the standard curve. The reading on the standard curve indicated the quantity of potassium solubilized by the isolates (Parmar and Sindhu, 2013).

8.4.3.2 Optimizing conditions of growth for maximum solubilization of potassium

Potassium solubilization was determined in Aleksandrov medium broth from mica powder at neutral pH and temperature of 28±2 °C. Various sugars, differential pH and temperatures were adjusted in order to arrive at the optimum conditions for maximum solubilisation. For carbon sources, isolate was inoculated into 25 ml of modified Aleksandrov medium broth with glucose replaced by arabinose, galactose and xylose respectively (Hu *et al.*, 2006). The flasks so inoculated were incubated for 10 days at 28±2°C. K released in inoculated and uninoculated (control) broths were estimated. Effect of incubation temperature was determined by inoculating the Aleksandrov broths with isolates, the flasks were then incubated at varying temperatures of 25, 35 and 45 °C along with 30°C for 10 days (Parmar and Sindhu, 2013). Effect of pH on K solubilisation was tested by preparing aleksandrov broths at varying of 6.5, 7.5 and 8.5 using N/10 HCl or N/10 NaOH, the broths were buffered with phosphate buffer. The broths were then inoculated with isolate and incubated 28±2°C for 10 days (Parmar and Sindhu, 2013). Study on the effect of potassium release from minerals was done by preparing aleksandrov broths to which were differentially added 3.0 g potassium solutions $AlK(SO_4)_2.12H_2O$, K_2SO_4 or KCl each along with mica powder. The isolates were then inoculated into the broths and at 28±2 °C for 10 days. Released K in each flask was now determined (Parmar and Sindhu, 2013).

8.4.4 Isolation of KSMs from ceramic industry soils

8.4.4.1 Collection of samples

The ceramic industries use feldspar (insoluble form of potassium) as raw material. For that soil samples were collected around ceramic industries.

8.4.4.2 Adaptation and enrichment

Feldspar (the insoluble potassium) was added to the soil samples collected and mixed. The mixture was incubated at room temperature for a week. Thereafter, 1 gm of the incubated soil was added to 100ml liquid which contained 0.05% yeast extract, 1% glucose, and 0.5% feldspar, incubated on 120rpm at 37°C for 1 week.

8.4.4.3 Isolation and screening of potassium solubilizing bacteria

Soil samples enriched serially diluted up to 10^{-6} and inoculated on Aleksandrov agar medium of the following composition 0.05% $MgSO_4.7H_2O$, 1% glucose, 0.01% $CaCO_3$, 0.0005% $FeCl_3$, 0.5% potassium aluminium silicate, 0.2% $CaPO_4$, and 3% agar at pH 6.5 and incubated at 37°C for 1 week (Sugumaran and Janartham, 2007). Colonies selected were those that showed clear zone around them indicating potassium solubilization. Such selection of colonies was from plates that had 10^{-4}, 10^{-5} and 10^{-6} dilutions (**Table 8.1**).

Zone of activity of the different isolates was calculated by using Khandeparkar's ratio (Prajapati and Modi, 2012).

Ratio = D/d = Diameter of zone of clearance / Diameter of growth

Table 8.1: A summary on various researches on isolation and study of KSMs, adopted from Shanware *et al.* (2014)

Experimental findings	Reference
Isolated *Bacillus* sp. as silicate solubilizing bacteria from rice ecosystem in a medium containing 0.25 insoluble magnesium tri silicate	Raj, 2004
Isolated KSB from the roots of cereal crops with using specific potassium bearing minerals	Mikhailouskaya and Tcherhysh, 2005
Reported a bacterium capable of dissolving silicate minerals from feldspar samples	Bardar, 2006
Reported K solubilizing strains from the soil on Aleksandrov medium and they were found to dissolve mineral potassium effectively	Hu *et al.*, 2006
Isolated, characterized and identified *Bacillus mucilaginosus* which solubilizes silicon from illite at 30°C.	Zhou, 2006
Isolated Potassium solubilizing bacteria from Orthoclase, muscovite mica. Among the isolates, *B. mucilaginosus* solubilized more potassium by producing slime in muscovite mica	Sugumaran and Janartham, 2007
Isolated fourteen potassium solubilizing microorganisms from ceramic industry soil. Among them, the best potassium solubilizing bacterial strain was identified as *Enterobacter hormaeche*. And fungi as *Aspergillus terreus* and *A. niger*.	Prajapati, 2012
Assess the potassium solubilization activity of *Bacillus* sp., *Burkholderia* sp. and *Pseudomonas* sp. from tea (*Camellia sinensis*). Among the various carbon sources the best carbon source for solubilization of muriate of potash was glucose at RT.	Bagyalakshmi, 2012
Isolated *Paenibacillus glucanolyticus*, a promising potassium solubilizing bacterium from rhizosphere of black pepper (*Piper nigrum* L.)	Sangeeth *et al.*, 2012
Isolated twenty five soil bacterial strains from Ha Tien Mountain, Kien Giang, Vietnam. Seven strains are related to *Bacillus megaterium* and *Bacillus coagulans*.	Diep and Hieu, 2013
Isolated alkaliphilic *Bacillus* species from mica mines of Nellore district of Andhra Pradesh, India. The 16S rDNA sequence showed 99% similarity with *Bacillus* sp.19 and closest relative is *Bacillus amyloliquefaciens*	Gundala *et al.*, 2013
Isolated a total of 30 KSB from rhizosphere of different crops from Dharwad and Belgaum districts. Morphological and Biochemical characterization upto genus level revealed 26 belongs to *Bacillus* and others from *Pseudomonas* genera and releases 2.41 to 44.49 g/mL of potassium.	Archana *et al.*, 2013

8.4.5 Characterization of potassium solubilizing bacterial isolates

Cultural, microscopical and biochemical methods were employed to characterize the bacteria isolated. More studies which include pattern of antibiogram, production of enzymes, optimal temperature, pH optima, production of organic acids and growth at different NaCl concentration (Osman, 2009) is also considered for the same.

8.4.6 Characteristics of bacterial strains

The isolates were inoculated on Aleksandrov's agar medium. Colony characteristics includes shape, size, texture, transparency and consistency, were checked. The chemical characteristics checked were Gram staining, capsule staining, endospore staining, Catalase test, Voges-Proskauer, Lysine utilization, Nitrate reduction, Ornithine utilization, H_2S production, Phenylalanine deamination and Carbohydrate utilization (Cappuccino, 1998; John *et al.*, 1997).

8.5 Potassium solubilizing fungi (KSF)

Like bacteria, *Aspergillus fumigatus*, *Aspergillus terreus*, *Aspergillus niger* fungi have been found to solubilize potassium.

8.5.1 Isolation of KSF

Following the methods of Sugumaran and Janartham (2007) and Prajapati and Modi (2012) with little-bit modification was followed for the isolation and growth of fungal strains respectively. Selected fungal isolates were also inoculated on Aleksandrov medium with pH indicator dye (0.025% bromothymol blue) for further studies (Prajapati *et al.*, 2012).

8.5.2 Selection of KSF

Fungal colonies that showing halozone on the agar plates meant have solubilized potassium. This consequently informed their choice as potassium solubilizers. Further screening was done consequent on the clear zone so formed by isolates by use of Khandeparkar's selection ratio.

8.5.3 Colonical and microscopic examination

Colony characteristics of the selected fungal strains were studied on PDA medium. They were stained with lactophenol cotton blue and observed under compound microscope for their morphologies (John *et al.*, 1997).

8.5.4 Potassium solubilization

The fungal isolates were grown in 100 ml broth that having 1% insoluble potassium, 0.05% yeast extract, 1% glucose, at pH-6.5 and incubated for 7 days at 30 °C on 120 rpm using two K supplements, feldspar and potassium aluminium silicate. Potassium released was determined after each 24 hours interval using Sodium cobaltinitrite and FolinCiocalteu Phenol reagent. Simultaneously, pH value of the broth was also measured after each 24 hours (Prajapati *et al.*, 2012).

8.6 Mechanism of potassium solubilization by microorganisms

Acids like malic acid, citric acid, oxalic acid and formic acid are produced when organic matter decomposes. The breakdown of pollution compounds into solution is enhanced the supply of protons and Ca^{2+} ions complex. Previous work has shown that microbes produce organic compounds that are able to enhance soil mineral dissolution. Such organic minerals produced include oxalate, citrate and acetate (Sheng *et al.*, 2003). Solubilization of potassium occurs by complex formation between metal ions (such as Ca^{2+}, Al^{3+} and Fe^{2+}) and organic acids (Styriakova *et al.*, 2003). Extracellular enzymes, metal chelaters, metabolic by products are some organic ligands produced by microorganisms. These ligands together with simple and complex organic acids promote aluminosilicate mineral or quartz to dissolve in experiments in the laboratory or on the field (Grandstaff, 1986; Surdam and MacGowan, 1987). Solubilization of illite and feldspar to release potassium is caused by organic acids (like oxalic and tartaric) and capsular polysaccharides which are produced by microorganisms (Sheng and He, 2006). The weathering ability of *Bacillus*, *Clostridium* and *Thiobacillus* is done by production of organic acids, protons, organic ligands and siderophores. It has been experimented that in *Cladosporoides*, *Cladosporium* and *Penicillium* sp. have the ability for production of appreciable quantity of gluconic, citric and oxalic acids in broth culture. The produced acids can cause dissolution of feldspar, mica and clay silicates (Shanware *et al.*, 2014) (**Table 8.2**).

8.7 Mass production and quality control

The Biomate, India in their project report for biofertilizer laboratory production unit outlined the following methodologies for mass production and quality control of biofertilizers.

8.7.1 Mass production of Biofertilizers

8.7.1.1 Media Preparation and Starter Culture

Bacteria require different nutrients for their growth. These include: (a) organic carbon source, (b) nitrogen source and (c) a variety of other elements dissolved

Table 8.2: Summary of work done on the effect of potassium solubilizing microorganisms on different crops. Adopted from Shanware et al., 2014

Work Done	Result	Reference
Experiments were conducted to evaluate the potentials of PSB *Bacillus megaterium* and KSB *Bacillus mucilaginosus* inoculated in nutrient limited soil planted with eggplant	Results showed that rock P and K materials either applied singly or in combination did not significantly enhance soil availability of P and K. PSB increased higher soil P availability than KSB, which was recommended as a K-solubilizer. Inoculation of these bacteria in conjunction with amendment of their respective rock P or K materials increased the availability of P and K in soil, enhanced N, P and K uptake, and promoted growth of eggplant	Han and Lee, 2005
The potential of PSB *Bacillus megaterium* var. *phosphaticum* and KSB *Bacillus mucilaginosus* inoculated in nutrient inadequate soil planted with pepper and cucumber.	Results showed that co-inoculation of PSB and KSB resulted in consistently higher P and K availability than in the control having without bacterial inoculum and without rock material fertilizer. Both bacterial strains consistently increased mineral availability, uptake and plant growth of pepper and cucumber, suggesting its potential use as biofertilizer	Han, 2006
In this study the dynamics of K released from waste mica inoculated with KSM *Bacillus mucilaginosus* and its effectiveness as potassic-fertilizer was studied using sudan grass *Sorghum vulgarepers* var *sudanensis* as test	Significant correlation between biomass yield, K uptake by sudan grass and different pools of K in soils were observed. X-ray diffraction analysis indicated greater dissolution of mica is due to inoculation of *B. mucilaginosus* strain in the soil.	Basak and Biswas, 2009
Bacterium possessing high ability to solubilize potash was isolated from the rhizosphere of black pepper and identified as *Paenibacillus glucanolyticus* IISRBK2. The strain was evaluated for plant growth and potassium (K) uptake in soil artificially treated with 0.5, 1 and 1.5g K kg-1 soil. In this study, wood ash was used as a source of K which contained 53.1g K Kg^{-1}	Inoculation with strain *P. glucanolyticus* was found to increase tissue dry mass (ranging from 37.0% to 68.3%) of black pepper in g K Kg^{-1} wood ash amended soil. In the soil treated with 0.5 - 1.5g K Kg^{-1}, K uptake in live bacterium inoculated black pepper plants increased by 125.0184% compared to uninoculated control.	Sangeeth et al., 2012

(Contd.)

In a greenhouse experiment, the nematicidal effect of some bacterial biofertilizers including the nitrogen fixing bacteria (NFB) *Paenibacillus polymyxa*, PSB *Bacillus megaterium* and KSB *B. circulans* were evaluated individually on tomato plants infested with the root-knot nematode *Meloidogyne incognita* in potted sandy soil	The results indicated that these bacterial biofertilizers were helpful in providing soil nutrients (nitrogen, phosphate and potassium) and for the biological control of *M. incognita*	El-Hadad et al., 2011
A potassium-releasing bacterial *Enterobacter hormaechei* and fungal *Aspergillus terreus* strains were examined for PGP effects and nutrient uptake on Okra (*Abelmoscus esculantus*) in K-deficient soil in pot experiments	Inoculation *E. hormaechei* was found to increase root and shoot growth of Okra and both microorganisms were able to mobilize potassium efficiently in plant when feldspar was added to the soil. In okra growing in soils treated with insoluble potassium and inoculated with strain *E. hormaechei* and *A. terreus* the potassium content was increased. Among all the three applications of microbial inoculants, combined application of seed and soil was found to be more effective.	Prajapati et al., 2013
Isolation, Characterization and Identification of KSF from rhizospheric soil in Bangalore, India.	*Aspergillus terreus* showed the ability to solubilize feldspar (an insoluble form of potassium) in Aleksandrov broth and encouraging solubilization activity under varied growth parameters.	Matthew and Tallapragada, 2017

in water. Blue Green Algae (BGA) that can fix atmospheric carbon dioxide, does not require any carbon source. Similarly, the nitrogen-fixing bacteria, can fix atmospheric nitrogen and not requires any nitrogen source. A medium is an aquatic solution of a variety of organic and inorganic compounds that can supplement the above requirements for the growth of different microorganisms. According to the physical appearance, media are of two types: (a) Liquid media and (b) Solid media. The liquid medium is solidified by the addition of agar-agar as solidifying agent. Liquid medium can harbour bacterial growth in submerged condition in the media, whereas solid medium harbours microbial growth on the surface.

8.7.1.2 Sterilization of medium and preparation of slants in autoclave

The prepared media is then autoclaved at 121°C temperature and 15lbs pressure. Then slant media tubes were prepared for necessary culture storage following standard protocol of preparation.

8.7.1.3 Preparation of Starter Culture

In the starter culture a small quantity of bacterial culture inoculum is added to the medium. For the preparation of starter culture, twin flask is used which is a pair of flasks of identical size joined together by a latex tube. The benefit of the use of the twin flask is to avoid contamination. Fermenter is used for the production of microbial metabolites such as antibiotics or enzymes; it may also be used for the growth and multiplication of potent microorganisms. A lowcost production unit is developed to produce microbial inoculants for use as biofertilizers. These include *Azotobacter*, *Rhizobium*, and other bacteria that solubilizes phosphate and potassium.

8.7.1.4 Sterilization of the fermenter

Fermenter is a metallic vessel for which moist sterilization is the appropriate method. The principle of moist sterilization lies in the fact that when water is boiled in a closed system, the water vapour produced being accumulated within the vessel and increases the inside pressure. Thus, the boiling point of water increases beyond 100°C and the steam, released from the boiling water is of both higher temperature and creates pressure above than the normal which easily destroys the microorganisms present in or on the article.

8.7.1.5 Inoculation, growth, quality testing and termination of growth

Inoculation means addition of starter culture to the medium in the fermenter. For the production of microbial biofertilizers a small amount of suspension of the desired bacterium in pure form is inoculated to the medium. Care should be taken to maintain the quality of starter culture, as extent of purity (no contaminants should be allowed), size of the starter culture (in terms of culture volume and density of cell) and stage of growth. Maintenance of proper physical and chemical environment inside the fermenter is essential for proper growth of microorganism. When the cell density reaches the desired level, growth is terminated and the culture is ready for mixing with carrier. The time period required for optimum cell density is thus standardized.

8.7.1.6 Carrier Preparation

Carrier is a medium, which can carry the microorganisms in sufficient quantities and keep them viable under specified conditions and easy in supply to the farmers (**Figure 8.1**). A good carrier should have the following qualities:

- Highly absorptive (water holding capacity) and easy to process.
- Non-toxic to microorganisms.
- Easy to sterilize effectively.
- Adequately available and low in cost.
- Provide good adhesion to seed.
- Has good buffering capacity.

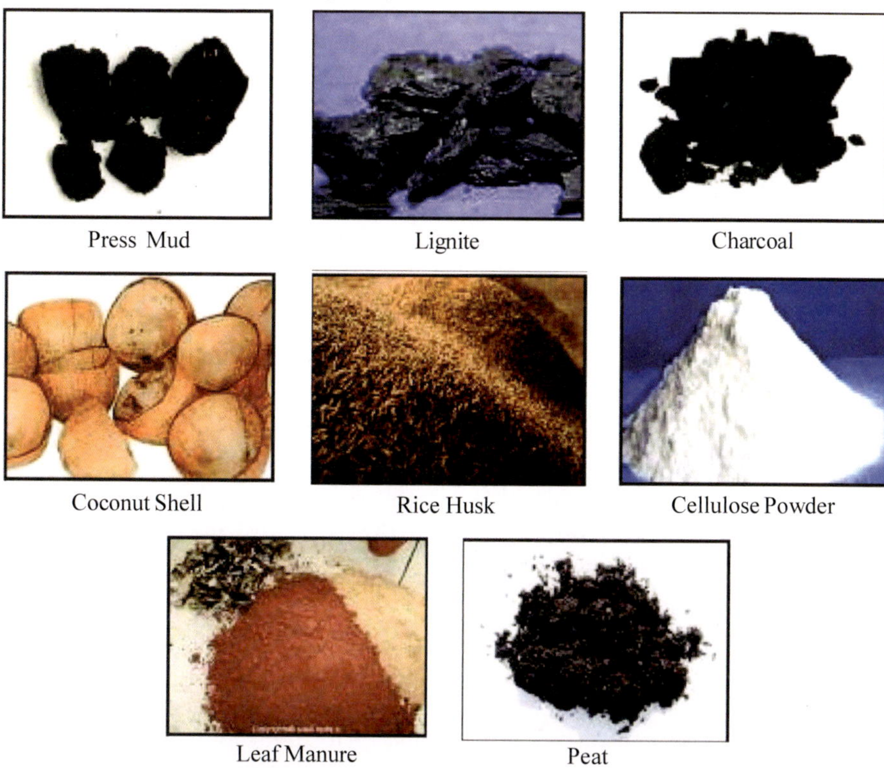

Fig. 8.1: Carriers for Bacterial inoculants

Different carriers are available in the market like, Charcoal, Peat, Lignite, Rice husk etc. But considering all the above qualities *Azolla* powder is the most suitable carrier in Indian context. This is due to:

- It has high water holding capacity (360%).
- It has good pH buffering capacity.
- It contains nutrient so bacteria can remain viable for a long period.
- It is easily available in this region.

8.7.1.7 Formulation

Inoculation of the carrier with the culture broth means the mixing of broth and carrier. This operation must be done in aseptic conditions to avoid any contamination.

8.7.2 Quality control of formulation

The viable cell count and the presence or absence of contaminants determines the quality of the carrier-based inoculum. A culture having$10^7 – 10^8$ viable cells or CFUs per g is considered as of good quality. No contaminants are permissible at $10^{-5} – 10^{-6}$dilutions. These critical values differ according to the type of biofertilizer.

8.7.2.1 How to Apply Microbial Biofertilizer

One important step in biofertilizer technology is how microbial biofertilizer is applied after mass production (**Figure 8.2**). The benefit obtained from its application is dependent on proper application of microbial inoculant. For proper application, the fact that microbial biofertilizers are mostly heterotrophic. They are dependent on organic carbon in the soil for supply of energy and growth. This makes them either colonizers of rhizosphere or symbionts inside the roots of higher plants. It is from the root exudates of higher plants that the rhizosphere colonizing bacteria derive organic carbon compounds while the symbionts derive directly from the roots. Microbial inoculants therefore should be added in methods that will cause the bacteria to adhere to the root surface. Inoculants are applied on roots for transplanting crops and on seed for seed sown crops. On the basis of the above principle, the following inoculation methods have been developed:

1. Inoculation of the seeds by slurry inoculating technique
2. Inoculation of seeds by seed pelleting technique
3. Inoculation of the seedlings
4. Inoculation of the soil by solid inoculation technique
5. Inoculation of soil by liquid inoculation technique

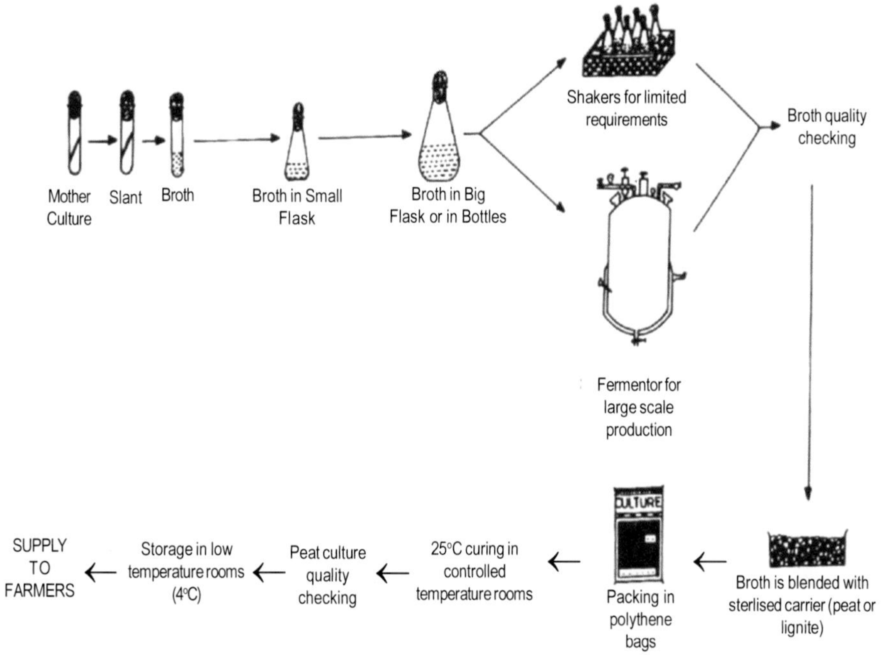

Fig. 8.2: Schematic diagram for mass scale production of bacterial Biofertilizers (retrieved from: https://www.slideshare.net/roshnimohan4/biofertilizers-production-and-their-applications)

8.8 Examples of researched application of KSM

1. Growth promoting effect of potassium solubilizing microorganisms on okra (*Abelmoschus esculentus* L.) (Prajapati *et al.*, 2013).

8.8.1 Increased K Uptake by Plants

Experiments to study how KSMs affect plant growth and uptake of potassium of Okra plant carried out in pots (15 cm diameter) which contained 5 kg of soil from Kalol (Gujarat, India). The experiment was designed thus:

treatment 1: soil not inoculated (control soil); treatment 2: soil not inoculated + soluble K (50 kg/ hec K_2O); treatment 3: soil not inoculated + soluble K (35 kg/ hec K_2O) + insoluble K (15 kg/ hec K_2O); treatment 4: inoculated soil + KSB strain + soluble K(35 kg/ hec K_2O)+ insoluble K (15 kg/ hec K_2O); treatment 5: inoculated soil + KSF strain + soluble K (35 kg/ hec K_2O) + insoluble K (15 kg/ hec K_2O); treatment 6: inoculated soil + KSB + KSF strains + soluble K (35 kg/ hec K_2O) + insoluble K (15 kg/ hec K_2O).

Soluble and insoluble potassium were added to soil in a plastic bag and mixed well. The mixture was added to pots and to each pot were sown 5 seeds at a depth of 2 cm. Each treatment was done in triplicate. Germinated plants were thinned to three plants in each pot. Log phase bacterial cells in broth culture was centrifuged for 10 minutes at 7000 rpm at 6°C. Then washed with sterile distilled water and centrifuged again. Pelleted cells were again suspended in sterile distilled water to obtain the bacterial inoculum. 1ml of bacterial suspension was used to inoculate Okra plant seedlings. The 1ml suspension was equivalent of an inoculum density of 10^8 cfu/ ml. In the case of fungal strain, suspension of spores in distilled water was done to give 10^9 spores/ml of inoculum density.

The three different experiments were planned for microbial application.

Experiment A: seed application - seeds were treated with microbial inoculants for 30 minutes before sowing.

Experiment B: soil application - microbial inoculants were inoculated in soil

Experiment C: seed + soil application - seeds were treated with microbial inoculants for 30 minutes before sowing and microbial inoculants were inoculated in soil.

The growth of plants in pots was for 55 days in greenhouse at a temperature of 30–34°C. The pots were watered to maintain 60% of soil water holding capacity. At the end of 55 days after sowing, number and weight of pods, shoot height, root length, fresh and dry weights were taken. The shoots and roots were separated and dried at 105°C. Their fresh and dry weights were also noted down.

Results

Shoot and root dry weights of inoculated plants showed significant increases better than uninoculated plants. Plants in soil with strain KSB-8 and rock K showed most increase in comparison to the KSF-1 and uninoculated soil.

Soil inoculation with KSB-8 and KSF-1 significantly increased plant growth in terms of length and weight of shoot and root, so also the height of plant. Total Plant weight and Dry weight especially when the potassium rock was added. Further soil co-inoculated with bacteria, soluble K and rock K showed significant increases in available K and chlorophyll content of the plants. **Figure 8.3** shows examples of tomato plants to which some KSMs were applied. It can be visibly seen that the inoculated plants exhibited very significant increase in height and fruiting than the control plants.

2. *Paenibacillus glucanolyticus*, a promising potassium solubilizing bacterium isolated from black pepper (*Piper nigrum* L.) rhizosphere (Sangeeth *et al.*, 2012).

8.8.2 Evaluation of growth promotion and K uptake in black pepper by K solubilizer

These experiments were conducted as: soil + farm yard manure + sand in a ratio 1:1:1 were mixed, sterilized and placed in polythene bags of size 15 × 21 cm. The mixture was added to pots in which black pepper cuttings were raised. The cuttings used for the experiment was Sreekara variety. Wood ash was used independently and in combination with the isolated organisms to study the release of K by the organisms and how it promoted growth and uptake of K by the plant. The design of the experiment was Completely Randomized Design (CRD) made of seven treatments in triplicate each. The treatments were (1) Control (2) Potting mixture + wood ash 0.5% (3) Wood ash 1.0% (4) Wood ash 1.5% (5) Potting mixture + wood ash 0.5% + BK1 (6) Wood ash 1.0% + BK2 and (7) Wood ash 1.5% + BK3.

Isolates used for inoculation were cultured differently in their media respectively as: Mc1, Mc2 and Mc3. To harvest cells, the broth was by centrifuged for 10 mins at 5000 rpm, sterile distilled water was used to wash them. The pellets obtained after repeated centrifugation were suspended in distilled water to prepare inoculum. This gave inoculum containing cfu of 10^7 mL^{-1}.

Roots of plants were dipped in inoculum for 15 minutes to have them inoculated at the point of planting. This was followed by pouring respective inoculum at 50 mL $plant^{-1}$ to drench the soil. Plants were allowed to grow for four months and were watered once in two days interval. Growth parameters which include number of leaves, root length and height of the plant were noted. Fresh and dry weights of plants were taken. N, P and K content of plants and soil were analysed as well as the soil pH.

Results: The experiment proved that there was release of insoluble K from ash into the soil by the isolates and it improved the content of K in the soil with consequent enhancement of plant growth. At the end of four months dry matter of the plants was between 3.31 and 5.57 g $plant^{-1}$. BK2(code name) isolate grown in 1.0% showed higher and significant increases in growth parameters and soil N and K than other isolates.

Fig. 8.3: Tomato Pants treated with Potassium Solubilizing Microorganisms. **Control** (no microbe), **A** (potassium solubilizing bacterial culture added), **B** (potassium solubilizing fungal culture added), **C** (Reference culture added), **D** (Arbuscular Mycorrhizal fungi added).

Figure 8.3 shows how effective KSM are in promoting plant growth and productivity. Pots A, B, C and D shows a luxuriant growth of the tomato plants which have some KSM added to them as against the control (first) pot to which no KSM was added.

8.9 Production of biofertilizer in India

Important biofertilizer producing states in India include Andhra Pradesh, Uttar Pradesh, Madhya Pradesh, Maharashtra, Kerala, Gujarat, Karnataka and Tamil

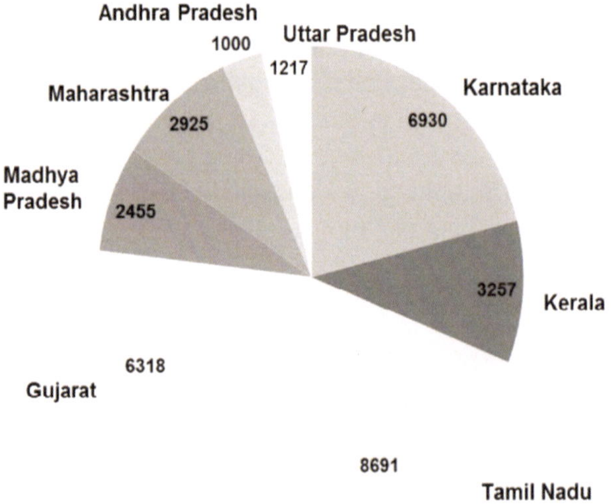

Fig. 8.4 Important Biofertilizer Producing States in India

Nadu with production of 1000, 1217, 2455, 3257, 6318, 6930 and 8691 metric tonnes respectively (**Fig. 8.4**). This indicates higher production in the southern part of India with 54%, followed by the West at 34%, then North 7%, North East 3% and lastly the East at 2% (**Fig. 8.5**).

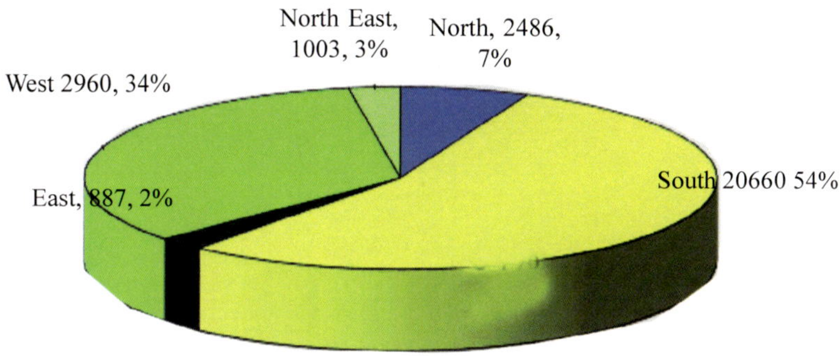

Fig. 8.5: Production of Biofertilizers in different regions of India

There is also a remarkable uplift of biofertilizer production in India from year 2005/6 to 2010/11 i.e. 100,000 metric tonnes to 400,000 metric tonnes respectively (**Fig. 8.6**). Biofertilizer production in total is shared by various microbes (**Fig. 8.7**). The bacteria are playing a major role followed by fungus and others. Phosphorus solubilizing bacteria (PSB) tops the list at 50%, Azospirillum 16%, Rhizobium 12%, Azotobacter 11%, Mycorrhiza 7% and others 4%.

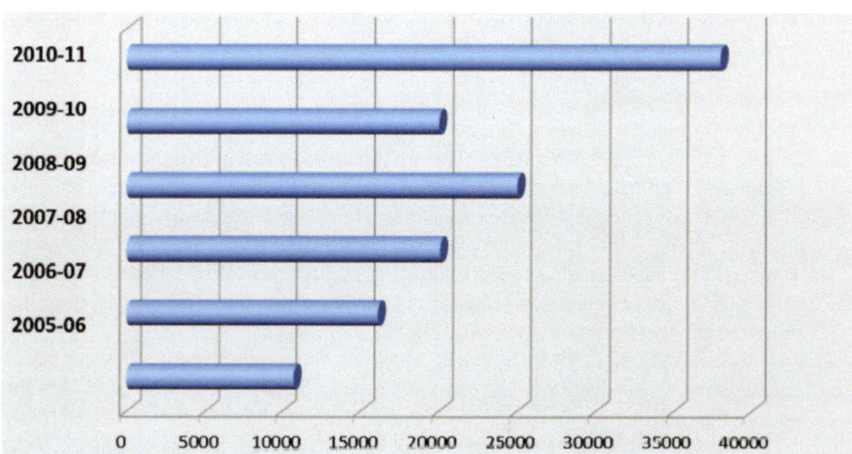

Fig. 8.6: Production Scenario of Biofertilizer in India

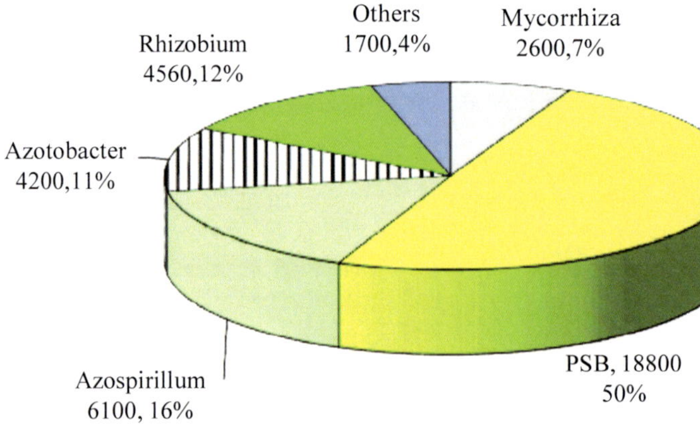

Fig. 8.7: Share of different biofertilizers to total production (2010-11)

The above statistics, though way below, desired production shows a great leap to meet up with the need. More of concerted and coordinated research, production and distribution efforts by researchers, government and industries will serve a great deal in meeting up demand and sufficiency of biofertilizer. This will in turn impact greatly on the human health, economic and ecological benefits.

In conclusion, biofertilizer is the way to go because of its several advantages like ecofriendly properties, affordability, plant health and yield. Several KSM have been unearthed and are in use in the formulation of biofertilizers as already shown. They are an invaluable component in the formulation of the biofertilizer. Hence the need to continue to explore and exploit their use in order to meet the growing food demand.

References

Archana, D.S., Nandish, M.S., Savalagi, V.P. and Alagawadi, A.R. 2013.Characterization of potassium solubilizing bacteria (KSB) from rhizosphere soil. Bioinfolet. vol. 10; 248-257

Badar, M. A., 2006. Efficiency of K feldspar combined with organic material and silicate dissolving bacteria on tomato yield. J. App. Sci. Res., 2(12): 1191-1198

Bagyalakshmi, B., Ponmurugan, P. and Balamurugan, A., 2012. Impact of different temperature, carbon and nitrogen sources on solubilization efficiency of native potassium solubilizing bacteria from tea (Camellia sinensis). J Biol Res. vol. 3(2); 36-42.

Basak, B.B. & Biswas, D.R., 2009.Influence of potassium solubilizing microorganism (*Bacillus mucilaginosus*) and waste mica on potassium uptake dynamics by sudan grass (Sorghum vulgare Pers.) grown under two Alfisols. Plant Soil. vol. 317; 235-255.

Billore, S, D., Ramesh, A., Vyas, A. K. and Joshi, O. P. (2009). Potassium use efficiencies and economic optimization as influenced by levels of potassium and soybean (Glycine max) genotypes under staggered planting. Ind. J. Agric. Sci. 79: 510–514.

Cakmak, I. (2005). The role of potassium in alleviating detrimental effects of abiotic stresses in plants. J. Plant. Nutri. Soil Sci. 168: 521–530. Doi:10.1002/jpln.200420485

Cappuccino, J.G. (1998). Microbiology – A Laboratory Manual 7th Edition Pearson Education. Dorling Kindersley (India) Pvt. Ltd.

Diep, C.N. and Hieu, T.N. 2013. Phosphate and potassium solubilizing bacteria from weathered materials of denatured rock mountain, Ha Tien, Kiên Giang province Vietnam. Am. J Life Sci. vol. 1(3); 88-92.

El-Hadad M.E. *et al.* 2011. The nematicidal effect of some bacterial biofertilizers on *Meloidogyne incognita* in sandy soil. Brazilian Journal of Microbiology 42; 105-113. Doi: 10.1590/S1517-83822011000100014

George, R. and Michael, S. (2002). Potassium for crop production. Communication and Educational Technology Services, University of Minnesota Extension.

Goldstein, A.H. (1994). Involvement of the Quino Protein Glucose Dehydrogenase in the Solubilization of Exogenous Mineral Phosphates by Gram Negative Bacteria. In Phosphate in Microorganisms: Cellular and Molecular Biology. Cell. Mol. Biol., Eds., pp. 197-203.

Grandstaff, D.E. (1986). The dissolution rate of foresteritic olivine from Hawaiian beach sand. In: Rates of Chemical Weathering of Rocks and Minerals. Eds. Colman, S.M., Dethier, D.P., Academic Press, New York, pp.41-60.

Gundala, P.B., Chinthala, P. and Sreenivasulu, B., 2013. A new facultative alkaliphilic, potassium solubilizing, Bacillus Sp. SVUNM9 isolated from mica cores of Nellore District, Andhra Pradesh, India. Research and Reviews. J Microbiol Biotechnol. 2(1); 1-7

Han, H.S. & Lee, K.D., 2005. Phosphate and potassium solubilizing bacteria effect on mineral uptake, soil availability and growth of eggplant. Res J Agric Boil Sci. 1(2): 176-180.

Han, H.S., Supanani S., Lee, K.D., 2006.Effect of co-inoculation with phosphate and potassium solubilizing bacteria on mineral uptake and growth of pepper and cucumber. Plant Soil Environ, 52: 130-136.

https://www.slideshare.net/gunjan_rjt/biofertilizer-18234085.

https://www.slideshare.net/roshnimohan4/biofertilizers-production-and-their-applications

Hu, X.F., Chen, J. and Guo, J.F. (2006). Two phosphate and potassium solubilizing bacteria isolated from Tiannu mountain, Zhejiang, China. World Journal of Microbiology and Biotechnology, 22: 983-990. Doi:10.1007/s11274-006-9144-2

John, G.H., Noel, R.K., Peter, H.S., James, T.S. and Stanley, T.W.s (1997). Bergey′s Manual of Determinative Bacteriology (In Russian) 9th Edition. Moscow, Mir Publishers.

Manib, M., Zahra, M.K., Abdel-Al, S.H.I. and Heggo, A. (1986). Role of silicate bacteria in releasing K and silicone from biotite and orthoclase. In: Soil biology and consevation of the biosphere. Szegi, J. Ed. Akademiai Kiado, Budapest. pp. 733-743.

Matthew, T. and Tallapragada, P. 2017. Isolation, characterization and identification of potassium solubilizing fungi from rhizosphere soil in Bangalore. IJBPAS 6(5): 931-940

Mikhailouskaya, N. and Tcherhysh, A., 2005. K-mobilizing bacteria and their effect on wheat yield. Latnian J. Agron., 8: 154-157

Osman, A.G. (2009). Study of Some Characteristics of Silicate Bacteria. Journal of Science and. Technology. 27-35.

Parmar, P. and Sindhu, S. S. (2013). Potassium Solubilization by Rhizosphere Bacteria: Influence of Nutritional and Environmental Conditions. Journal of Microbiology research, 3(1): 25-31. Doi:10.5923/j.microbiology.20130301.04

Prajapati, K. B. and Modi, H. A. (2012). Isolation and Characterization of Potassium Solubilizing Bacteria from Ceramic Industry Soil. CIBTech Journal of Microbiology. Vol. 1(2-3), 8-12.

Prajapati, K., Sharma, M. C. and Modi, H. A. (2012). Isolation of Two Potassium Solubilizing Fungi from Ceramic Industry Soil. Life Sciences Leaflets 5:71-75. Culled from http://lifesciencesleaflets.ning.com/

Prajapati, K., Sharma, M. C. and Modi, H. A. (2013). Growth Promoting Effect of Potassium Solubilizing Microorganisms on Okra (*Abelmoscus esculantus*). International Journal of Agricultural Science and Research (IJASR), 3(1):181-188.

Raj, S. A., 2004. Solubilization on a silicate in concurrent release of phosphorus and potassium in rice ecosystem. In: Biofertilizer technology for rice based cropping system, pp. 372-378.

Rana, R., Ramesh and Kapoor, P. (2013). Biofertilizers and Their Roles in Agriculture. Popular Kheti. 1(1)56-61. Culled from http://www.popularkheiti.com

Rehm, G. and Schmitt, M. (2002). Potassium for crop production. Retrieved February 2, 2011, from Regents of the University of Minnesota website: http://www.extension.umn.edu/distribution/cropsystems/dc6794.html.

Rogers, J.R., Bennett, P.C. and Choi, W.J. (1998). Feldspars as a source of nutrients for microorganisms. American Mineralogy, 83: 1532-1540. Doi:10.2138/am-1998-11-1241

Sahu, G.K. and Sindhu, S.S. (2011). Disease control and plant growth promotion of green gram by siderophore producing *Pseudomonas* sp. Journal of Microbiology Research. 3(1): 25-31. DOI: 10.3923/jm.2011.735.749

Sangeeth, K. P., Bhai, S. R and Srinivasan, V. (2012). *Paenibacillus glucanolyticus*, a promising potassium solubilizing bacterium isolated from black pepper (*Piper nigrum* L.) rhizosphere. Journal of Spices and Aromatic Crops. 21(2): 118-124.

Shanware, A. S., Surekha, A. K. and Minal, M. T. (2014). Potassium Solubilizers: Occurrence, Mechanism and Their Role as Competent Biofertilizers. International Journal of Current Microbiology and Applied Sciences. 3(9):622-629.

Sheng, X. F. and He, L. Y. (2006). Solubilization of potassium bearing minerals by a wild type strain of *Bacillus edaphicus* and its mutants and increased potassium uptake by wheat. Can, J. Microbial., 52(1); 6672. DOI: 10.1139/w05-117

Sheng, X. F., Xia, J.J. and Chen, J. (2003). Mutagenesis of the *Bacillus edaphicus* strain NBT and its effect on growth of chili and cotton. Agric Sci China, 2:40-41.

Shri AMM Murugappa Chettiar Research Centre (SAMCRC) (2010). Booklet on Biofertilizer (Phosphobacteria).

Sindhu, S.S., Gupta, S.K. and Dadarwal, K.R. (1999). Antagonistic effect of Pseudomonas spp. on pathogenic fungi and enhancement of plant growth in green gram (*Vigna radiata*). Biology and Fertility of Soils, 29, 62-68.

Sivaramaiah, N., Malik, D.K. and Sindhu, S.S. (2007). Improvement in symbiotic efficiency of chickpea (Cicer arietinum) by coinoculation of *Bacillus* strains with *Mesorhizobium* sp. Cicer. Indian Journal of Microbiology, 47, 51-56. Doi:10.1007/s12088-007-0010-1

Styriakova, I., Styriak, I., Hradil, D. and Bezdicka, P., (2003). The release of iron bearing minerals and dissolution of feldspar by heterophic bacteria of *Bacillus species*. Ceramic. Silicaty, 47(1); 20-26.

Sugumaran, P. and Janartham, B. (2007). Solubilization of potassium minerals by bacteria and their effect on plant growth. World Journal of Agricultural Sciences 3(3): 350-355.

Surdam, R.C. and MacGowan, D.B. (1987). Oil field waters and sandstone diagenesis. Appl. Geo Chem., 2(5-6): 613-619. Doi:10.1016/0883-2927(87)90013-8

Zhou Hong, B. O., Zheng Xiao X. I., Liu Fei-Fei, Qiu Guan-Zhou and Hu Yue Hua. 2006. Screening, identification and desiccation of silicate bacterium. J. Cent. Sou. Univ tech., 13; 337- 341.

9

Advances in Potassium Nutrition in Crop Plants by Potassium Solubilizing Bacteria

Nur Uddin Mahmud, Musrat Zahan Surovy, Dipali Rani Gupta and Md Tofazzal Islam

Department of Biotechnology, Bangabandhu Sheikh Mujibur Rahman Agricultural University, Gazipur 1706, Bangladesh

Abstract

Potassium (K) is the third most essential macronutrient element for plants. More than 98% of K in agricultural soils are remained as insoluble organic and inorganic compounds or complexes that are unavailable for plant uptake. Nutrition of crop plants are heavily relied on application of expensive inorganic chemical fertilizers, which is a potential cause of environmental pollution. An eco-friendly alternative strategy for plant K nutrition could be the application of K solubilizing bacteria (KSB) as biofertilizer. Plant and soil-associated bacteria from diverse taxonomic genera such as Bacillus, Pseudomonas, Acidithiobacillus, Paenibacillus, Flectobacillus, Arthrobacter, Enterobacter, Sphingomonas, Paraburkholderia etc. have shown high K solubilizing activities and promote plant nutrition in K deficient soils. The mechanisms involved in their activities include acidolysis, chelation, exchange reactions, and enzymatic degradation of insoluble K minerals and other K complexes. However, inconsistency in the efficiency of KSB-based biofertilizer for crop production in diverse ecological settings is still a big challenge. Recent genomics, post-genomics and CRISPR-Cas genome editing approaches seem to be useful for a better understanding the underlying mechanisms of KSB and development of highly efficient K fertilizers for promoting sustainable agriculture. This chapter comprehensively reviews current updates of KSB-based biofertilizer and their potentials for promoting sustainable crop production. The role of genomics and post-genomics approaches for better understanding of the mechanisms of the functions of applied KSB-based biofertilizers in the crop field and mitigation of the existing inconsistency in the efficiency of KSB-based fertilizers are also discussed.

Keywords: Potassium biofertilizer; Rhizobacteria; *Bacillus* sp.; Genome editing; Sustainable agriculture

9.1. Introduction

Potassium (K) is an essential macronutrient, which is involved in several physiological and biochemical activities (Zhang and Kong 2014). Plants available form of potassium is K^+. It involves with over 60 different enzyme systems, production of starch, and synthesis of protein. The K^+ assists in the growth of root, administers cell stomatal alteration and, contributes plants resistance to diseases, insects, and other biotic and abiotic stresses (Rehm and Schmitt 2002; Sangeeth *et al*., 2012; Mahmud *et al*., 2017). Besides, it carries a vital role in the photosynthetic system and triggers enzymes in carbohydrates metabolism that leads to the production of amino acids and proteins. Deficiency of potassium causes a reduction in photosynthesis and scorching or burning of small grains. The adverse effects of soil K deficiency in plants include poorly developed roots, stunted growth, tiny seeds and decreased yields of crops and increased susceptibility to biotic and abiotic stresses (Meena *et al*., 2013, 2014a; Bakhshandeh *et al*., 2017). Symptoms of K deficiency include yellowing or schorching of leaf edge and falling out the leaf. Sometimes yield and quality loss can be occurred long before the appearance of deficiency symptoms (Khanwilkar and Ramteke 1993).

Although K content in most of the soils of tropical and sub-tropical regions is abundant, only 1-2% of the soil K is available as K^+ for plant uptake. About 98% of K remains in the soil as unavailable forms as K-feldspars, phyllosilicates and organic K (Andrist-Rangel *et al*., 2007). Besides, soil erosion, leaching and run off, imbalance fertilizer application and use of modern crop varieties for higher crop production decrease the content of available K in soils (Sheng and Huang 2002a; Maurya *et al*., 2014; Singh *et al*., 2015). The major potassic fertilizer, muriate of potash (MoP) is highly expensive and not affordable by the resource-poor farmers in the developing countries. These worsen situations demand the search for an innovative alternative to the commercial potassium fertilizer for better K nutrition in crop plants in tropical and sub-tropical areas for ensuring sustainable agriculture. Plant-associated potassium solubilizing bacteria (KSB) is considered one of the viable alternatives and a potential biotechnology for sustainable K nutrition in crop plants. A large body of literature indicates that soil rhizospheric and plant endophytic bacteria can release K^+ by degradation of silicate minerals and decomposition of the organic matter in soils. In fact, these beneficial bacteria increase the available K^+ and other essential nutrient element in soil that promote plant growth and contributes to ion cycling (Lian *et al.*, 2008). Although the mechanism of functions of these bacteria are not fully understood, however, KSB can also secrete secondary metabolites, low molecular weight organic acids, hormones, enzymes and suppress pathogens that confer advantages to plant. The KSB also weather silicate minerals and make free K^+, Si^{+4}, Al^{+3} ions that improve soil structure and nutrients status (Lian *et al.*, 2002; Bosecker 1997). The history of studying KSB and their roles in soils and plant is not new. For example, Muentz in the year 1890 experimented the microbial association in potassium solubilization from rocks. Aleksandrov *et al.*, (1967) reported the solubilization of potassium from silicate and aluminosilicates by bacteria from agricultural soils. The KSB diffuse K^+ from different K-bearing minerals such as muscovite, biotite, illite, and potash feldspar through synthesis and release of organic acids (Zhang and Kong 2014). Therefore, applying the KSB as biofertilizer can lessen the use of chemical fertilizer and support environmentally safe and sustainable agriculture (Sindhu *et al.*, 2010). The KSB based biofertilizers are widely used in many countries like India, Korea and China due to deficiency in potassium availability for crops in the agricultural lands (Xie 1998).

This chapter reviews advances of K nutrition in plants and updates our understanding of the use of KSB-based biofertilizers and molecular mechanism of the functions of KSB. The formulations of effective biofertilizer from KSB, field application of KSB, efficiency of these biofertilizers in crop plants and the future prospects of KSB-based biofertilizers are also discussed.

9.2. Discovery of the elite strains of KSB

Bacteria from diverse sources and taxonomic classes can solubilize insoluble and fixed forms of K into the available form (K^+) for plant uptake (Zarjani *et al.*, 2013; Gundala *et al.*, 2013). Several studies reported that various genera of bacteria such as *Bacillus*, *Acidithiobacillus*, *Pseudomonas*, *Enterobacter*, *Burkholderia* and *Paenibacillus* bring potassium into K^+ from insoluble potassium-minerals (Sheng 2005; Liu *et al.*, 2012; Maurya *et al.*, 2014; Kumar *et al.*, 2015). Many elite strains of KSB are isolated from the root and rhizosphere soil of cultivated land with different plants which can solubilize potassium from muscovite, microcline, orthoclase, mica or potassium muriate (Saiyad *et al.*, 2015; Maurya *et al.*, 2014; Bagyalakshmi *et al.*, 2012a; Sugumaran and Janarthanam 2007). Mica mine and ceramic industries are also the great sources of potential KSB that solubilize K from potassium aluminum silicate minerals (Gundala *et al.*, 2013, Prajapati and Modi 2012; Prajapati *et al.*, 2012). The

KSB were isolated mainly from different rhizospheric soils of cultivated plants such as wheat (Parmar and Sindhu 2013; Zhang *et al.*, 2013), maize (Abou-el-Seoud and Abdel-Megeed 2012), potato-soybean cropping pattern (Biswas 2011), pepper and cucumber (Han and Supanjani 2006), common bean (Kumar *et al.*, 2012), oak-mycorrhizosphere (Uroz *et al.*, 2007), cotton (Sheng and He 2006), Sudan grass (Basak and Biswas 2009), and sorghum and maize (Archana *et al.*, 2013).

Generally, the KSB were isolated from soil using serial dilution technique and plating in Aleksandrov medium composed of 1g glucose, 0.05M $MgSO_4.7H_2O$, 0.0005M $FeCl_3$, 0.01M $CaCO_3$, 0.2M $CaPO_4$ and 0.5M potassium aluminum silicate, agar 3 % (pH 7.6) and incubated at 37°C for one week (Aleksandrov *et al.*, 1967; Sugumaran and Janartham 2007; Meena *et al.*, 2014). Isolated bacteria are screened for solubilization of potassium in modified Aleksandrov broth medium containing an insoluble source of potassium minerals such as muscovite, biotite and wood ash (Sangeeth *et al.*, 2012; Meena *et al.*, 2013). Colonies exhibiting clear halo zones **(Figure 9.1)** indicate the potential strains of KSB that are screened on the basis of their potassium solubilizing index (KSI). KSI=A/B, where A= Diameter of the halo zone of, B= Diameter of colony growth. Quantitative estimation of K in modified Aleksandrov broth (MAB) medium is done following a standard protocol using a flame photometer or atomic absorption spectrophotometer (Nath *et al.*, 2017). Considering the high K solubilization capacity, efficient potassium-solubilizing strains are selected by *in vitro* seedling assay, *in vivo* pot experiment and field authentication using various crops (Archana *et al.*, 2013).

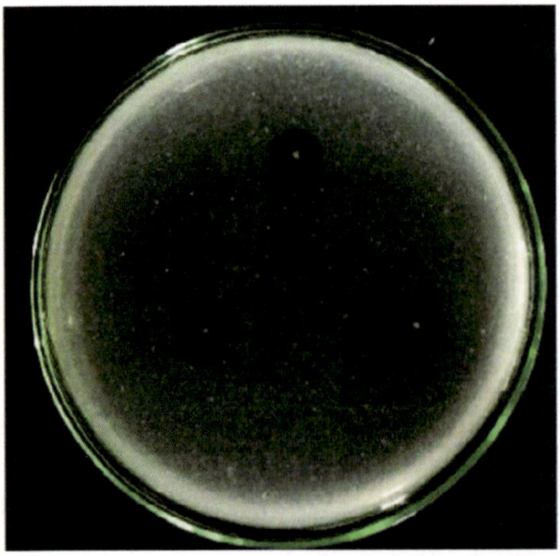

Fig. 9.1 Halo zone formation by KSB in muscovite containing Aleksandrov media. Whitish dot at the center of the halo zones are the colony of a KSB isolated from agricultural soils of Bangladesh.

Routine bacteriological test including morphological, physiological, biochemical test and 16S rRNA gene sequencing provide the identification and characterization of the bacteria. Bacteriological test included color, shape, colony margin, elevation, the zone of solubilization, optical density, biofilm formation, slime production, gram staining, KOH, citrate, indole, catalase, oxidase and starch production, siderophore production, endospore production, Voges-Proskauer (VP) and Methyl red (MR) reactions, citrate utilization, H_2S production, utilization of organic acids, anaerobic growth, production of acid from carbohydrates, and antibiotic assay (Claus and Berkeley 1986; Meena *et al*., 2013). Xiufang *et al.* (2006) described the molecular identification of potassium solubilizing bacteria by 16S rRNA gene sequence analysis. Molecular identification of KSB using 16S rRNA includes several steps such as DNA extraction, PCR amplification using universal primer, gel electrophoresis, PCR product purification, sequencing of the amplified DNA, construction of phylogenetic tree using software such as CLC, clustalW2, Mega 4, and finally submission of the aligned sequence to the NCBI GenBank for obtaining accession number of a particular strain. Advanced molecular techniques such as metagenomics provide a precise characterization of bacteria including unculturable ones in soils or rhizosphere settings such as mechanism of potassium solubilization, exploring new enzymes, organic acid and plant growth-promoting activity. Genomic studies of the KSB are a time demanding approach for obtaining potential KSB for the development of a biofertilizer (**Fig. 9.2**).

9.3. Mechanisms of plant K nutrition by bacteria

Considerable information is available on mechanism of potassium solubilizing plant-associated and soil bacteria, and their effect on plant K nutrition (**Table 9.1**). Various factors including strains of bacteria, bacterial exudates, soil environment such as aerobic condition, pH, temperature, sources and nature of potassium-containing minerals affect the solubilization of potassium by the KSB (Sheng and Huang 2002; Uroz *et al*., 2009; Chen *et al*., 2008). For instance, *Bacillus edaphicus* can solubilize higher amount of potassium from illite than feldspar in the liquid medium (Sheng and He 2006). Badr (2006) recorded 4.29mg/L potassium solubilization by silicate-solubilizing bacteria at pH 6.5–8.0. Similarly, *B. mucilaginosus* solubilized 4.29 mg K/L in muscovite supplemented media (Sugumaran and Janarthanam 2007).

Several mechanisms are known to mobilize and solubilize structurally unavailable form of potassium compounds by the KSB. These mechanisms include production of different inorganic and organic acids that are accompanied by acidolysis and complexolysis exchange reactions, lowering the pH or enhance the chelation of K-bound cations, production of capsular polysaccharides (CPS), exopolysaccharides (EPS), siderophores, and formation of bacterial biofilm.

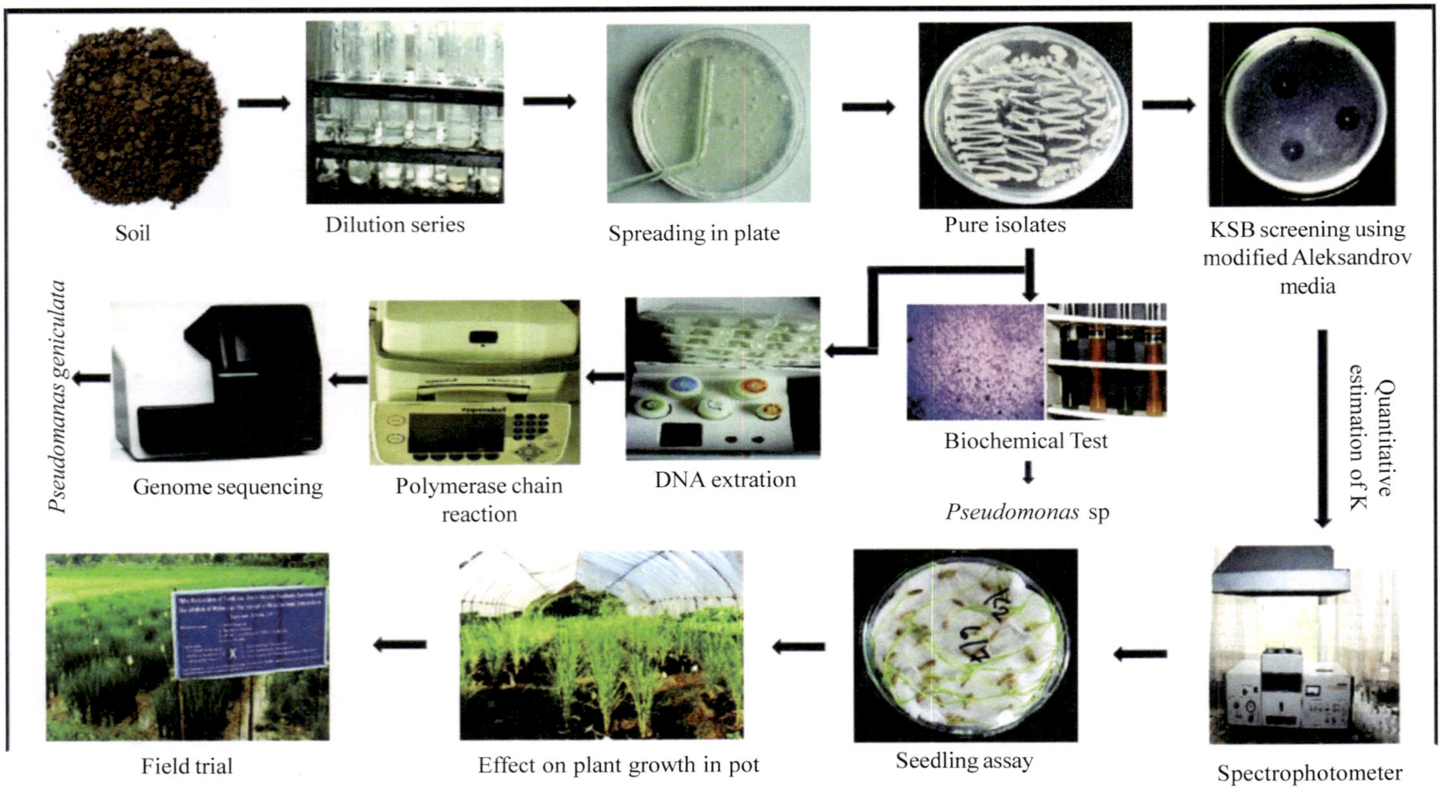

Fig. 9.2: Flow chart showing the major steps of research for discovery of KSB from rhizosphere soil to develop a K-biofertilizer.

Table 9.1 Diversity of potassium solubilizing bacteria with their source of isolation and their mechanism of actions for K solubilization

Bacterial species	Source of bacteria	Mechanism of K solubilization	Reference
Pseudomonas sp.	Wheat rhizosphere	Excretion of organic acids	Sheng and Huang (2002)
Bacillus sp.	Granite crusher yard	Lipo-chitooligosaccharides	Raj (2004)
Bacillus mucilaginosus	Plastic film house area	Production of IAA	Han and Lee (2005)
B. edaphicus	Cotton and rapeseed	Production of organic acids	Sheng (2005)
B. mucilaginosus	Illite	ND	Zhou et al. (2006)
B. mucilaginosus	Pepper and cucumber	Organic acids	Han and Supanjani (2006)
B. mucilaginosus	Tianmu mountain	ND	Hu et al. (2006)
B. edaphicus	Wheat	Tartaric acid, oxalic acid	Sheng and He (2006)
B. glathei	Mycorrhizosphere, bulk soil	Siderophores, organic ligands	Calvarusoet al. (2007)
Sphingomonas sp., *Burkholderia* sp.	Oak-mycorrhizosphere	Acidification, complexation	Urozet al. (2007)
B. mucilaginosus	Muscovite mica	ND	Sugumaran and Janarthanam (2007)
B. globisporus	Weathered feldspar	Gluconic and acetic acid production	Sheng et al. (2008)
B. mucilaginosus	Sudan grass	Organic acids	Basak and Biswas (2009)
B. subtilis	Rhizospheric soil of agricultural fields in Thailand	ND	Leaungvutivirojet al. (2010)
B. mucilaginosus	Maize	Acidification	Abou-el-Seoud and Abdel-Megeed (2012)
Enterobacter hormaechei	Ceramic industry soil	Organic acids	Prajapati and Modi (2012)
Bacillus sp., *Azotobacter* sp., and *Microbacterium* sp.	Potassium mines	ND	Lang-bo et al. (2012)
Paenibacillus glucanolyticus	Black pepper rhizosphere	Organic acids	Sangeeth et al. (2012)
P. mucilaginosus	Silicate minerals	Tartaric, citric, oxalic acids	Liu et al. (2012)
Frateuria aurantia	Banana rhizosphere	Release of hormone	Pindi and Satyanarayana (2012)
Pseudomonas putida	Southern Indian tea soils	Production of phytohormones	Bagyalakshmi et al. (2012)
B. megaterium, *B. coagulans*	Soils/weathered rocks	ND	Diep and Hieu (2013)

B. amyloliquefaciens	Mica mines	Gluconic acid	Gundala et al. (2013)
B. megaterium, *Arthrobacter* sp.	Iranian soils	ND	Zarjani et al. (2013)
Bacillus sp., *Pseudomonas* sp.	Rhizosphere soil	Organic acids	Archana et al. (2013)
E. hormaechei	Okra	Organic acids	Prajapati et al. (2013)
Klebsiella sp., *Enterobacter* sp.	Tobacco rhizosphere soil	ND	Zhang and Kong (2014)
Pseudomonas sp., *Bacillus* sp.	Loamy sand soil	ND	Syed and Patel (2014)
Pantoea ananatis, *Rahnella aquatilis*, *Enterobacter* sp.	Rice paddy soil	Organic acids	Bakhshandeh et al. (2017)

ND, not determined.

Production of various types of low molecular weight organic acids is the well-understood and widely studied mechanisms of potassium solubilization by the KSB. Various kinds and levels of organic acids such as citric, tartaric, 2-ketogluconic, oxalic, gluconic, malic, propionic, fumaric, glycolic, and succinic acids are produced by the KSB to solubilize K from various insoluble forms (Zarjani *et al.*, 2013; Prajapati and Modi 2012; Prajapati *et al.*, 2012; Wu *et al.*, 2005).

Romheld and Kirkby (2010) observed the direct dissolution of minerals or chelation of Al and Si-bound minerals by organic acid to slowly release available K^+. Production of organic acids by the KSB can also be detected and quantified through enzymatic and high performance liquid chromatographic (HPLC) method (Archana *et al.*, 2013; Zhang *et al.*, 2013). The organic acid enhances dissolution directly by H^+ or ligand-mediated mechanism and/or indirectly by forming complexes in solution with reaction products (Ullman and Welch 2002; Meena *et al.*, 2014a; Kumar *et al.*, 2015). For example, *B. mucilaginosus* and *B. edaphicus* produce carboxylic acids like citric, tartaric, and oxalic acids that are associated with solubilization of K from feldspar (Malinovskaya *et al.*, 1990; Sheng and Huang 2002a). Silicate solubilizing bacteria produce phosphoric acid which dissolves potassium, silica, and aluminum from the insoluble mineral, apatite (Heinen 1960). Hutchens *et al.*, (2003) documented that some strains of *B. megaterium* and *Arthrobacter* sp. produce organic acids and mono-hydroxamate siderophores and solubilize K, Si, and Fe from the liquid medium supplemented with acid-leached soil, muscovite, and biotite. It is also reported that *Thiobacillus* sp. and *Bacillus* sp. produce inorganic acids that can dissolve potassium from K-containing minerals (Berthelin and Belgy 1979). However, organic acids can dissolve more K than inorganic acids produced by the bacteria (Welch and Ullman 1993).

Bacteria colonized the sub-surface of plant roots produce slime, polysaccharide and exopolysaccharides (EPS).These bacteria are also known to promote dissolution of silicates minerals and contribute to release K from silicate minerals (Duff *et al.*, 1963; Groudev 1987; Surdam and MacGowan 1988; Vandevivere *et al.*, 1994). A high concentration area of organic acids near the minerals occur due to the adsorption of organic acids by polysaccharides that adhere to the surface of the mineral causing dissolution of that mineral (Liu *et al.*, 2006). The EPS affects the equilibrium between the mineral and fluid phases by absorbing SiO_2 which results the release of K solubilization. Naturally-occurring fresh microbial EPS obtained from the subsurface microorganisms may form complexes with framework ions in solution and increase the dissolution rate of feldspars (Welch and Vandevivere 2009). A wild-type strain of *B. edaphicus* and its four mutants produce oxalic, tartaric acids, and capsular polysaccharides

(CPS) that can release potassium from illite and feldspar minerals (Sheng and He 2006).

Production of biofilms by bacteria is another potential mechanism to mobilize soil K (Balogh-Brunstad *et al.*, 2008). Bacteria form biofilm on the surface of mineral that release K, Si, and Al by accelerating the corrosion of potassium-rich shale in the bacteria-mineral contact model (Li-yang *et al.*, 2014). Higher weathering rates were supposed to be caused by biofilms and biocrusts in the process of biodegradation (Warscheid and Braams 2000).

9.4. Formulation and practical delivery of KSB to the crop field

Soilborne bacteria and other microorganisms play a vital role in several biogeochemical cycles and cycling of essential plant nutrients for plant uptake. They directly involve in maintaining soil nutrient status and productivity (Wall and Virginia 1999). They can solubilize the important macronutrients such as nitrogen, phosphorus, and potassium etc. from their unavailable and insoluble sources by the biological mechanisms (Islam *et al.*, 2017). Rhizospheric topsoils contain earthworms (ca. 1000 kg/ha), fungi (ca. 2700kg/ha), bacteria (ca. 1600 kg/ha), protozoa (ca. 150 kg/ha) and arthropods (ca. 995 kg/ha) (Verma *et al.*, 2009). Several lines of evidence indicate that rhizosphere bacteria possess the capability to confer plant growth promotion and enhance biotic and abiotic resistance of plants (Dimkpa *et al.*, 2009; Sindhu *et al.*, 2014b; Islam *et al.*, 2005; 2017).

Potassium-solubilizing bacteria, an integral part of soil microorganisms play a vital role in K cycle by dissolving the unavailable K to available form for plants uptake (Marques *et al.*, 2010; Sindhu *et al.*, 2014a). The KSB based biofertilizer is an alternative to the environmentally hazardous chemical fertilizers as they can colonize inthe plant roots and rhizosphere. Sheng *et al.* (2003) suggested using of KSB as biofertilizers for obtaining safe and eco-friendly agriculture.

Formulation of an effective KSB biofertilizer is a prerequisite for sustainable use of these bioresources. At present, biofertilizers are found in carrier-based solid or powder form or in liquid form. Liquid biofertilizers achieve greater acceptance by the farmers than conventional carrier-based technology. In case of biofertilizer formulation, potential strains of bacteria can be used either in single or in a consortium with other plant growth promoting bacteria such as phosphate solubilizing bacteria and nitrogen-fixing bacteria to promote effectivity of the biofertilizer. In fact, application of a biofertilizer as a bacterial consortium may be a more ecologically viable approach for its better colonization and adaptation to the soils. *Frateuria aurantia*, a new bacterial species belonging to the family Pseudomonadaceae can reduce 50–60% usage of potassic chemical fertilizers. This K biofertilizer can be also applied in consortium with *Rhizobium*, *Azospirillum*, *Azotobacter*, *Acetobacter* and with phosphate solubilizing bacteria. The *F. aurantia* is a highly adaptive to almost all types of soils having low K content and variable soil pH ranging 5 to11 and withstand in a temperature of up to 42°C. This strain of KSB promotes plant growth and enhances the efficiency of chemical fertilizer (Patel 2011).

Potassium solubilization is a complex phenomenon which is influenced by several factors such as available essential nutrient elements in soil, soil microorganisms, plant-microbe interactions, soil mineral type, amount of mineral, size of mineral, and other environmental factors. To formulate a biofertilizer, selection of a potential strain of KSB is critical (Kirk *et al.*, 2004; Naik *et al.*, 2008). Selection should be done based on *in vitro* and *in vivo* assay towards growth promoting efficiency, adaptation of the strain to a highly buffering condition (Sindhu *et al.*, 2002, 2014b). To prepare a consortium formulation, all member bacteria must be compatible to each other.

The efficacy of KSB can be improved by developing a better cultural conditions and delivery system for their survivability in the rhizosphere (Sindhu and Dadarwal 2000). Bacterial formulation can be done using two different approaches such as the single-culture approach (SCA) and the multiple or mixed culture approach (MCA). In SCA, K solubilizer is used alone. In contrast, in MCA, which is often called co-inoculation, one or more KSB are used combined with other plant growth promoting microorganisms (Bahadur *et al.*, 2014). Seed treatment with KSB biofertilizer is a cheap and the easiest means of inoculation. Proper application of this method ensures receiving the introduced bacteria in each seed. The KSB fertilizer can also be used as a treating agent of plant seedlings. The bacterial inoculant is mixed with 500 ml of sterilized water to make slurry (Bahadur *et al.*, 2014) and gum acacia can be used to improve the adherence of the inoculant KSB on the seed. The seeds are poured into the slurry for uniform coating of the inoculant over the seeds following shade dried and then sown the seeds within 24 h in the seed bed. For coating 10 kg of seeds, about 200 g inoculants is enough (Subba 2001). Seedlings treatment is practiced for transplanted crops. In case of seedling treatment, seedling should be washed with water and root priming should be done overnight in the above-mentioned slurry and then transplant in the field. Another method of biofertilizer application is mixing the bioinoculants with compost manure and applied in the main field just before seed sowing or seedling transplanting. This method is recommended where seed application may be an ineffective way of application, such as treatment of seeds with chemical pesticides which is incompatible with the KSB. This method has several advantages i.e. greater population of KSB per unit area and reduced direct contact with chemically treated seeds. Proper inoculation technique enhances the survivability of KSB. The growth and multiplication of KSB is greatly affected by the soil condition such as low quality or poor organic matter content, low moisture condition, and also by the seasonal variation. Soil-inhabiting indigenous microbes compete with applied bacterial inoculant for nutrition and moisture, thus inoculated bacteria cannot show their productive establishment in the soil. Thus, application of potential KSB in soil with proper environment for inoculant establishment is a prerequisite for maintaining better plant nutrition. The KSB are sophisticated biofertilizer which need careful handling, maintenance, storage, and transportation facility. The quality and shelf-life of the population of commercial formulations should be maintained carefully to get the desired benefit (Lian *et al.*, 2008; Sindhu *et al.*, 2014).

9.5. Performances of KSB in promoting growth and yield of crop plants

Plant probiotic bacteria can exert beneficial effects to plants through helping better intake of necessary nutrients by N_2 fixation, solubilization of P, K and Zn production of phytohormones and siderophores, and transformation of macro-nutrients such N, P, K, Fe, and others (Islam *et al.*, 2007; Herridge *et al.*, 2008; Sindhu *et al.*, 2010; Islam and Hossain 2012; Islam *et al.*, 2017). Treatments of seeds and seedling with KSB significantly enhanced seed germination, seedling vigor, K uptake, growth, and yield of crop plants (Singh *et al.*, 2010; Awasthi *et al.*, 2011; Zhang *et al.*, 2013; Zhang and Kong 2014). Numerous researchers demonstrated that the KSB can remarkably reduce the usage of chemical fertilizers. The first report of the application of silicate bacteria with the combination of organo-minerals for the enhancement of growth of wheat and maize was published by Aleksandrov (1985). Since then a large body of literature is available on beneficial effects of KSB and K-biofertilizer for growth and yield of crop plants. Some important literatures are briefly reviewed in following sub-sections.

9.5.1. Rice

A good number of reports are available on the promotion of growth and yield of rice by the inoculation with silicate-solubilizing bacteria (Muralikannan 1996; Kalaiselvi 1999). For example, application of *B. circulans* solubilizes Si and K from various minerals. It increases a significant amount of organic matter in the soils and also increased the grain yield (17%) of rice (Zahra *et al.*, 1984). Similarly, Raj (2004) and others demonstrated that grain yield, and silica contents in rice and soils were increased by the application of silicate-solubilizing *Bacillus* species (Balasubramaniam and Subramanian 2006). Bakhshandeh *et al.* (2017) conducted both pot and field trials of three potential strains viz. *Pantoea ananatis*, *Rahnella aquatilis* and *Enterobacter* sp. on rice. In the pot experiment, it has been found that plant height, stem diameter, root length, leaf area and plant biomass were increased by 4.09–10.8%, 4.07–10.4%, 8.0–13.1%, 19.8–21.4% and 7.53–15.7%, respectively compared to untreated control. In the field experiment, plant height, biomass dry weight, Soil Plant Analysis Development (SPAD) value, K uptake in the leaves, stem, and root of rice seedlings were also augmented by 10.8–15.1%, 27.4–65.3%, 8.64–12.0%, 38.5–76.9%, 17.6–52.9% and 25.0–75.0%, respectively (Bakhshandeh *et al.*, 2017). Mahmud *et al.* (2017) screened 11 potential KSB isolated from the rhizosphere soils of rice in Bangladesh. Seed priming or root dipping of rice seedlings with those KSBs significantly enhanced growth and yield of rice compared to untreated control under low K soils.

9.5.2. Wheat and maize

Beneficial effects of KSB were also observed in wheat and maize. Wheat yield was increased up to 1.04 tons/ha by inoculating KSB on several eroded soils (Mikhailouskaya and Tcherhysh 2005). In a pot experiment, inoculation of wheat with *B. edaphicus* enhanced growth of roots and shoots as well as increased NPK contents in plants as compared to control plants in yellow-brown soils containing low K (Sheng and He 2006). Almost similar positive effects of KSB on plant growth and nutrient contents in soils have been described by many other researchers (Archana *et al.*, 2008; Parmar 2010). The inoculation of maize with *B. subtilis* KT7/2 increased the dry weight of shoots and roots and K content in soils (Leaungvutiviroj *et al.*, 2010). Singh *et al.* (2010) carried out a hydroponics study using *B. mucilaginosus*, *Rhizobium* sp., and *A. chroococcum* on wheat and maize under a phytotron growth chamber and observed the capacity of those bacteria to solubilize K from potassium minerals as a sole source of K. They found significant biomass accumulation, K content, and acquisition of crude protein and chlorophyll content in the tissues of wheat and maize plants. *B. mucilaginosus* showed the highest performance in mobilizing the potassium than *Rhizobium* sp. and *A. chroococcum.* In another study, *B. mucilaginosus* and *B. subtilis* were co-inoculated in K deficient calcareous soil supplemented with rock K materials to evaluate the synergistic effects using maize as a test crop. These two KSB increased available potassium in soil, promoted plant shoot and root growth, and enhanced potassium uptake by maize (Abou-el-Seoud and Abdel-Megeed 2012).

9.5.3. Sorgum and green gram

The beneficial effects of KSB inoculation were also reported in sorghum (Sheng and Huang 2002; Zhang *et al.*, 2013). The inoculation of KSB in sorghum plants supplemented with potassium minerals enhanced uptake of K by 79%, 41%, and 93% and increased dry matter content by 58%, 48%, and 65% in calcareous, clay, and sandy soils, respectively (Badr 2006). Basak and Biswas (2009) inoculated K-solubilizing *B. mucilaginosus* combined with waste mica and observed an increased biomass yield and uptake of K by Sudan grass (*Sorghum vulgare* var. *sudanensis*) grown under two Alfisols. Like sorgum, inoculation of KSB also improved the growth of green gram seedlings (Khudsen *et al*., 1982). The KSB strain isolated from the rhizosphere of mung bean and rock were found efficient in K solubilization in acidic soils and improved the growth and yield of green gram (Archana *et al*., 2008).

9.5.4. Groundnut and tobacco

The KSB application also enhanced growth and yield of groundnut and tobacco. For example, application of a KSB strain *P. mucilaginosus* MCRCp1 increased the dry matter (ca. 25%) and oil contents (ca. 35%) of groundnut compared to untreated control (Sugumaran and Janarthanam 2007). Inoculation of KSB in combination with a soluble NPK chemical fertilizer increased the yield of tobacco green leaf and K content both in crop shoots and cultivated soils. The inoculation of the KSB strains on growth of tobacco and improvement of cultivated soils have been reported by many other investigators (Zhang and Kong 2014; Subhashini 2015).

9.5.5. Cotton, rapeseed and sunflower

Similar to previously described crops, the growth and yield of cotton, rapeseed and sunflower were also increase by the application of KSB (Sheng 2005; Kammar *et al*., 2015). For example, Sheng (2005) performed in a pot experiment in K deficient soil, *B. edaphicus* NBT enhanced root and shoot growth of both cotton and rapeseed. The potassium content in both plants were also significantly increased by the application of this KSB. Kammar *et al*. (2015) evaluated 6 efficient strains of KSB namely, KSBL-5, KSBL-9, KSBL-32, KSBL-41, KSBD58 and KSBD-67 on growth and yield of sunflower and found that strain KSBD-58 is the best in promoting growth and seed yield of sunflower.

9.5.6. Tea and pine

Both tea and pine growth were also promoted by the application of KSB. Bagyalakshmi *et al*. (2012a) isolated a promising K-solubilizing strain of *P. putida* from the rhizosphere of tea and used as biofertilizer in tea plantation. For example, a KSB strain of *P. putida* from the rhizosphere of tea improved the productivity, quality of tea leaves, and nutrient uptake by tea plants. Similarly, inoculation of *Burkholderia glathei* strains PMB (7) and PML1 (12) on pine roots significantly increased the weathering of biotite and better K nutrition and growth in pine (Christophe *et al*., 2006).

9.5.7. Okra and eggplant

The KSB, *Enterobacter hormaechei* was applied to okra (*Abelmoschus esculantus*) in a K deficient soil (Prajapati *et al.*, 2013). It significantly enhanced root and shoot growth as well as increased K use efficiently in okra plant. The available K content in the cultivated soils was also increased by the application of this KSB. Similarly, application of KSB increased the uptake of K and biomass of brinjal plant as compared to the untreated control (Nayak 2001). In another study, a significant increase in brinjal yield, plant height, and K uptake was recorded in *F. aurantia* treated okra plants compare to the control (Ramarethinam and Chandra 2006). Co-inoculation of a PSB (*B. megaterium*) and a KSB (*B. mucilaginosus* strain KCTC3870) mixed with of rock P and K minerals in the soils enhanced N, P, and K uptake, improved photosynthesis, and promoted growth of eggplant in nutrient-limited soil (Han and Lee 2005).

9.5.8. Pepper, cucumber and tomato

Co-inoculation a PSB (*B. megaterium* var. phosphatic) and a KSB (*B. mucilaginosus*) consistently improved soil P and K availability and improved the growth of pepper (*Capsicum annuum* L.) and cucumber (*Capsicum annum* L.) (Han and Supanjani 2006). In tomato, inoculation of a silicate-dissolving bacterium, *B. mucilaginosus* strain RCBC13 increased biomass (125%) and K and P uptake compared to uninoculated plants (Lin *et al.*, 2002). Badr (2006) reported that inoculation of silicate-dissolving bacteria *B. cereus* along with feldspar and rice straw increased yield in tomato crop.

9.5.9. Yam and Tapioca

Co-inoculation of KSB, *Rhizobium*, *Azospirillum*, *Azotobacter*, *Acetobacter* and PSB increased K^+ uptake as well as increased yield of tapioca (*Manihot esculenta*) and yam (*Dioscorea* sp.) (Clarson 2004).

9.6. Molecular genetics of KSB and mechanism of their action

Potassium is an essential mineral nutrient element for all organisms including the KSB. K^+ uptake is required for the homeostatic processes of turgor pressure regulation and maintenance of the cytoplasmic pH of the bacterial cells (Csonka and Epstein 1996; Stumpe *et al.*, 1996). In KSB, generally three different K transporters viz. *Trk*, *Kdp*, and *Kup* are involved in the uptake of K from the soils. However, two major types of K^+ uptake systems (*Trk* and *Kdp*) and one minor K^+ uptake system (*Kup*) are present in *Escherichia coli* K-12 (Schleyer and Bakker 1993). The K transporter, *Trk* is a multicomponent complex which is widespread in both archaea and bacteria. It consists of a *trans*-membrane protein named *TrkH* or *TrkG* that are rapid K^+-uptake systems with a relatively low affinity for K^+ (Dosch *et al.*, 1991). The transporter *Trk* is a NAD-binding protein which required for the system's activity (Sleator and Hill, 2002). The Kdp which was found in *E. coli* and many other bacteria, is an inducible system belongs to the family of P-type ATPases with high affinity and specificity for K. The Kdp is a unique bacterial K acquisition system. Its expression is strongly regulated at the transcriptional level. The transporter Kdp is linked with the activity of KdpD sensor kinase and the KdpE response regulator (Epstein, 2003; Domínguez-Ferreras *et al.*, 2009). The genes of KUP/HAK/KT family are homologous to bacterial KUP (TrkD) potassium transporters. The KUP transporter was characterized from *E. coli* by a midrange (0.37 mM) KM for K^+ as well as a similar affinity for Rb^+ and Cs^+ (Bossemeyer *et al.*, 1989). Researchers transformed AtKUP1gene in a mutant *E. coli* TK2463 cell using a plasmid containing the AtKUP1 gene or with an empty vector (Epstein and Kim 1971). It was confirmed that the cell is defective in three K1 uptake transporters (Trk, Kup, and Kdp). In *Vibrio alginolyticus*, the *ktrA* and *ktrB* are two adjacent genes that encode a new type of bacterial K^+-uptake system. Nakamura *et al.* (1998) confirmed that the *ktrA* gene is peripheral and ktrB is integral membrane proteins. The expression of *Trk* and *Kdp* genes in the *Mycobacterium tuberculosis* (*Mtb*), wildtype and a *trk*-gene knockout strains were evaluated at various stages of logarithmic growth in relation to extracellular K^+ concentrations and pH (Cholo *et al.*, 2015). It was found that the Trk and Kdp systems of *Mtb* function at high extracellular K^+ concentration and neutral pH that favour membrane potential-driven uptake of K^+. Genes and primers used for the expression studies of TrK and Kdp K^+ transporter systems are tabulated (Cholo *et al.*, 2015) **(Table 9.2)**.

9.7. Challenges and opportunities

The plant associated KSB have beneficial effects to growth and yield of crops. Discovery of the elite strains of KSB are prerequisite for the commercial production of K-biofertilizer. Applications of KSB have several advantages over conventional chemical K fertilizers in crop production. The KSB are eco-friendly, self-multiplying ability (renewable) and are safe to the ecological processes. For commercialization of KSB, a consortium formulation is desirable for sustainable crop production system. In-consistency in the efficacy at field application and poor shelf-life under varying environmental conditions are two major challenging issues for commercialization of any KSB (Lugtenberg *et al.*, 2001). Another important issue is the ability of root colonization of the KSB under varying crop cultivars and environmental settings (Lin *et al.*, 2002; Sugumaran and Janarthanam 2007). Better understanding of the biology of the potential KSB through genomic and post-genomic approaches are needed for utilizing them effectively as biofertilizers for sustainable K nutrition in soils with high insoluble K contents (Bahadur *et al.*, 2016).

Table 9.2: Genes and primers for expression study of Trk and Kdp K^+ transporter systems (modified from Cholo et al., 2015)

Gene name	Forward primer	Reverse primer	Fragment length/ amplicon size (bp)	Tm(°C)
ceoB[a]	CGG CGA CAA CTC CAA CAT	CGG CAC ACC GAA GGT TTC	58	86.2
ceoC[a]	CTG CTG GAG TCG ATT CAC CT	ACC GCG AAT TCG GTC TTG	107	88.9
kdpA[b]	GAT TGA ACG GCC TAC TGG TC	CTG GAT CTT CTT GCC GAG AT	95	87.5
kdpB[b]	CTG GGC TGA CGA TCA TCT TT	CGT GGT CGG AAT GAG ACA C	122	90.6
kdpC[b]	TAC CGC AAG GAA AAC AAT CTG	GGC ATT GAC CAC CGA TATG	102	90.2
Kdpc[b]	AAC ATC GTC GGG TTG GTG	GCG AAT AGG AAC GCC ATT AG	50	83.7
kdpD[c]	AGT CCA TCG ACC AAC TCA CC	CAC CGC TTC CTC CAG GTA T	110	91.4
kdpE[c]	TGG AAT GGA CGA GTT TCT GG	GAC GGT GAA TGA ATC GGT TT	103	91.4
sigA[d]	CAA GGA CGC CGA ACT CAC	CTT GCC GAT CTG TTT GAG GT	64	87.4

[a]Gene encoding Trk system; [b] Gene encoding Kdp system; [c]Component response regulator protein; [d]House keeping gene

Studies are needed to understand the natural diversity and genetic manipulation of the elite strains of KSB for augmenting their efficiency of K solubilization in soils (Ryan *et al.*, 2009; Sindhu *et al.*, 2014). Other important aspects are effective formulation with suitable carrier materials and efficient delivery of the KSB to the crop plants. Molecular biological studies are obviously needed to characterize genes enzymes, and other metabolites involved in potassium solubilization by the bacteria. As CRISPR-Cas system is becoming a powerful tool for creating genetic diversity and genome editing of any organism (Haque *et al.*, 2018), the greater application of the revolutionary technology would obviously help development of powerful KSB fertilizer for ensuring sustainable K nutrition in crop production (Zhang *et al.*, 2013; Sindhu *et al.*, 2014).

9.8. Conclusion and perspective

To promote sustainable agricultural system and productivity for the increasing population of the world, judicious utilization of the natural resources is badly needed (Rahman *et al.*, 2018). The KSB are promising natural resources that could effectively be utilized for the development of eco-friendly biofertilizer for K nutrition in crop plants. To overcome the existing inconsistency in efficacy of biofertilizers in the field application and poor shelf-life under varying environmental settings, genomic and post-genomics studies are needed for elucidating the underlying molecular mechanisms of K solubilization by the plant-associated bacteria. The revolutionary CRISPR-Cas genome editing tool could be utilized for the development of highly efficient KSB with multiple plant growth promoting traits potential for commercialization as biofertilizers. It is expected that a combination of classical microbiological, genomics, post-genomics and genome editing approaches would help us to overcome existing limitations and develop highly efficient biofertilizers for K nutrition in crop plants in a sustainable manner.

References

Abou-el-Seoud, Abdel-Megeed A (2012) Impact of rock materials and biofertilizations on P and K availability for maize (*Zea mays*) under calcareous soil conditions. Saudi J BiolSci 19:55–63. Doi: 10.1016/j.sjbs.2011.09.001

Aleksandrov VG (1985) Organo-mineral fertilizers and silicate bacteria. DoklAkadNauk 7:43-48

Aleksandrov VG, Blagodyr RN, Ilev IP (1967) Liberation of phosphoric acid from apatite by silicate bacteria. Microchem J 29(2):111–114

Andrist-Rangel Y, Edwards AC, Hillier S, Oborn I (2007) Long-term K dynamics in organic and conventional mixed cropping systems as related to management and soil properties. AgricEcosyst Environ 122:413–426. doi.org/10.1016/j.agee.2007.02.007

Archana DS, Nandish MS, Savalagi VP, Alagawadi AR (2013) Characterization of potassium solubilizing bacteria (KSB) from rhizosphere soil. Bioinfolet 10:248–257

Archana DS, Savalgi VP, Alagawadi AR (2008) Effect of potassium solubilizing bacteria on growth and yield of maize. Soil Biol Ecol 28:9–18

Awasthi R, Tewari R, Nayyar H (2011) Synergy between plants and P-solubilizing microbes in soils: effects on growth and physiology of crops. Int Res J Microbiol 2:484–503

Badr MA (2006) Efficiency of K-feldspar combined with organic materials and silicate dissolving bacteria on tomato yield. J Appl Sci Res 2(12):1191–1198

Bagyalakshmi B, Ponmurugan P, Balamurugan A (2012) Impact of different temperature, carbon and nitrogen sources on solubilization efficiency of native potassium solubilizing bacteria from tea (*Camellia sinensis*). J Biol Res 3(2):36–42

Bagyalakshmi B, Ponmurugan P, Marimuthu S (2012a) Influence of potassium solubilizing bacteria on crop productivity and quality of tea (*Camellia sinensis*). Afr Agric Res 7:4250–4259

Bahadur I, Maurya BR, Kumar A, Meena VS, Raghuwanshi R (2016) Towards the soil sustainability and potassium-solubilizing microorganism. In: Meena VS, Maurya BR, Verma JP, Meena RS (eds) Potassium solubilizing microorganism for sustainable agriculture, Springer, India 2016, p 254 DOI 10.1007/978-81-322-2776-2

Bahadur I, Meena VS, Kumar S (2014) Importance and application of potassic biofertilizer in Indian agriculture. Int. Res. J. Biol. Sci 3, 80-85

Bakhshandeh E, Pirdashti H, Lendeh KS (2017) Phosphate and potassium-solubilizing bacteria effect on the growth of rice. Ecol. Eng. 103 (2017): 164-169. Doi:10.1016/j.ecoleng.2017.03.008

Balasubramaniam P, Subramanian S (2006) Assessment of soil test based potassium requirement for low land rice in udic haplustalf under the influence of silicon fertilization. Tamil Nadu Agricultural University, Tiruchirapalli, pp 621–712

Balogh-Brunstad Z, Keller CK, Gill RA, Bormann BT, Li CY (2008) The effect of bacteria and fungi on chemical weathering and chemical denudation fluxes in pine growth experiments. Biogeochemistry 88: 153–167. Doi:10.1007/s10533-008-9202-y

Basak BB, Biswas DR (2009) Influence of potassium solubilizing microorganism (*Bacillus mucilaginosus*) and waste mica on potassium uptake dynamics by Sudan grass (*Sorghum vulgare* Pers.) grown under two Alfisols. Plant Soil 317(1-2):235–255. Doi:10.1007/s11104-008-9805-z

Berthelin J, Belgy G (1979) Microbial degradation of phyllosilicates during simulate podzolization. Geoderma 21(4):297–310.Doi:10.1016/0016-7061(79)90004-1

Biswas DR (2011) Nutrient recycling potential of rock phosphate and waste mica enriched compost on crop productivity and changes in soil fertility under potato soybean cropping sequence in an Inceptisol of IndoGangetic Plains of India. Nutr Cycl Agroecosyst 89:15–30

Bosecker K (1997) Bioleaching: metal solubilization by microorganisms. FEMS Microbiol Rev 20(3-4): 591–604. Doi: 10.1016/S0168-6445(97)00036-3

Bossemeyer D, Schlosser A, Bakker EP (1989) Specific Cesium Transport via the *Escherichia coli* Kup (TrkD) K+ uptake system. J Bacteriol 171(4):2219-2221. Doi:10.1128/jb.171.4.2219-2221.1989

Calvaruso C, Turpault MP, Leclerc E, Frey-Klett P (2007) Impact of ectomycorrhizosphere on the functional diversity of soil bacterial and fungal communities from a forest stand in relation to nutrient mobilization processes. Microb Ecol 54:567–577

Chen S, Lian B, Liu CQ (2008) *Bacillus mucilaginosus* on weathering of phosphorite and primary analysis of bacterial proteins during weathering. Chin J Geochem 27:209–216

Cholo MC, Rensburg EJV, Osman AG, Anderson R (2015) Expression of the genes encoding the Trk and Kdp potassium transport systems of *Mycobacterium tuberculosis* during growth *in vitro*. BioMed Research International 2015, 11 pages.doi.org/10.1155/2015/608682

Christophe C, Turpault MP, Freyklett P (2006) Root associated bacteria contribute to mineral weathering and to mineral nutrition in trees and budgeting analysis. Appl Environ Microbiol 72:258–1266. Doi: 10.1128/AEM.72.2.1258-1266.2006

Clarson D (2004) Potash biofertilizer for ecofriendly agriculture. Agro-Clinic and Research Centre, Kottayam, pp 98–110

Claus D, Berkeley CW (1986) The genus *Bacillus*. In: PHA Sneath (ed) Bergey's manual of systematic bacteriology, vol 2. Williams, Wilkins, Baltimore. 34, 1105–1139

Csonka LN. and Epstein W. (1996) Osmoregulation. In: F. C. Neidhardt, Ed., *Escherichia coli* and *Salmonella*: Cellular and Molecular Biology. ASM Press, Washington DC, 1210-1223

Diep CN, Hieu TN (2013) Phosphate and potassium solubilizing bacteria from weathered materials of denatured rock mountain, Ha Tien, Kie^nGiang province Vietnam. Am J Life Sci 1:88–92

Dimkpa C, Weinand T, Asch F (2009) Plant-rhizobacteria interactions alleviate abiotic stress conditions. Plant Cell Environ 32:1682–1694. Doi: 10.1111/j.1365-3040.2009.02028.x.

Domínguez-Ferreras A, Munoz S, Olivares J, Soto MJ, and Sanjua J (2009) Role of potassium uptake systems in *Sinorhizobium meliloti* osmoadaptation and symbiotic performance. J Bacteriol, 191(7):2133–2143. Doi: 10.1128/JB.01567-08

Dosch DC, Helmer GL, Sutton SH, Salvacio FF, Epstein W (1991) Genetic Analysis of potassium transport loci in *Escherichia coli*: evidence for three constitutive systems mediating uptake of potassium. J Bacteriol, 173(2):687-696. Doi: 10.1128/jb.173.2.687-696.1991

Duff RB, Webley DM, Scott RO (1963) Solubilization of minerals and related materials by 2-ketogluconic acid producing bacteria. Soil Sci 5:105–114

Epstein W (2003) The roles and regulation of potassium in bacteria. Prog Nucleic Acid Res MolBiol 75: 293–320. Doi: 10.1016/s0079-6603(03)75008-9

Epstein W, Kim BS (1971) Potassium transport loci in *Escherichia coli* K-12. J Bacteriol 108(2): 639-44.

Groudev SN (1987) Use of heterotrophic microorganisms in mineral biotechnology. ActaBiotechnol 7:299–306

Gundala PB, Chinthala P, Sreenivasulu B (2013) A new facultative alkaliphilic, potassium solubilizing *Bacillus* sp. SVUNM9 isolated from mica cores of Nellore District, Andhra Pradesh, India: research and reviews. J Microbiol Biotechnol 2:1–7

Han HS, Lee KD (2005) Phosphate and potassium solubilizing bacteria effect on mineral up take, soil availability and growth of eggplant. J Agric Biol Sci 1:176–180

Han HS, Supanjani LKD (2006) Effect of co-inoculation with phosphate and potassium solubilizing bacteria on mineral uptake and growth of pepper and cucumber. Plant Soil Environ 52(3):130–136. DOI: 10.17221/3356-PSE

Haque E, Taniguchi H, Hassan MM, Bhowmik P, Karim MR, Smiech M, Zhao K, Rahman M and Islam T (2018) Application of CRISPR/Cas9 Genome Editing Technology for the Improvement of Crops Cultivated in Tropical Climates: Recent Progress, Prospects, and Challenges. Front. Plant Sci. (2018) 9:617. doi: 10.3389/fpls.2018.00617

Heinen W (1960) Silicon metabolism in microorganisms. Arch Microbiol 37:199–210

Herridge DF, Peoples MB, Boddey RM (2008) Global inputs of biological nitrogen fixation in agricultural systems. Plant Soil 311:1–18. doi:10.1007/s11104-008-9668-3

Hu X, Chen J, Guo J (2006) Two phosphate and potassium solubilizing bacteria isolated from Tianmu Mountain, Zhejiang, China. World Journal Microbiology and Biotechnology 22, 983-990. Doi:10.1007/s11274-006-9144-2

Hutchens SE, Valsami JE, Eldowney MS (2003) The role of heterotrophic bacteria in feldspar dissolution. Minerol Mag 67:1151–1170

Islam MT, Deora A, Hashidoko Y, Rahman A, Ito T, Tahara S (2007) Isolation and identification of potential phosphate solubilizing bacteria from the rhizoplane of *Oryza sativa* L. cv. BR29 of Bangladesh. Z. Naturforsch. 62c: 103-110

Islam MT, Hashidoko Y, Deora A, Ito T, Tahara S (2005) Suppression of damping-off disease in host plants by the rhizoplane bacterium *Lysobacter* sp. strain SB-K88 is linked to plant colonization and antibiosis against soilborne Peronosporomycetes. Appl. Environ. Microbiol. 71: 3786-3796

Islam MT, Hossain MM (2012) Plant probiotics in phosphorus nutrition in crops, with special reference to rice. In Bacteria in Agrobiology: Plant Probiotics, Maheshwari DK ed., Springer Berlin-Heidelberg, pp. 325-363. DOI: 10.1007/978-3-642-27515-9_18

Islam MT, Rahman MM, Pandey P, Jha CK, Aeron A(eds.) (2017) Bacilli and Agrobiotechnology. Springer International Publishing, pp. 1-416.

Kalaiselvi P (1999) Influence of silicate solubilizing bacteria on dissolution of silicate and potassium in rice soil added with graded levels of potassium and rice residues. M. Sc. (Agric) thesis, Tamil Nadu Agricultural University, Coimbatore

Kammar SC, Gundappagol RC, Santhosh GP, Shubha S, Ravi MV (2015) Influence of Potassium Solubilizing Bacteria on Growth and Yield of Sunflower (*Helianthus annuus* L.). Environ. & Ecol. 34 (1): 33—37

Khanwilkar SA, Ramteke JR (1993) Response of applied K in cereals in Maharashtra. Agriculture 4:84–96

Khudsen D, Peterson GA, Prov PF (1982) Lithium, sodium and potassium. In: Page AL (ed) Methods of soil analysis part (2) agronomy monograph 9, 2nd edn. ASA and SSSA, Madison. Doi:10.2134/agronmonogr9.2.2ed.c1.

Kirk JL, Beaudette LA, Hart M, Moutoglis P, Klironomus JN, Lee H, Trevors JT (2004) Methods of studying soil microbial diversity. J Microbiol Methods 58:169–188

Kumar A, Bahadur I, Maurya BR, Raghuwanshi R, Meena VS, Singh DK, Dixit J (2015) Does a plant growth-promoting rhizobacteria enhance agricultural sustainability. J Pure Appl Microbiol 9: 715-724

Kumar P, Dubey RC, Maheshwari DK (2012) *Bacillus* strains isolated from rhizosphere showed plant growth promoting and antagonistic activity against phytopathogens. Microbiol Res 167(8): 493–499. Doi: 10.1016/j.micres.2012.05.002

Lang-bo YI, Qing-zhong P, Qi-zhuang HE, Qing-jing P (2012) Isolation and identification of potash feldspar solubilizing bacteria and their potassium-releasing activities. Chin J Microecol 24:773–776

Leaungvutiviroj C, Ruangphisarn P, Hansanimitkul P, Shinkawa H, Sasaki K (2010) Development of a new biofertilizer with a high capacity for N2 fixation, phosphate and potassium solubilization and auxin production. BiosciBiotechnolBiochem 74:1098–1101

Lian B, Fu PQ, Mo DM, Liu CQ (2002) A comprehensive review of the mechanism of potassium release by silicate bacteria. Acta Mineral Sin 22:179–182

Lian B, Wang B, Pan M, Liu C, Tang HH (2008) Microbial release of potassium from K-bearing minerals by thermophilic fungus *Aspergillus fumigatus*. Geochem Cosmochim Acta 72:87–98. Doi:10.1016/j.gca.2007.10.005

Lin Q, Rao Z, Sun Y, Yao J, Xing L (2002) Identification and practical application of silicate-dissolving bacteria. AgricSci China 1:81–85

Liu D, Lian B, Dong H (2012) Isolation of *Paenibacillus* sp. and assessment of its potential for enhancing mineral weathering. Geomicrobiol J 29:413–421

Liu W, Xu X, Wu S, Yang Q, Luo Y, Christie P (2006) Decomposition of silicate minerals by *Bacillus mucilaginosus* in liquid culture. Environ Geochem Health 28:133–140

Li-yang M, Xiao-yan CAO, De-si S (2014) Effect of potassium solubilizing bacteria-mineral contact mode on decomposition behavior of potassium-rich shale. C J Non Ferr Metal 24:1099–1109

Lugtenberg BJJ, Dekkers L, Bloemberg CV (2001) Molecular determinants of rhizosphere colonization by *Pseudomonas*. Annu Rev Phytopathol 39:461-490. Doi: 10.1146/annurev.phyto.39.1.461

Mahmud NU,Surovy MZ, Gupta DR, Rahman MM, Islam MT (2017) Search for new biologicals from native environmental bacteria for potassium nutrition in rice. Paper presented at SQUARE-ACI International Conference on Biotechnology in Health and Agriculture, University of Dhaka, Bangladesh 29-30 December, 2017, GNOBB abstract book p.27

Malinovskaya IM, Kosenko LV, Votselko SK, Podgorskii VS (1990) Role of *Bacillus mucilaginosus* polysaccharide in degradation of silicate minerals. Microbiology 59:49–55

Marques APGC, Pires C, Moreira H, Rangel AOSS, Castro ML (2010) Assessment of the plant growth promotion abilities of six bacterial isolates using *Zea mays*as indicator plant. Soil Biol Biochem42:1229–1235. Doi:10.1016/j.soilbio.2010.04.014

Maurya BR, Meena VS, Meena OP (2014) Influence of Inceptisol and Alfisol's potassium solubilizing bacteria (KSB) isolates on release of K from waste mica. Vegetos 27:181-187.

Meena OP, Maurya BR, Meena VS (2013) Influence of K- solubilizing bacteria on release of potassium from waste mica, Agriculture for Sustainable Development, 1(1):1-5, 2.

Meena VS, Maurya BR, Bahadur I (2014) Potassium solubilization by bacterial strain in waste mica. Bangladesh J Bot 43:235–237

Meena VS, Maurya BR, Verma JP (2014a) Does a rhizospheric microorganism enhance K+ availability in agricultural soils? Microbiol. Res. 169(5-6):337-347. Doi:10.1016/j.micres.2013.09.003

Mikhailouskaya N, Tchernysh A (2005) K–mobilizing bacteria and their effect on wheat yield. Latv J Agron 8:147–150

Muralikannan M (1996) Biodissolution of silicate, phosphate and potassium by silicate solubilizing bacteria in rice ecosystem. M. Sc. (Agric) thesis, Tamil Nadu Agricultural University, Coimbatore

Naik PR, Raman G, Narayanan KB, Sakthivel N (2008) Assessment of genetic and functional diversity of phosphate solubilizing fluorescent pseudomonads isolated from rhizospheric soil. BMC Microbiol 8:230–243. Doi: 10.1186/1471-2180-8-230

Nakamura T, Yuda R, Bakker EP, Unemoto T (1998) Ktr AB, a New Type of Bacterial K Uptake System from *Vibrio alginolyticus*. J Bacterial. 180(13):3491-3494

Nath D, Maurya BR, Meena VS (2017) Documentation of five potassium- and phosphorus-solubilizing bacteria for their K and P solubilization ability from various minerals. Bicat & Agric Biotech, volume 10, April 2017, P 174-181

Nayak B (2001) Uptake of potash by different plants with the use of potash mobilizing bacteria (*Frateuria aurantia*). M.Sc. (Agric) thesis, OUAT, Bhubaneswar

Parmar P (2010) Isolation of potassium solubilizing bacteria and their inoculation effect on growth of wheat (*Triticum aestivum* L. em. Thell.). M.Sc. thesis submitted to CCS Haryana Agricultural University, Hisar

Parmar P, Sindhu SS (2013) Potassium solubilization by rhizosphere bacteria: influence of nutritional and environmental conditions. J Microbiol Res 3(1):25–31. DOI: 10.5923/j.microbiology.20130301.04

Patel BC (2011) Advance method of preparation of bacterial formulation using potash mobilizing bacteria that mobilize potash and make it available to crop plant. WIPO Patent Application WO/2011/154961

Pindi PK, Satyanarayana SDV (2012) Liquid microbial consortium- a potential tool for sustainable soil health. J Biofertil Biopesticide 3(4):1–9. DOI: 10.4172/2155-6202.1000124

Prajapati K, Sharma MC, Modi HA (2012) Isolation of two potassium solubilizing fungi from ceramic industry soils. Life Sci Leaflets 5:71–75

Prajapati K, Sharma MC, Modi HA (2013) Growth promoting effect of potassium solubilizing microorganisms on *Abelmoscusesculantus*. Int J AgricSci 3:181–188

Prajapati KB, Modi HA (2012) Isolation and characterization of potassium solubilizing bacteria from ceramic industry soil. CIB Tech J Microbiol 1:8–14

Rahman M, Sabir AA, Mukta JA, Khan MMA, Mohi-Uddin M, Miah MG, Rahman M, Islam MT(2018) Plant probiotic bacteria *Bacillus* and *Paraburkholderia* improve growth, yield and content of antioxidants in strawberry fruit. Scientific Reports 8:2504. Doi:10.1038/s41598-018-20235-1

Raj SA (2004) Solubilization of silicate and concurrent release of phosphorus and potassium in rice ecosystem. In: Kannaiyan S, Kumar K, Govindarajan K (eds) Biofertilizer technology for rice based cropping system. Scientific Book Center, pp 372–378

Ramarethinam S, Chandra K (2006) Studies on the effect of potash solubilizing bacteria *Frateuria aurantia* (Symbion-K- liquid formulation) on Brinjal (*Solanum melongena* L.) growth and yield. Pestology 11:35–39

Rehm G, Schmitt M (2002) Potassium for crop production. University of Minnesota Extension, www.extension.umn.edu/distribution/cropsystems. 46, pp 229–236

Romheld V, Kirkby EA (2010) Research on potassium in agriculture: needs and prospects. Plant Soil 335:155–180. Doi:10.1007/s11104-010-0520-1

Ryan PR, Dessaux Y, Thomashow LS, Weller DM (2009) Rhizosphere engineering and management for sustainable agriculture. Plant Soil 321:363–383. doi:10.1007/s11104-009-0001-6

Saiyad SA, Jhala YK, Vyas RV (2015) Comparative efficiency of five potash and phosphate solubilizing bacteria and their key enzymes useful for enhancing and improvement of soil fertility. Int J Sci Res Pub 5:1–6

Sangeeth KP, Bhai RS, Srinivasan V (2012) *Paenibacillus glucanolyticus*, a promising potassium solubilizing bacterium isolated from black pepper (*Piper nigrum* L.) rhizosphere. J Spic Aromat Crops 21(2):118–124

Schleyer M, Bakker EP (1993) Nucleotide sequence and 30-end deletion studies indicate that the K+-uptake protein Kup from *Escherichia coli* is composed of a hydrophobic core linked to a large and partially essential hydrophilic C terminus. J Bacteriol 175(21):6925–6931. Doi: 10.1128/jb.175.21.6925-6931.1993

Sheng XF (2005) Growth promotion and increased potassium uptake of cotton and rape by a potassium releasing strain of *Bacillus edaphicus*. Soil Biol Biochem 37:1918–1922

Sheng XF, He LY (2006) Solubilization of potassium bearing minerals by a wild type strain of *Bacillus edaphicus* and its mutants and increased potassium uptake by wheat. Can, J. Microbial. 52(1), 66-72. DOI: 10.1139/w05-117

Sheng XF, Huang WY (2002) Study on the conditions of potassium release by strain NBT of silicate bacteria. SciAgric Sin 35:673–677

Sheng XF, Huang WY (2002a) Mechanism of potassium release from feldspar affected by the strain NBT of silicate bacterium. ActaPedol Sin 39:863–871

Sheng XF, Xia JJ, Chen J (2003) Mutagenesis of the *Bacillus edaphicus* strain NBT and its effect on growth of chilli and cotton. Agric Sci China 2: 400–412

Sheng XF, Zhao F, He H, Qiu G, Chen L (2008) Isolation, characterization of silicate mineral solubilizing *Bacillus globisporus* Q12 from the surface of weathered feldspar. Can J Microbiol 54:1064–1068

Sindhu SS, Dadarwal KR (2000) Competition for nodulation among rhizobia in Rhizobium-legume symbiosis. Indian J Microbiol 40:211–246

Sindhu SS, Dua S, Verma MK, Khandelwal A (2010) Growth promotion of legumes by inoculation of rhizosphere bacteria. In: Khan MS, Zaidi A, Musarrat J (eds) Microbes for legume improvement. Springer Wien, New York, pp 195–235

Sindhu SS, Parmar P, Phour M (2014a) Nutrient cycling: potassium solubilization by microorganisms and improvement of crop growth. In: Parmar N, Singh A (eds) Geomicrobiology and biogeochemistry. Springer, Berlin/Heidelberg, pp 175–198. Doi:10.1007/978-3-642-41837-2_10

Sindhu SS, Parmar P, Phour M, Kumari K (2014b) Rhizosphere microorganisms for improvement in soil fertility and plant growth. In: Nagpal R, Kumar A, Singh R (eds) Microbes in the service of mankind: tiny bugs with huge impact. JBC Press, New Delhi, pp 32–94

Sindhu SS, Phour M, Choudhary SR, Chaudhary D (2014) Phosphorus cycling: prospects of using rhizosphere microorganisms for improving phosphorus nutrition of plants. In: Parmar N, Singh A (eds) Geomicrobiology and biogeochemistry. Springer, Berlin/Heidelberg, pp 199–237. Doi:10.1007/978-3-642-41837-2_11

Sindhu SS, Suneja S, Goel AK, Parmar N, Dadarwal KR (2002) Plant growth promoting effects of *Pseudomonas* sp. on co-inoculation with *Mesorhizobium* sp. Cicerstrain under sterile and wilt sick soil conditions. Appl Soil Ecol 19:57–64

Singh G, Biswas DR, Marwah TS (2010) Mobilization of potassium from waste mica by plant growth promoting rhizobacteria and its assimilation by maize (*Zea mays*) and wheat (*Triticum aestivum* L.). J Plant Nutr 33:1236–1251

Singh NP, Singh RK, Meena VS, Meena RK (2015) Can we use maize (*Zea mays*) rhizobacteria as plant growth promoter? Vegetos. 28(1): 86-99

Sleator RD, Hill C (2002) Bacterial osmoadaptation: the role of osmolytes in bacterial stress and virulence. FEMS Microbiol Rev 26:49–71. Doi: 10.1111/j.1574-6976.2002.tb00598.x

Stumpe S, Schlösser A, Schleyer M, Bakker EP (1996) Handbook of Biological Physics (Konings W N, Kaback HR, Lolkema JS, eds), Vol 2, p. 473, Elsevier Science B.V., Amsterdam

Subba RNS (2001) An appraisal of biofertilizers in India. In: Kannaiyan S (ed) The biotechnology of biofertilizers. Narosa Pub House, New Delhi

Subhashini DV (2015) Growth promotion and increased potassium uptake of tobacco by potassium-mobilizing bacterium *Frateuria aurantia* grown at different potassium levels in Vertisols. Commun Soil Sci Plant Anal 46:210–220. Doi:10.1080/00103624.2014.967860

Sugumaran P, Janarthanam B (2007) Solubilization of potassium containing minerals by bacteria and their effect on plant growth. World J AgrSci 3:350–355

Surdam RC, MacGowan DB (1988) Oil field waters and sandstone diagenesis. Appl Geo Chem 2:613–620

Syed BA, Patel B (2014) Investigation and correlation of soil biotic and abiotic factors affecting agricultural productivity in semi-arid regions of north Gujarat, India. Int J Res Stud Biosci 2:18–29

Ullman WJ, Welch SA (2002) Organic ligands and feldspar dissolution. Geochem Soc 7:3–35

Uroz S, Calvaruso C, Turpault MP, Pierrat JC, Mustin C, Frey-Klett P (2007) Effect of the mycorrhizosphere on the genotypic and metabolic diversity of the bacterial communities involved in mineral weathering in a forest soil. Appl Environ Microbiol 73:3019–3027

Uroz S, Calvaruso C, Turpault P, Frey-Klett P (2009) Mineral weathering by bacteria: ecology, actors and mechanisms. Trends Microbiol 17:378–387

Vandevivere P, Welch SA, Ullman WJ, Kirchman DJ (1994) Enhanced dissolution of silicate minerals by bacteria at near neutral pH. Microbiol Ecol 27: 241–251

Verma JP, Yadav J, Tiwari KN (2009) Effect of *Mesorhizobium* and plant growth promoting rhizobacteria on nodulation and yields of chickpea. Biol Forum AnInt J 1(2):11–14

Wall DH, Virginia RA (1999) Control of soil biodiversity- in sight from extreme environments. Appl Soil Ecol 13:137–150

Warscheid T, Braams J (2000) Biodeterioration of stone: a review. Int Biodeterior Biodegrad 46:343–368

Welch SA, Ullman WJ (1993) The effect of organic acids on plagioclase dissolution rates and stoichiometry. Geochim Cosmochim Acta 57:2725–2736

Welch SA, Vandevivere P (2009) Effect of microbial and other naturally occurring polymers on mineral dissolution. Geomicrobiol J 12:227–238

Wu SC, Cao ZH, Li ZG, Cheung KC, Wong MH (2005) Effects of biofertilizer containing N-fixer, P and K solubilizers and AM fungi on maize growth: a greenhouse trial. Geoderma 125:155–166. Doi: 10.1016/j.geoderma.2004.07.003

Xie JC (1998) Present situation and prospects for the world's fertilizer use. Plant Nutri Fertil Sci 4:321–330

Xiufang H, Jishuang C, Jiangfeng G (2006) Two phosphate and potassium solubilizing bacteria isolated from Tianmu Mountain Zhejiang, China. World J Microbiol Biotechnol 22:983–990.Doi:10.1007/s11274-006-9144-2

Zahra MK, Monib MS, Abdel-Al I, Heggo A (1984)Significance of soil inoculation with silicate bacteria. ZentralblMikrobiol 139(5):349–357. Doi:10.1016/S0232-4393(84)80013-X

Zarjani JK, Aliasgharzad N, Oustan S, Emadi M, Ahmadi A (2013) Isolation and characterization of potassium solubilizing bacteria in some Iranian soils. Arch Agron Soil Sci 59:1713–1723. Doi:10.1080/03650340.2012.756977

Zhang A, Zhao G, Gao T, Wang W, Li J, Zhang S *et al* (2013) Solubilization of insoluble potassium and phosphate by *Paenibacillus kribensis* CX-7: a soil microorganism with biological control potential. Afr J Microbiol Res 7:41–47. DOI: 10.5897/AJMR12.1485

Zhang C, Kong F (2014) Isolation and identification of potassium-solubilizing bacteria from tobacco rhizospheric soil and their effect on tobacco plants. Appl Soil Ecol 82:18–25

Zhou H, Zeng X, Liu F, Qiu G, Hu Y (2006) Screening, identification and desilication of a silicate bacterium. J Cent South Univ Technol 13:337–341.Doi:10.1007/s11771-006-0045-1

10

Nano Biofertilizer for Sustainable Agriculture

S. Pati[1], A. Dash[1], S. Maity[2], S. Pattnaik[1], S. Mohapatra[3] and D. P. Samantaray*[1]

[1]*Department of Microbiology, CBSH, OUAT, Odisha, India*
[2]*Department of Biotechnology, UIC, CCT, University of Rajasthan, Jaipur Rajasthan, India*
[3]*Department of Microbial Technology, AMITY University, Noida, New Delhi, India*

Abstract

Agriculture provides food, feed, fiber and many other desired products for survival of human beings in the world. Modern agriculture emphasizes on use of hybrid seeds, high yielding varieties and chemical fertilizer to increase the crop productivity within a short period. Indiscriminate use of synthetic fertilizers has led to environmental pollution and deterioration of soil health, which results in depletion of essential plant nutrients and organic matter in soil. On account of the environmental hazard and soil infertility, biofertilizer concept was developed. Biofertilizer are natural fertilizers including bacteria, fungi and algae alone or in combination that augment availability of nutrients to the plants. However, due to its short shelf life period it may not be a suitable candidate for modern agriculture. Thus, it is the need of the hour to develop an economic and eco-friendly nano-biofertilizer for sustainable agriculture.

Keywords: Agriculture, Synthetic fertilizer, Biofertilizer, Nano-biofertilizer

10.1 Introduction

Agriculture is highly indispensable for existence human beings in the world as it continues to play a crucial role for production of food, feed, fiber and many other desired products. Agriculture is the backbone of the Indian economy and this sector provides 50% employment to the country. Besides this, India is world's largest producer of pulses, rice, wheat, spices followed by fruits and vegetables. According to the Department of Economics and Statistics (DES) the food grain production in India increased from 264 to 257million tons by 2012-14. However, loss of biodiversity, climate change, land degradation, decrease in rural population and soil erosion, compaction & pollution are the barriers in for progress of agriculture (Velten *et al.*, 2015). Initially, these threats pave the way of scientists towards chemical fertilizer to increase the crop yield as well as global food supply. Though regular land use practices have established a pilot scale market of chemical fertilizer but, its consumption deteriorate the environment by ground water pollution, eutrophication, loss of biodiversity, soil acidification, declination of soil health, reduction of soil microflora & pollinators and birds (Tomer *et al.*, 2016; Mahanty *et al.*, 2017; Kourgialas *et al.*, 2017; Duhan *et al.*, 2017).

Thus, it is the need of the hour to replace chemical fertilizer by biofertilizer for sustainable agriculture. It refers to an economic, eco-friendly and site-specific agriculture system that will fulfil the global food demand thereby, enhancing the quality of life and society for a long period of time (Velten *et al.*, 2015). Several researchers develop new approaches such as biofertilizer and nano-biofertilizer for sustainable agriculture in different agro-climatic zones. Biofertilizers are the formulation of solid or liquid products containing latent or live microbes that improves the nutrient accessibility to plant and soil health in a sustainable manner (El-Ghamry *et al.*, 2018). Though, biofertilizers offer better agricultural practices but it's different limitations pave the way for nano-biofertilizer (Chen, 2006). On the other hand, nano-biofertilizers are the biomaterials (>100 nm size) extracted either from plants or microbes by different physio-chemical or biological methods with the help of nano-technology. Nano-biofertilizer have a key role in improving soil fertility and crop productivity as well as quality of agricultural products (Brunnert *et al.*, 2006). Scientists are exploiting nano-biofertilizer as an effective tool in agriculture due to its high penetration capacity and surface area that provides novel physical and chemical properties to the nano-biofertilizer (Singh *et al.*, 2017). Both biofertilizers and nano-biofertilizer have effective role for sustainable agriculture. Thus, the present chapter provides insight on comparative account of bio & nano-biofertilizer with respect to its mechanism, biosynthesis, types and advantages.

10.2 Biofertilizer in agriculture

The growing need of agricultural products pave the way for biofertilizers. Biofertilizers are known as the living microbial inoculants, containing one or more microbial strains that directly or indirectly enhances crop productivity. Generally, biofertilizers are classified into several categories such as nitrogen fixing, phosphate solubilising, phosphate mobilizing, potash solubilising, vesicular arbuscular mycorrhiza, plant growth promoting rhizobacteria and biofertilizer for micronutrients (Jehangir *et al.*, 2017). The biofertilizer effectively enhances the final crop yield by providing required nutrients to the plant and soil, thereby giving a better health to the plant. The biofertilizer not only acts as a sustainable alternate to chemical fertilizers and organic manures but also provides natural resistance to plants against pests and soil borne diseases. In addition to that the beneficial microbes also increase the soil fertility (Board, 2004). The biofertilizers not only replace the synthetic one for economic crop production but also, intensify the crop yield by producing growth hormones (auxin, cytokinin and gibberellins) (Yasin *et al.*, 2012). Besides this, the non-toxic biofertilizers also encourage the microbes to enrich the land which eventually improve the soil quality (Dumitrescu *et al.*, 2009).

The assets and utility of biofertilizers galvanize the government and private bodies to go for large scale field trials. Presently, India Government has focused on promoting biofertilizer application through the following steps: farm level promotion programmes, financial assistance to farmers and direct production in public, cooperative, research & educational organizations. Despite of all the advantages the short shelf life, lack of storage facilities, higher implementation cost, high volume requirement and low transfer of micro & macronutrients holds back its thriving commercialization (Mazid & Khan, 2014; Carvajal-Muñoz & Carmona-Garcia, 2012). Hence, the above drawbacks of biofertilizer trigger the researchers to focus on nano-biofertilizer for sustainable agriculture.

10.3 Nano-biofertilizer

Agriculture, the cornerstone of developing countries is now adversely affecting the farmer community due to extensive use of toxic chemical fertilizer and high implementation cost of biofertilizer. To overcome the above drawbacks in a different way, nanotechnology is envisioned to create revolution for a sustainable agriculture. Nano-biofertilizer is defined as the bio-synthesis of fertilizer from microbes or plant extracts at nano scale to revitalize the plant growth and soil quality (El-Ghamry *et al.*, 2018; Qureshi *et al.*, 2018). Zinc oxide, titanium dioxide, silver are the popular examples of nano-particles biosynthesized to be used for precision farming (Bansal *et al.*, 2014).

The basic principle of nano-fertilizer synthesis includes wet and dry method. The wet method includes sol-gel, homogeneous precipitation, hydrothermal, protein template, biosynthesis using enzyme and reversed micelle approach, while the aerosol-based processes are included in dry method (Kaul *et al.*, 2012; Raliya *et al.*, 2017). In general nano-particles are synthesized biologically or by using microorganisms and during that time they exhibit some properties with unique applications. For synthesis of nano-nutrients, the microbes are grown on nutrient source and the biomass is collected by centrifugation. The biomass is then used for extraction of specific proteins from which nano-particles are synthesized. After synthesis; seed treatment, soil amendment and foliar spray, the three approaches for application of these fertilizer to plant (Raliya *et al.*, 2017).

The nano-particles enter the plant cell through vascular or stomatal system and increase the crop yield by boosting the metabolic activity of plant. Out of the two system, stomatal system has gained the higher attention because of its large size exclusion limit and high transport velocity. These tiny, economic, stable nano-biofertilizers are eco-friendly as they are encapsulated by hydrophilic fungal proteins (Patel & Krishnamurthy, 2015). The small size, exceptional optical properties and high surface to volume ratio make the nano-biofertilizer vital in the field of agriculture as it helps in management of farm pesticides, growth & germination of plants and plant protection (Duhan *et al.*, 2017). Nano-biofertilizer is the emerging star in the world of agriculture and is highly active in fertilizer delivery, micronutrient supply, insect pest management, nano-sensors, nano-herbicides and nano-fungicides production. A list of nano-biofertilizer along with their role in agriculture is given in **Table 10.1.**

10.4 Advantages of Nano-biofertilizer

Nanotechnology recreates a sustainable manner of crop management through nano-fertilizer, which results in effective absorption of nutritional elements for plant growth and metabolism (El-Ramady *et al.*, 2017, 2018; Ditta *et al.*, 2015; Morteza *et al.*, 2013). Besides this, the nano-fertilizers are also act as a potential replacement of chemical fertilizer which is not only harmful to us but also make undesirable effects on plant when applied in more amount. In addition to that chemical fertilizers on over application cause environmental pollution due to surface run-off, which can be overcome by using nano-fertilizer because of its controlled release property (Wilson *et al.*, 2008). The controlled release of nutrients from nano-fertilizer also increases the nutrient availability there by prevent the loss of nutrient from denitrification, volatilization and leaching. Furthermore, the nano-coatings on fertilizer have high surface tension thus give better surface protection to plant (Solanki *et al.*, 2015; Subramanian *et al.*, 2015). The small size and high surface area of nano-biofertilizer, boost photosynthate production as well as solubility, which facilitate penetration of micronutrients from the applied surface to other parts of plant (Lin *et al.*, 2007). Furthermore, application of nano-biofertilizer enhances the quality parameter of plant which involves oil, protein and sugar content. The nano-nutrients make the plant disease & stress resistant and also sustain in nutrient deficient condition (Naderi *et al.*, 2012). Reports are also available in favor of application of nano-

Table 10.1: Nano-biofertilizers and their role in agriculture

Sl.No	Nano-biofertilizer	Application in agriculture	References
1	Cu-chitosan	Active against fungal pathogens such as, *Alternaria alternata, Macrophomina phaseolina* and *Rhizoctonia solani*	Saharan *et al.*, 2013
2	Silver nanoparticles chitosa nencapsulated paraquate	Active against weeds *Eichhornia crassipes*	Namasivayam and Aruna 2014
3	Silver nanoparticle- PGPR	Active against plant pathogens	Duhan *et al.*, 2017
4	Biodegradable thermoplastic starch (TPS)	Good tensile strength and lowered water permeability	Park *et al.*, 2002
5	Macronutrient fertilizers coated with zinc oxidenanoparticles	Enhancement of nutrients absorption by plants and the delivery of nutrients to specific sites	Milani *et al.*, 2015
6	Acetamprid loaded alginate-chitosan nano-capsules	Improved delivery of agrochemicals in the field, better efficacy, better control of application/dose.	Kumar *et al.*, 2015
7	Chitosan nanoparticle	Used as herbicide carrier	Ghaly, 2009
8	Nano-silica (Plant origin)	Insectcontrol *Artemisia arborescens*	Barik *et al.*, 2008

biofertilizer in seed germination and growth by positively affecting the seed and seed vigor. These nano-nutrients enhance root and shoot length of plant by increasing nutrient availability to the growing seedling (Qureshi *et al.*, 2018).

10.5 Future prospects of nano-biofertilizer

Though, nano-biofertilizers have lots of advantages because of its unique physico-chemical properties, but this can't make us to ignore its liabilities. Studies reported that direct exposure of plant to nano-particle can cause phyto-toxic effect on plant (Boonyanitipong *et al.*, 2011). ZnO, SiO_2, TiO_2 and CeO_2 nano-particles adversely affect the functional microbes and metabolic profiles in soil making the soil infertile. The nano-biofertilizer sometimes negatively affects the enzymatic activity of soil (Raliya *et al.*, 2017). Nano-biofertilizer also creates obstacles in seed germination and growth of seedling when applied in high concentration (Qureshi *et al.*, 2018). Despite of so many advantages very few drawbacks place the nano-biofertilizer in a rudimentary stage. Thus, researchers are now focusing for making it safer and environmental friendly. Though reports are available in support of the non-cytotoxic nature of nano-biofertilizer but further investigation is still required to know their behavior inside the living body.

Apart from this, scientists should concentrate on following aspects of exploiting nano-biofertilizer for sustainable agriculture: formulation & characterization of nano-structure size, shape, chemical composition, crystal phase, porosity, hydrophilicity, surface charge, stability, dissolution, agglomeration/aggregation and valence of the surface layer, precise delivery pathway for proper management, understanding the fate of nanoparticles in plants and food chain, response of nano-biofertilizer towards crop productivity and its pilot scale production (Raliya *et al.*, 2017). Nano-technology is the foremost approach for bringing the concept of smart agriculture in developing countries like India. Thus, it is highly essential to develop an economic and ecofriendly nano-biofertilizer and transfer of technology from laboratory to field for sustainable agriculture.

References

Bansal, P., Duhan, J.S. and Gahlawat, S.K. (2014). Biogenesis of nanoparticles: A review, African Journal of Biotechnology, 13: 2778-2785.

Barik, T.K., Sahu, B. and Swain, V. (2008). Nano-silica from medicine to pest control, Parasitology Research, 103: 253-258.

Board, NIIR. (2004). The complete technology book on bio-fertilizer and organic farming, National Institute of Industrial Research.

Boonyanitipong, P., Kositsup, B., Kumar, P., Baruah, S. and Dutta, J. (2011). Toxicity of ZnO and TiO2 nanoparticles on germinating rice seed Oryza sativa L, International Journal of Bioscience, Biochemistry & Bioinformatics, 1: 282.

Brunnert, I., Wick, P., Manserp, Spohnp, Grass, R.N., Limbach, L.K., Bruinink, A .and Stark, W.J. (2006). Environmental Science & Technology, 40: 4374-4381.

Carvajal-Muñoz, J.S. and Carmona-Garcia, C.E. (2012). Benefits and limitations of biofertilization in agricultural practices, Livestock Research for Rural Development, 24(3): 1-8.

Chen, J. (2006). The combined use of chemical and organic fertilizer and or biofertilizer for crop growth and soil fertility, International Workshop on Sustained Management of the Soil-Rhizosphere System for Efficient Crop Production and Fertilizer Use, Thailand: 16-20.

Ditta, A., Arshad, M. and Ibrahim, M. (2015). Nanoparticles in sustainable agricultural crop production: applications and perspectives, In: M.H. Siddiqui et al. (eds.), Nano-technology and Plant Sciences, 55-75.

Duhan, J.S., Kumar, R., Kumar, N., Kaur, P., Nehra, K., Duhan, S. (2017). Nanotechnology: the new perspective in precision agriculture, Biotechnology Reports, 15: 11-23.

Dumitrescu, Manciulea, Sauciuc, A. and Zaha, C. (2009). Obtaining biofertilizer by composting vegetable waste, sewage sludge and sawdust, Bulletin of the Transilvania University of Brasov, 2(51).

El-Ghamry, A.M., Mosa, A.A., Alshaal, T.A. and El-Ramady, H.R. (2018). Nano-fertilizers vs. biofertilizers: New insights, Environment, Biodiversity & Soil Security, 2: 1-22.

El-Ramady, H., Alshaal, T., Abowaly, M., Abdalla, N., Taha, H.S., Al-Saeedi, A.H., Shalaby, T., Amer, M., Fári, M., Domokos-Szabolcsy, E., Sztrik, A., Prokisch, J., Selmar, D., Pilon Smits, E.A.H. and Pilon, M. (2017). Nano-remediation: towards sustainable crop production, In: S. Ranjan et al. (eds.), S. Ranjan et al. (eds.), Nanoscience in Food and Agriculture 5, Sustainable Agriculture Reviews, 26, DOI 10.1007/978-3-319-58496-6_12.

Ghaly, A.E.. (2009). The Black Cutworm as a Potential Human Food. American Journal of Biochemistry & Biotechnology, 5: 210-220.

Jehangir, I.A., Mir, M.A., Bhat, M.A. and Ahangar, M.A. (2017). Biofertilizers an approach to sustainability in agriculture: A review, International Journal of Pure & Applied Bioscience, 5(5): 327-334.

Kaul, R., Kumar, P., Burman, U., Joshi, P., Agrawal, A., Raliya, R. and Tarafdar, J. (2012). Magnesium and iron nanoparticles production using microorganisms and various salts, Materials Science-Poland, 30: 254-258.

Kourgialas, N.N., Karatzas, G.P. and Koubouris, G.C. (2017). A GIS policy approach for assessing the effect of fertilizers on the quality of drinking and irrigation water and well head protection zones (Crete, Greece), Journal of Environmental Management, 189: 150-159.

Kumar, S., Bhanjana, G., Sharma, A., Sarita, Sidhu, M.C. and Dilbaghi, N. (2015). Herbicide loaded carboxymethyl cellulose nano-capsules as potential carrier in agri-nanotechnology, Science of Advanced Materials, 7: 1143-1148.

Lin, D. and Xing, B. (2007). Phytotoxicity of nanoparticles: inhibition of seed germination and root growth. Environmental, Pollution, 150: 243-250.

Mahanty, T., Bhattacharjee, S., Goswami, M., Bhattacharyya, P., Das, B., Ghosh, A. and Tribedi, P. (2017). Biofertilizers: a potential approach for sustainable agriculture development, Environmental Science & Pollution Research, 24(4): 3315–3335., doi:10.1007/ s11356-016-8104-0.

Mazid, M. and Khan, T.A. (2014). Future of bio-fertilizers in Indian agriculture: An overview, International Journal of Agricultural and Food Research, 3(3): 10-23.

Milani, N., Hettiarachchi, G.M., Kirby, J.K., Beak, D.G., Stacey, S.P. and McLaughlin, M.J. (2015). Fate of zinc oxide nanoparticles coated onto macronutrient fertilizers in an alkaline calcareous soil, PLoS One, 10, doi: org/10.1371/journal. pone.0126275.

Morteza, E., Moaveni, P., Farahani, H.A., Kiyani, M. (2013). Study of photosynthetic pigments changes of maize (*Zea mays* L.) under nano TiO2 spraying at various growth stages, Springerplus, 2: 247.

Naderi, M.R. and Abedi, A. (2012). Application of nanotechnology in agriculture and refinement of environmental pollutants Journal of Nano-technology, 11(1): 18-26.

Namasivayam, S.K.R. and Aruna, A. (2014). Gokila evaluation of silver nanoparticles-chitosan encapsulated synthetic herbicide paraquate (AgNp-CS-PQ) preparation for the controlled release and improved herbicidal activity against Eichhornia crassipes, Research Journal Biotechnology, 9: 19-27.

Park, H.M., Li, X., Jin, C.Z., Park, C.Y., Cho, W.J. and Ha, C.S. (2002). Preparation and properties of biodegradable thermoplastic starch/clay hybrids, Macromolecular Materials & Engineering, 287: 553-558.

Patel, H.R. and Krishnamurthy. (2015). Antimicrobial efficiency of biologically synthesized nanoparticles using root extract of Plumbago zeylanica as bio-fertilizer application, International Journal of Bioassays, 4(11): 4473-4475.

Qureshi, A., Singh, D.K. and Dwivedi, S. (2018). Nano-fertilizers: A novel way for enhancing nutrient use efficiency and crop productivity, International Journal of Current Microbiology and Applied Sciences, 7(2): 3325-3335.

Raliya, R., Saharan, V., Dimkpa, C. and Biswas, P. (2017). Nano-fertilizer for precision and sustainable agriculture: current state and future perspectives, Journal of Agricultural and Food Chemistry, 66: 6487-6503, doi: 10.1021/acs.jafc.7b02178.

Saharan, V., Mehrotra, A., Khatik, R., Rawal, P., Sharma, S.S. and Pal, A. (2013). Synthesis of chitosan-based nanoparticles and their *in vitro* evaluation against phytopathogenic fungi, International Journal of Biological Macromolecule, 62: 677-683.

Singh, M.D., Chirag, G., Prakash, P.O,. Mohan, M.H., Prakasha, G. and Vishwajith. (2017). Nano-fertilizers is a new way to increase nutrients use efficiency in crop production, International Journal of Agriculture Sciences, 9(7): 3831-3833.

Solanki, P., Bhargava, A., Chhipa, J., Jain, N. and Panwar, J. (2015). Nano-technologies in food and agriculture, Cham: Springer International Publishing: 81-101.

Subramanian, K., Manikandan, A., Thirunavukkarasu, M. and Rahale, C. (2015). Nano-fertilizers for balanced crop nutrition, In: Rai, M., Ribeiro, C., Mattoso, L., Duran, N. (eds), Nanotechnologies in food and agriculture. Springer International Publishing, Cham: 69-80.

Tomer, S., Suyal, D.C. and Goel, R. (2016). Biofertilizers: a timely approach for sustainable agriculture, In: D.K. Choudhary et al. (eds.), Plant-Microbe Interaction: An Approach to Sustainable Agriculture, doi:10.1007/978-981-10-2854-0_17.

Velten, S., Leventon, J., Jager, N. and Newig, J. (2015). What is sustainable agriculture? A systematic review, Sustainability, 7: 7833-7865, doi:10.3390/su7067833.

Wilson, M.A., Tran, N.H., Milev, A.S., Kannangara, G.S.K., Volk, H. and Lu, G.H.M. (2008). Nanomaterials in soils, Geoderma, 146: 291-302.

Yasin, M., Ahmad, K., Mussarat, W. and Tanveer, A. (2012). Bio-fertilizers, substitution of synthetic fertilizers in cereals for leveraging agriculture, Crop & Environment, 3(1-2): 62-66.

11

Role of Phosphorus Solubilizing Microorganisms for Sustainable Crop Production

Sushanta Saha[1], Bholanath Saha[2], Ayon Alipatra[2], Biplab Pal[1], Partha Deb Roy[3] and Abhas Kumar Sinha[4]

[1]Directorate of Research, Bidhan Chandra Krishi Viswavidyalaya, Kalyani Nadia, West Bengal-741235, India
[2]Dr. Kalam Agricultural College, Bihar Agricultural University, Kishanganj Bihar-855107, India
[3]ICAR-Indian Institute of Water Management, Bhubaneswar, Odisha -751023 India
[4]Uttar Banga Krishi Viswavidyalaya, Pundibari, Cooch Behar West Bengal 736165, India

Abstract

Phosphorus is the second important key element after nitrogen as a mineral nutrient in terms of quantitative plant requirement. Although abundant in soils, in both organic and inorganic forms, its availability is restricted as it occurs mostly in insoluble forms. The dynamics of phosphorus in soil is closely related to the dynamics of the biological cycle in which microorganisms play a central role. They also participate in the cycles of the most important macro and microelements such as phosphorus besides their major role in the decomposition of plant residues, creation of humus and maintenance of stable soil structure. Microorganisms affect the amount of phosphorus accessible to plants by means of mineralization of organic phosphorus compounds, immobilization of available phosphorus and solubilization of non-soluble phosphorus minerals such as tricalcium phosphate. Mineralization is catalyzed by microbiological enzymes phosphatases secreted by a diverse group of bacteria, fungi and other groups of microorganisms. Organic phosphorus mineralization in soil depends on several soil and environmental factors such as temperature, pH, soil moisture, degree of aeration. In this chapter the major emphasis is given on the transformation of P through microbiologically involving different enzymes and also the microorganisms involved in the process. Emphasis is also given on the factors that have profound influence on the microbial transformation of P in soil and ultimately making it available for plant uptake. This chapter also focuses on the mechanism of P solubilization, role of various phosphatases, impact of various factors on P solubilization and potential for application of this knowledge in managing a sustainable environmental system.

Keywords: Phosphorus, Mineralization, Immobilization, Solubilization, Phosphatases

11.1. Introduction

In terms of quantitative plant requirement, phosphorus (P) is the second most important element as a mineral nutrient after nitrogen. It is found abundantly both in organic and inorganic forms in soils. But due to its occurrence mostly as insoluble forms, availability is less. Among the macro-elements, phosphorus is a very essential which is required for plant nutrition. P participates in metabolic processes such as photosynthesis, energy transfer & synthesis and breakdown of carbohydrates. An adequate supply of phosphorus during early phases of plant development is important for laying down the primordia of plant reproductive parts. It plays very vital role in increasing root proliferation and strengthening, thereby imparting vitality and disease resistance capacity to plant. It also helps in seed formation and in early maturation of crops like cereals and legumes.

Phosphorus is found in the soil in organic compounds and in minerals. Nevertheless, the amount of readily available phosphorus is very low compared with the total amount of phosphorus in the soil. Therefore, in many cases phosphorus fertilizers should be applied in order to meet crop requirements.

11.2. Functions of phosphorus in plants

- Phosphorus plays vital role in plant growth processes like nitrogen, potassium etc. But energy storage and transfer is the most important function of P. Within plants, ADP (Adenosine di-phosphate) and ATP (Adenosine tri-phosphate) acts as energy currency. A huge quantity of energy is released when phosphate molecule is split off from ADP or ATP. Generally, this splitting off occurs in the form of $H_2PO_4^-$ molecules. When this transfer occurs from ATP to other materials inside plant, this process is known as phosphorylation. This process involves the transformation of ATP to ADP. During this process, a bulk amount of energy (12,000 cal/mol) is released.

- Phosphorus is a vital component of DNA, the genetic "memory unit" of all living things in the earth. It is also a component of RNA, the compound that reads the DNA genetic code to build proteins and other compounds essential for plant structure, seed yield and genetic transfer. The structures of both DNA and RNA are linked together by phosphorus bonds.
- Phosphorus stimulates root development in plants which is very essential so that the plants *get all* the required nutrients from the soil. The roots also provide necessary support for the plant in the soil. Due to well developed roots, they are able to penetrate the soil and extract all essential nutrients required for development of the plant.
- It also helps in boosting up the development of plant. For the development, plants require proper nutrition. The processing of nutrition is generally occurs in the leaves and after that, they are stored or transferred to other parts inside the plants. Phosphorus plays a major role in photosynthesis as well as in the storage and transportation of the nutrients within plant body.
- Under the favorable circumstances, the fruit production in plants is expected after a certain period of time. The maturity of crop at the right time depends upon the phosphorous. When the plants lacking phosphorous, it takes longer time to mature and after maturity, the number of fruits or seeds are very few and poor in quality.
- Legumes help in fixing atmospheric nitrogen in the soil through their roots. This process would not be possible well without phosphorous which is responsible for boosting up the development of plant roots.
- Without the availability of phosphorous, the materials required for formation as well as development of genes cannot perform well. The transfer of the genes from one generation to the next is only possible when phosphorous is available.
- Plants that have access to enough phosphorous have the ability to resist diseases due to their well developed and quickly growing parts. Plants grown using hydroponics are supplied with enough phosphorous to ensure they grow well.

11.3. Deficiency of phosphorus in plants

If phosphorus deficiency occurs in plants, they becomes stunted in growth and develops an abnormal dark-green color. Accumulation of sugars causing development of anthocyanin pigments which in turn produces a reddish-purple color. On extremely P deficient soils, these symptoms usually persist. It should be noted that these are severe phosphorus deficiency symptoms and crops may respond well to phosphorus fertilization without showing characteristic deficiencies. The reddish-purple colouration does not always indicate phosphorus deficiency. Red coloring may be induced by other factors such as insect damage which causes interruption of sugar transport to the grain. In small plants, phosphorus deficiencies may even look somewhat similar to nitrogen deficiency. Under cold temperatures, phosphorus deficient plants may become yellow, unthrifty owing to small root extension and low uptake of soil phosphorus. These deficiencies may disappear under heated soil conditions.

11.4. Sources of phosphorus in soils

Soils generally contains 500-1000 parts per million (ppm) of total phosphorus (inorganic and organic), but most of this is in a "fixed" form that is unavailable for plant use. In addition to this, soluble phosphorus in fertilizer or other sources is quickly converted to less available forms when added to the soil. Although some soils may require large phosphorus additions for best yield, the past use of phosphorus fertilizer and application of manure have led to high phosphorus levels in many soils.

The two main categories of phosphorus (P) in soils are organic and inorganic. The organic form is found in humus and other organic materials. The inorganic portion occurs in various combinations with iron, aluminium, calcium and other elements, most of which are not very soluble in water. Both organic and inorganic forms of phosphorus are important sources of phosphorus for plant growth, but their availabilities are controlled by soil characteristics and environmental conditions.

11.5. Soil solution phosphorus

Plants generally take phosphorus in the form of $H_2PO_4^-$ and HPO_4^{-2} ions. Depending upon the soil pH, the amount of these two ions varies in the soil solution. At neutral pH range, these two ions are in equivalent amount. Under acidic condition, the concentration of $H_2PO_4^-$ is higher than HPO_4^{-2} whereas under alkaline condition the concentration of HPO_4^{-2} is higher than $H_2PO_4^-$. Plant uptakes more $H_2PO_4^-$ in comparison with the HPO_4^{-2} ion.

The young tissues present in the root tips are very active absorbent. That's why these root tips has the highest concentration of accumulated phosphorus. Generally, plant roots absorbs phosphorus from the soil solution whereas, mass flow and diffusion process provides additional phosphorus to the active surface of plant roots.

11.6. Inorganic soil phosphorus

Contribution of inorganic form of phosphorus to growing plants is of prime importance to the investigators. Generally, various compounds of calcium, iron and aluminium as well as zinc and copper etc. constitutes the major portion of inorganic phosphorus sources in the soil. Most of these compounds are insoluble in neutral to alkaline soil reactions. But they are soluble under acidic soil reactions. For this reason, under acidic conditions availability of inorganic phosphorus is much higher.

Since the availability of iron and aluminium is more under acidic soil reactions, there is very high chance of re-fixed or re-precipitation of released phosphorus. Thus release of P from insoluble calcium, iron and aluminium phosphates and its re-fixation is a continuous reversible process. Aluminium and iron phosphate are mostly present in Ultisols and Oxisols of southern India and north-east India whereas calcium phosphate is mostly found in calcareous soils of northern India.

Certain portion of phosphorus is also adsorbed as anions on the surface of clay particles. When oxides and hydroxides of iron precipitate on phosphorus compounds, they form coatings that make phosphorus unavailable to crop plants. This type of P is also known as occluded P. In low-land rice eco-system, ferric ion is reduced to soluble ferrous ion due to water submergence. This incidence releases P occluded by iron oxides and increases P availability.

11.7. Organic soil phosphorus

Generally phosphorus in inorganic forms is mostly taken up by crop plants. That's why organic P has not been given much importance but it also plays an integral part of total P in the soils. Inositol phosphates are the major constituents of organic form of P, which contributes little to even up to two-third of organic P. Among these, most common is phytic acid. The other common organic forms of P are nucleic acids and phospholipids, which contributes 5-6% of organic phosphorus.

11.8. Phosphorus cycle

The process associated with transformation of phosphorus occurring in soil is governed by the principles of soil-based biogeochemical cycles. **Figure 11.1** illustrates the relationships between various forms of P in soils. Plant roots absorbs P from soil solution, which is later replenished by both the inorganic and organic P in soils. Primary and secondary P minerals then dissolve to resupply $H_2PO_4^-$ and HPO_4^{2-} in soil solution. Inorganic P adsorbed on minerals and clay surfaces as $H_2PO_4^-$ and HPO_4^{2-} (labile inorganic P) can also desorb to buffer solution P. Soil microorganisms digest plant residues containing P and produce organic P compounds that are mineralized through microbial activity to supply solution P.

When phosphatic fertilizers are applied to soil which is water soluble, they readily dissolve and increase the P in soil solution. In addition to P uptake by plant roots, inorganic and organic P fractions buffer the increased solution P through P adsorption on mineral surfaces, precipitation as secondary P minerals, and immobilization as microbial or organic P. Maintaining solution P concentration (intensity) for adequate P nutrition depends on the ability of labile P (quantity) to replace soil solution P taken up by the plant. The ratio of quantity to intensity factors defines buffer capacity (BC) or the relative ability of the soil to buffer changes in soil solution P. The larger the BC, the greater the ability to buffer solution P.

Labile P is the readily available portion of the quantity factor that exhibits a high dissociation rate and rapidly replenishes solution P. Depletion of labile P causes some non-labile P to become labile, but at a slow rate. Thus, the quantity factor comprises both labile and non-labile P fractions. Understanding the dynamics of P transformations in soils will provide the basis for sound management of soil and fertilizer P to ensure adequate P availability to plants.

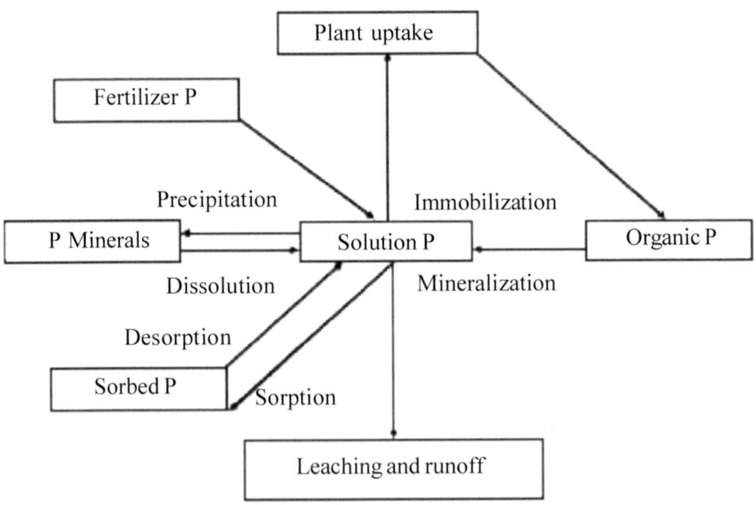

Fig. 11.1: Phosphorus cycle in soil

11.9. Phosphate Solubilizing Microorganisms (PSM)

A certain number of microorganisms possess P solubilization capacity which includes bacteria, fungi, actionomycetes and even algae. Among the bacteria, besides *Pseudomonas* and *Bacillus*, other phosphate solubilizing bacteria are *Rhodococcus*, *Arthrobacter*, *Chryseobacterium* etc. (Wani *et al.*, 2005), *Azotobacter* (Kumar *et al.*, 2001), *Xanthomonas* (De Freitas *et al.*, 1997), *Enterobacter* (Chung *et al.*, 2005), *Xanthobacter agilis* (Vazquez *et al.*, 2000).

In addition to that, *Rhizobium* sp., which fix atmospheric nitrogen, also shown the phosphate solubilizing activity (Zaidi *et al.*, 2009). As for example, *Rhizobium leguminosarum* bv. *trifolii* (Abril *et al.*, 2007) and Rhizobium species which nodulates in sunnhemp (*Crotalaria* sp.) (Sridevi *et al.*, 2007) also mobilizes inorganic and organic phosphorus.

Among the total fungal population, only 0.1-0.5% contributes in P-solubilization (Kucey, 1983). Fungi are more important than bacteria with respect to P-solubilization in soils owing to their long distance traversing capacity (Kucey, 1983). Another reason for better P-solubilizing activity of the fungi is due to their more acid producing capacity (Venkateswarlu *et al.*, 1984). Among fungi, most common P-solubilizers are the genera *Aspergillus* and *Penicillium* which are filamentous in nature (Fenice *et al.*, 2000; Khan and Khan, 2002; Reyes *et al.*, 1999, 2002). The strains of *Trichoderma* (Altomare *et al.*, 1999) and *Rhizoctonia solani* (Jacobs *et al.*, 2002) have been also reported as phosphate solubilizers.

Only in the recent years, the P-solubilizing ability of actinomycetes has attracted interest. Actinomycetes not only capable of surviving in extreme environments (e.g. drought, fire.) but also possess other potential benefits (e.g. production of antibiotics and phytohormone-like compounds) that could simultaneously benefit plant growth (Fabre *et al.*, 1988; Hamdali *et al.*, 2008a, 2008b). A study by Hamdali *et al.* (2008a) has indicated that approximately 20% of actinomycetes can solubilize P, including those in the common genera *Streptomyces* and *Micromonospora*.

Besides bacteria, fungi and actinomycetes, algae such as cyanobacteria and mycorrhiza also have P solubilization capacity. The interactive effects of arbuscular mycorrhizal fungi (AMF) and rhizobacteria on the growth and nutrients uptake of *Sorghum bicolor* were studied in acid and low availability phosphate soil. The microbial inocula consisted of the AMFs *Glomus manihotis* and *Entrophospora colombiana*, PSB *Pseudomonas* sp., results indicated that the interaction of AMF and the selected rhizobacteria has a potential to be developed as biofertilizers in acid soil. The potential of dual inoculation with AMF and rhizobacteria needs to be further evaluated under different crop and agroclimatic conditions, particularly in the field (Widada *et al.*, 2007).

11.9.1. Mycorrhizae

A mycorrhiza (fungus-root) is a mutualistic and symbiotic association between a fungus and cortical and epidermal root cells. The fungi receive organic nutrients (sugar) from the plant and, in turn, help the roots in the absorption of water and plant nutrients especially P.

There are broadly two groups of mycorrhizae, the ectomycorrhizae and the endomycorrhizae, although a group with intermediate properties, the ectendotrophic mycorrhizae is also reported. In ectomycorrhizae, the fungal hyphae form a mantle outside the root and in the intercellular spaces of the epidermis and cortex (but no intracellular penetration of fungal hyphae). On the other hand endomycorrhizae enter the cortical and epidermal cells and form vesicles and arbuscules. The most important subgroup of endomycorrhizae is vesicular arbuscular mycorrhizae popularly known as VAM.

While the ectomycorrhizae are generally found in the roots of trees such as pine, fair oak, beech, chestnut etc, the VAM are present on the roots of several dicot annual and perennial crops. Without the nutrient absorbing properties of mycorrhizae, many communities of trees could not exist. As regards VAM considerable research has been done on the help that these can provide in making available native soil P and applied ground rock phosphate. A partial list of PSM including various groups is given in **Table 11.1.**

Table 11.1: Biodiversity of Phosphate Solubilizing Microorganisms

Bacteria	*Alcaligenes* sp., *Aerobactor aerogenes*, *Achromobacter* sp., *Actinomadura oligospora*, *Agrobacterium* sp., *Azospirillum brasilense*, *Bacillus* sp., *Bacillus circulans*, *B. cereus*, *B. fusiformis*, *B. pumils*, *B. megaterium*, *B. mycoides*, *B. polymyxa*, *B. coagulans*, *B. chitinolyticus*, *B. subtilis*, *Bradyrhizobium* sp., *Brevibacterium* sp., *Citrobacter* sp., *Pseudomonas* sp., *P. putida P. striata*, *P. fluorescens*, *P. calcis*, *Flavobacterium* sp., *Nitrosomonas* sp., *Erwinia* sp., *Micrococcus* sp., *Escherichia intermedia*, *Enterobacter asburiae*, *Serratia phosphoticum*, *Nitrobacter* sp., *Thiobacillus ferroxidans*, *T. thioxidans*, *Rhizobium meliloti*, *Xanthomonas* sp.
Fungi	*Aspergillus awamori*, *A. niger*, *A. tereus*, *A. flavus*, *A. nidulans*, *A. foetidus*, *A. wentii*, *Fusarium oxysporum*, *Alternaria teneius*, *Achrothcium* sp., *Penicillium digitatum*, *P lilacinium*, *P balaji*, *P. funicolosum*, *Cephalosporium* sp., *Cladosprium* sp., *Curvularia lunata*, *Cunnighamella*, *Candida* sp., *Chaetomium globosum*, *Humicola inslens*, *Humicola lanuginosa*, *Helminthosporium* sp., *Paecilomyces fusisporous*, *Pythium* sp., *Phoma* sp., *Populospora mytilina*, *Myrothecium roridum*, *Morteirella* sp., *Micromonospora* sp., *Oideodendron* sp., *Rhizoctonia solani*, *Rhizopus* sp., *Mucor* sp., *Trichoderma viridae*, *Torula thermophila*, *Schwanniomyces occidentalis*, *Sclerotium rolfsii*.
Actinomycetes	*Actinomyces*, *Streptomyces*.
Cyanobacteria	*Anabena* sp., *Calothrix braunii*, *Nostoc* sp., *Scytonema* sp.
VAM	*Glomus fasciculatum*

11.9.2. Mechanisms and phosphate-solubilization by Phosphate Solubilizing Microorganisms (PSM)

Sims and Pierzynski (2005) reported that, the major processes which affect soil solution P concentrations as (1) dissolution–precipitation (mineral equilibria), (2) sorption–desorption (interactions between P in solution and soil solid surfaces), and (3) mineralization–immobilization (biologically mediated conversions of P between inorganic and organic forms).

The soil microorganisms employed three main P solubilization mechanisms which includes: (1) release of complexing or mineral dissolving compounds e.g. organic acid anions, siderophores, protons, hydroxyl ions, CO_2, (2) liberation of extracellular enzymes (biochemical P mineralization) and (3) the release of P during substrate degradation (biological P mineralization) (McGill and Cole, 1981). Therefore, microorganisms play an important role in all three major components of the soil P cycle (i.e. dissolution–precipitation, sorption–desorption, and mineralization–immobilization) (**Figure 11.2**). Additionally, these microorganisms in the presence of labile C serve as a sink for P, by rapidly immobilizing it even in low P soils; therefore PSM become a source of P to plants upon its release from their cells. Release of P immobilized by PSM primarily occurs when cells die due to changes in environmental conditions, starvation or predation. Environmental changes, such as drying–rewetting or freezing–thawing, can result in so-called flush-events, a sudden increase in available P in the solution due to an unusually high proportion of microbial cell lysis (Butterly *et al*., 2009). Grierson *et al*., 1998 found that about 30–45% of microbial P (0.8–1 mg kg^{-1}) was released in a sandy Spodosols in an initial flush after drying–rewetting cycles within the first 24 hour **(Figure 11.3)**.

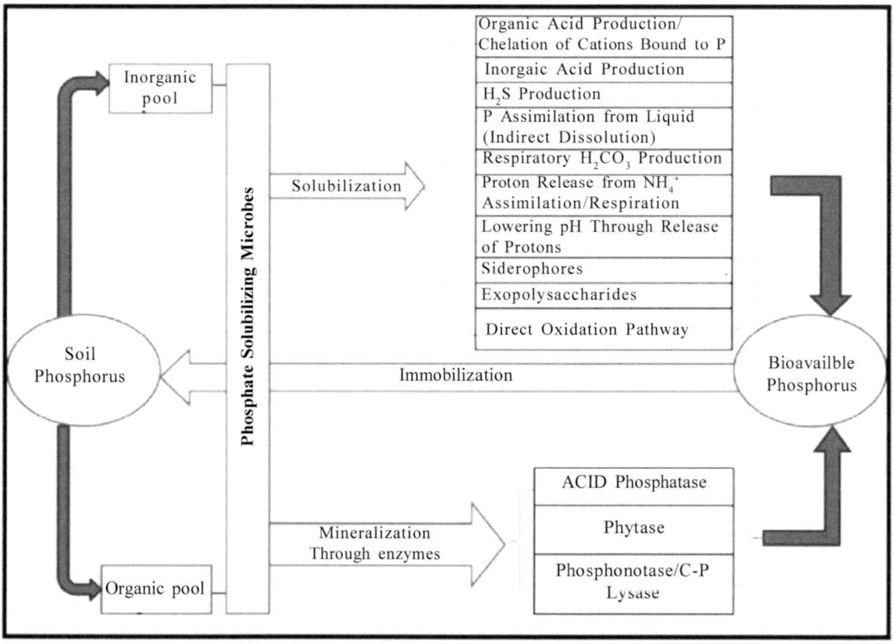

Fig. 11.2: Schematic representation of mechanism of soil P solubilization/ mineralization and immobilization by PSM

Soil sample collection

↓

Serial dilution

↓

Inoculation on media (pour plale/streaking) having different sources of insoluble P, depending on soil type

↓

Clear zone around colony growth indicates PSM activity

↓

Additional test in liquid media to assay P dissolution

↓

Test isolates for abundant production of organic acids

↓

Pure cultures by reinoculation

↓

Study of the morphology and colony characteristics and biochemical tests

↓

Screening of best inoculants in terms of P solubilizing activity

↓

Identification at genetic level (molecular characterization)

↓

Test on a model plant

↓

Development of microphos

↓

Green house trials

↓

Field trials

↓

Standardization (quality control)

↓

Commercial biofertiliser

Fig. 11.3: Protocol for isolation and development of effective inoculants of PSM based biofertilizer

11.10. Conclusion

The role of phosphorus solubilizing microorganisms (PSM) in enhancing dissolution of insoluble and sparingly-soluble phosphorus was known as far back as 1960s. The interest in PSM has revived in recent years with a view to enhancing the availability of native soil-P and also to explore the possibilities of direct application of rock phosphate in neutral to mildly alkaline soils. Bacterial cultures namely *Bacillus megaterium* and *Pseudomonas striata* or fungal species namely *Aspergillus awamori* is reported to add 30-35 kg P_2O_5/ha under ideal conditions. The performance of PSM is, however, constrained in semi-arid subtropical regions as those in north-west India. Reports on effectiveness of PSM under field conditions are scarce, yet the available information reveals a greater crop response to PSM inoculation on high organic matter soils or with the application of FYM on low organic matter soils. Hence, the possibility of poor colonization of PSM in rhizosphere at the high P-demanding vegetative growth stages of crops cannot be denied. Nonetheless, it is an interesting area that needs to be explored.

References

Abril A, Zurdo-Pineiro, J.L., Peix, A., Rivas, R., Velazquez, E. (2007). Solubilization of phosphate by a strain of *Rhizobium leguminosarum* bv. *trifolii* isolated from Phaseolus vulgaris in El Chaco Arido soil (*Argentina*) In: Velazquez E, Rodriguez-Berrueco C, editors. Developments in Plant and Soil Sciences. The Netherlands: Springer; pp. 135–138.

Altomare C, Norvell WA, Bjorkman T, Harman GE. (1999). Solubilization of phosphates and micronutrients by the plant-growth-promoting and biocontrol fungus Trichoderma harzianum rifai 1295-22. Appl Environ Microbiol. 65(7):2926-33.

Butterly CR, Bunemann EK, McNeill AM, Baldock JA, Marschner P. (2009). Carbon pulses but not phosphorus pulses are related to decrease in microbial biomass during repeated drying and rewetting of soils. Soil Biol Biochem. (41): 1406–1416. doi: 10.1016/j.soilbio.2009.03.018.

Chung H, Park M, Madhaiyan M, Seshadri S, Song J, Cho H, Sa T. (2005). Isolation and characterization of phosphate solubilizing bacteria from the rhizosphere of crop plants of Korea. Soil Biol Biochem. 2005; 37:1970–1974. doi: 10.1016/j.soilbio.2005.02.025.

De Freitas JR, Banerjee MR, Germida JJ. (1997) Phosphate-solubilizing rhizobacteria enhance the growth and yield but not phosphorus uptake of canola (*Brassica napus* L.) Biol Fertil Soils. 24:358–364. doi: 10.1007/s003740050258.

Fabre B, Armau E, Etienne G, Legendre F, Tiraby G. (1988). A simple screening method for insecticidal substances from actinomycetes. J Antibiot (Tokyo). 41(2):212-219.

Fenice M, Seblman L, Federici F, Vassilev N. (2000). Application of encapsulated Penicillium variabile P16 in solubilization of rock phosphate. Bioresour Technol. 73:157–162. doi: 10.1016/S0960-8524(99)00150-9.

Grierson PF, Comerford NB, Jokela EJ. (1998). Phosphorus mineralization kinetics and response of microbial phosphorus to drying and rewetting in a Florida Spodosol. Soil Biol Biochem. 30:1323–1331. doi: 10.1016/S0038-0717(98)00002-9.

Hamdali H, Bouizgarne B, Hafidi M, Lebrihi A, Virolle MJ, Ouhdouch Y. (2008a). Screening for rock phosphate solubilizing Actinomycetes from Moroccan phosphate mines. Appl Soil Ecol. 38:12–19. doi: 10.1016/j.apsoil.2007.08.007.

Hamdali H, Hafidi M, Virolle MJ, Ouhdouch Y. (2008b). Growth promotion and protection against damping-off of wheat by two rock phosphate solubilizing actinomycetes in a P-deficient soil under greenhouse conditions. Appl Soil Ecol. 40:510–517. doi: 10.1016/j.apsoil.2008.08.001.

Jacobs H, Boswell GP, Ritz K, Davidson FA, Gadd GM. (2002). Solubilization of calcium phosphate as a consequence of carbon translocation by *Rhizoctonia solani*. FEMS Microbiol Ecol., 40(1):65-71. Doi: 10.1111/j.1574-6941.2002.tb00937.x

Khan MR, Khan SM. (2002). Effects of root-dip treatment with certain phosphate solubilizing microorganisms on the fusarial wilt of tomato. Bioresour Technol., 85(2):213-5. DOI: 10.1016/s0960-8524(02)00077-9

Kucey RMN. (1983). Phosphate solubilizing bacteria and fungi in various cultivated and virgin Alberta soils. Can J Soil Sci. 1983;63:671–678. doi: 10.4141/cjss83-068.

Kumar V, Behl RK, Narula N. (2001). Establishment of phosphate-solubilizing strains of *Azotobacter chroococcum* in the rhizosphere and their effect on wheat cultivars under green house conditions. Microbiol Res. 156(1):87-93. DOI: 10.1078/0944-5013-00081

McGill WB, Cole CV. (1981). Comparative aspects of cycling of organic C, N, S and P through soil organic matter. Geoderma. 26:267–268. doi: 10.1016/0016-7061(81)90024-0.

Reyes I, Bernier L, Antoun H. (2002). Rock phosphate solubilization and colonization of maize rhizosphere by wild and genetically modified strains of *Penicillium rugulosum*. Microb Ecol. 2002 Jul; 44(1):39-48.

Reyes I, Bernier L, Simard RR, Antoun H. (1999). Effect of nitrogen source on the solubilization of different inorganic phosphates by an isolate of Penicillium rugulosum and two UV-induced mutants. FEMS Microbiol Ecol. 128:281–290. doi: 10.1111/j.1574-6941.1999. tb00583.x.

Sims JT, Pierzynski GM. (2005). Chemistry of phosphorus in soil. In: Tabatabai AM, Sparks DL, editors. Chemical processes in soil, SSSA book series 8. Madison: SSSA; 2005. pp. 151–192.

Sridevi M, Mallaiah KV, Yadav NCS. (2007). Phosphate solubilization by Rhizobium isolates from Crotalaria species. J Plant Sci. 2:635–639. doi: 10.3923/jps.2007.635.639.

Vazquez P, Holguin G, Puente M, Lopez-cortes A, Bashan Y. (2002). Phosphate solubilizing microorganisms associated with the rhizosphere of mangroves in a semi-arid coastal lagoon. Biol Fertil Soils. 30:460–468. doi: 10.1007/s003740050024.

Venkateswarlu B, Rao AV, Raina P, Ahmad N. (1984). Evaluation of phosphorus solubilization by microorganisms isolated from arid soil. J Indian Soc Soil Sci. 32:273–277.

Wani PA, Zaidi A, Khan AA, Khan MS. (2005). Effect of phorate on phosphate solubilization and indole acetic acid (IAA) releasing potentials of rhizospheric microorganisms. Annals Plant Protection Sci. 2005;13:139–144.

Widada J, Damarjaya DI, Kabirun S. (2007).The interactive effects of arbuscular mycorrhizal fungi and rhizobacteria on the growth and nutrients uptake of sorghum in acid soil. In: Rodriguez-Barrueco C, editor. Velazquez E. Springer: First International Meeting on Microbial Phosphate Solubilization pp. 173–177.

Zaidi A, Khan MS, Ahemad M, Oves M, Wani PA, *et al.* (2009). Recent Advances in Plant Growth Promotion by Phosphate-Solubilizing Microbes. In: Khan MS, *et al.*, editors. Microbial Strategies for Crop Improvement. Berlin Heidelberg: Springer-Verlag; pp. 23–50.

12

Plant Growth Promoting Traits Exhibited by *Rhizobium* Isolated from Different Regions of Odisha

Aiswarya Panda* and Bibhuti Bhusan Mishra

Department of Microbiology, College of Basic Science and Humanities Odisha University of Agriculture and Technology, Bhubaneswar, Odisha, India

Abstract

Rhizobia are soil bacteria that fix atmospheric dinitrogen after establishing symbiotic relationship inside the root nodules of legumes (Fabaceae). Plant Growth Promoting Rhizobacteria are the group of bacteria which thrive in the rhizosphere region and enhance the plant growth through direct & indirect mechanisms. Pulses are being inoculated with specific rhizobial strains to have better nodulation in India. The native rhizobial strains from the root nodules of legumes from different agroclimatic zones of Odisha was isolated, characterised and screened for PGP activities. Out of all eleven isolates showed similar characteristics with the family Rhizobiacae. Among them six isolates contained nif-gene and PGPR activity (B2, B4, B5, B6, B7 and B8) were applied to seeds, that helped in faster germination and better elongation of root and shoot. The observations were statistically significant.

Keywords: Rhizobia; PGPR; Root nodules; Germination

12.1 Introduction

India is predominantly an agriculture-based country. To ensure food security to the growing population, by 2020 the targeted production of 321 million tonnes of food grain is being achieved, but nutrient deficiency of about 7.2 million tonnes still prevails (Arun 2007). Soil enriched with nutrient is essential for enhancing the crop productivity and sustainability. In Green Revolution-I, application of synthetic fertilizers and agrochemicals were advocated to replenish soil nutrients.

Over the time soil has been drastically affected due to excessive use of agrochemicals, fertilizers and now, organic farming is prioritized in Green Revolution-II. In organic farming, biofertilizers are considered as the future of agriculture and the best supplement to chemical fertilizers. Biofertilizers are efficient microbial formulations that can be applied to seed, soil or composting areas to enhance the number of such microorganisms to supplement the necessary nutrients for easy plant assimilation. Biofertilizers may be nitrogen fixers, phosphate solubilizers, sulphur oxidisers or organic matter decomposers. Bacterial biofertilizer can diminish the utilization of chemical fertilizers through fortification of nutrients in the soil and increasing its availability to the plants (Mia et al. 2005 and 2010; Pahari and Mishra 2017a).

Rhizobium is a group of soil bacteria that establish symbiotic association inside root nodules of leguminous plants and fix atmospheric nitrogen. This process is known as Biological Nitrogen fixation, in which the atmospheric dinitrogen is reduced to ammonia (NH_3) by rhizobia through symbiosis and significantly contributes to available nitrogen in the biosphere (Bhatt et al. 2013). The cultivation of leguminous plants in rotation with non leguminous plants help nitrogen enrichment in soil and also maintains soil fertility (Shahzad et al. 2012). Pulses are being inoculated with specific rhizobial strains to have better nodulation in India since 1934 (Deka and Azad 2006). Nevertheless, even in traditional legume growing areas, there is deficiency of specific and efficient Rhizobium strains (Tilak et al. 2006).

Bacteria that reside in the rhizosphere region and are known to enhance the plant growth are referred to as Plant Growth Promoting Rhizobacteria (PGPR). Even though microorganisms are found in all parts of the plant, but the plant beneficial microbes are exclusively present in the rhizosphere. Plant Growth Promoting Rhizobacteria promotes growth of plants through direct & indirect mechanisms. In direct mechanism it facilitates nutrient uptake, nitrogen fixation, phosphorus solubilization, siderophore production, growth hormone production (Glick 1995; Bhardwaj et al. 2014).

On account of this an attempt was made in this study to isolate the native Rhizobial strains from the root nodules of legumes collected from different regions of Odisha, a state of different agroclimatic zones, to characterise them using various cultural and biochemical tests and study their PGPR properties, so that in addition to nitrogen, *Rhizobium* with PGPR traits can further enhance growth and productivity of crop.

12.2 Materials and Methods

12.2.1 Collection of Samples (Soil and Nodules)

Rhizospheric soil and nodules were collected aseptically from different pulse crops (green gram, black gram, arhar) of different locations of the following districts, Ganjam, Boudh and Angul. From freshly uprooted plants, nodules were collected and washed thoroughly to remove soil from the roots. Healthy Pink nodules were carefully cut from the roots and were preserved after washing.

12.2.2 Physico-chemical properties of soil

The various Physico-chemical properties of soil like pH, Electrical Conductivity (EC), Oxidative Reductive potential (Eh), Total dissolved solid (TDS) were studied by taking soil samples with distilled water in 1:2 proportion.

12.2.3 Isolation of Rhizobacteria and Morphological characterization

Microorganisms were isolated from the rhizospheric sample by *in vitro* cultivation method of bacterial isolation in Yeast Extract Mannitol Agar medium. The sample was processed in the laboratory and enumerations of heterotrophic bacteria were conducted through serial dilution and spread plate technique. The CFUs of different morphology were selected and sub cultured on YEMA slants preserved at 4°C. Colony morphology and gram's reaction was studied. The isolates were streaked in different media like EMB, Glucose peptone agar+ Bromocresol purple, Keolactose Media, Starch media, YMA with bromothymol blue to observe their growth.

12.2.4 Biochemical tests for the generic level identification

Several Biochemical tests were performed in order to morphologically identify the isolates. The tests performed were IMVIC, ONPG, TTC, TSI etc. and the motility was studied by performing hanging drop experiment. Enzymatic tests like Oxidase, Urease, Catalase, Decarboxylation test and Starch, gelatin &casein hydrolysis were performed. Apart from these, sugar utilization and antibiotic sensitivity test was carried out. Effect of the pH and temperature (physical parameters) on the isolates was also studied.

12.2.5 Determination of PGPR traits

Plant growth promoting traits of the bacterial isolates were tested *in vitro*. The PGP activities, *viz*. IAA production, Phosphate solubilisation, Hydrogen cyanide production, Nitrate reduction, Antibiosis and ammonia production, which were determined following standard methods.

Seed germination was carried out using the isolates as inoculants and average root length & shoot length was measured. The total dry biomass of the germinated seeds was also measured and statistically analysed.

12.3 Results And Discussion

Enrichment of the soil was done in Yeast mannitol broth (YEMB), for *Rhizobium* isolation (Bromfield et al. 1994). Several physico-chemical parameters of the soil was observed, pH of the samples ranged from 5.72-7.02 (**Table 12.1**), Howieson and Ewing, (1986) isolated several strains of *Rhizobium melliloti* from acidic soil. Standard isolation method (Donadio et al. 2002) was followed by spread plate method from the serial dilutions and the colony morphology was studied and found to be oval, entire margin, raised elevation, translucent or opaque, mucoid surface colonies (**Table 12.1; Figure 12.1**). Gram reaction was performed and Gram negative rods were further purified. Out of 21 isolates, 11 were found to be gram negative (**Table 12.2**) and further work was carried on taking these isolates. Shahzad et al. (2012) isolated *Rhizobium meliloti* from root nodule of Alfalfa by also studying the colony morphology (circular, raised & smooth edge colony) and gram staining technique by choosing the gram negative pink rods concomitant to their findings.

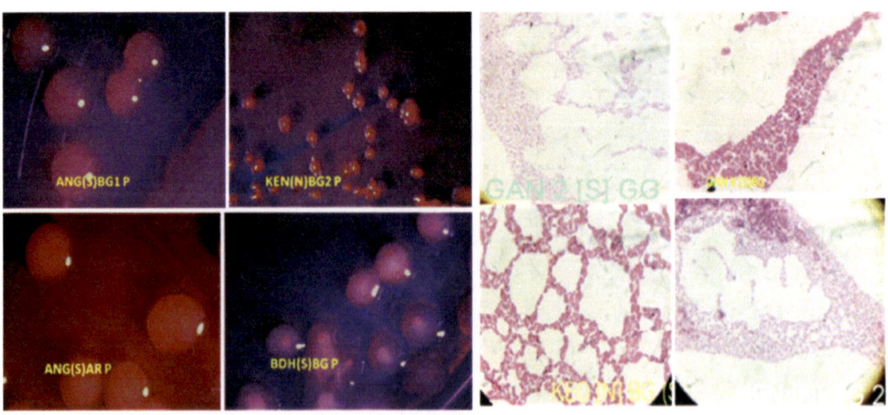

Fig.12.1: Colony morphology and Gram's reaction

For identification of the bacterial isolates biochemical and enzymatic activities of all the isolates showed different specific reactions. Bergey's manual of systematic bacteriology was used for identification of the isolates (**Table 12.3; Figure 12.2**). Hussain et al. (2002) had conducted biochemical tests for the generic level identification of their isolates from alfalfa and identified as *Sinorhizobium meliloti.*

Table 12.1: Colony characteristics and Gram's reaction of the bacterial isolates

Orgs	Shape	Margin	elevation	opacity	surface	colour	Motility	Gram staining
B1	Oval	Entire	raised	translucent	mucoid	red	Motile	Gram –ve rod
B2	irregular	undulated	flat	opaque	mucoid	red	Motile	Gram –ve rod
B3	Oval	Entire	convex	opaque	mucoid	pink	Motile	Gram –ve rod
B4	Oval	Entire	flat	translucent	mucoid	white	Motile	Gram –ve rod
B5	irregular	undulated	flat	opaque	mucoid	white	Motile	Gram –ve rod
B6	spherical	Entire	raised	translucent	mucoid	white	Motile	Gram –ve rod
B7	irregular	Lobate	raised	translucent	mucoid	white	Nonmotile	Gram –ve rod
B8	Oval	Entire	raised	translucent	mucoid	White	Motile	Gram –ve rod
B9	spherical	Entire	convex	opaque	mucoid	red	Motile	Gram –ve rod
B10	Oval	undulated	raised	opaque	mucoid	pink	Motile	Gram –ve rod
B11	irregular	undulated	flat	opaque	mucoid	red	Nonmotile	Gram –ve rod

Table 12.2: Biochemical characteristics of the isolates

Sl.no.	Biochemical test		B1	B2	B3	B4	B5	B6	B7	B8	B9	B10	B11
1.	Indole test		-	-	-	-	-	-	-	-	-	-	-
2.	Methyl red test		+	+	-	+	-	-	-	-	-	-	-
3.	Voges-Proskauer test		-	-	-	-	-	-	-	-	-	-	-
4.	Citrate utilization test		-	+	+	+	+	+	+	+	-	-	-
5.	TSI	Butt	Y	Y	Y	Y	Y	Y	Y	Y	Y	Y	R
		Slant	R	R	R	R	R	R	R	R	R	R	R
		Gas production	-	+	+	-	+	+	+	-	+	-	-
6.	Mannitol motility test	Mannitol	+	+	+	+	+	+	+	+	+	+	+
		Motility	+	+	+	+	+	+	-	+	+	+	-
7.	ONPG test		+	+	+	+	+	-	+	+	-	-	-
8.	Hanging drop experiment		+	+	+	+	+	+	-	+	+	+	-
9.	Catalase test		+	+	+	+	+	+	+	+	+	+	+
10.	Oxidase test		+	-	-	+	-	-	-	-	+	+	-
11.	Urease test		+	+	+	+	+	+	+	+	+	+	+
12.	GPA+BCP	(growth,color change)	-/-	+/+	+/+	+/+	+/+	+/+	+/+	+/+	-/+	-/-	-/-
13.	YMA+BTB	(growth, color change)	-/-	+/-	+/+	+/+	+/+	+/+	+/+	+/+	-/-	-/-	-/-
14.	YMB+2% NaCl		-	+	+	+	+	+	+	+	+	+	+
15.	EMB		-	Greenish	pink	pink	pink	pink	pink	pink	-	Pink	-
16.	Ketolactose test		-	-	-	-	-	-	-	-	-	-	-
17.	2,3,5, Triphenyl tetrazolium chloride (TTC) test		-	+	+	+	+	+	+	+	+	-	-

Table 12.3: Enzymatic characterization

Orgs	oxidase	catalase	Urease(24/72hr)	amylase	Lysine Decarboxylase	Arginine Decarboxylase	gelatinase	caseinase
B1	+	+	+/++	+	+	+	+	-
B2	-	+	-/++	+	+	+	-	-
B3	-	+	++/++	+	+	+	-	-
B4	+	+	-/+	+	-	+	-	-
B5	-	+	++/++	-	+	+	-	-
B6	-	+	+/++	+	-	+	-	-
B7	-	+	+/++	+	-	+	-	-
B8	-	+	-/++	+	+	+	-	-
B9	+	+	++/++	-	+	+	-	-
B10	+	+	-/++	-	+	+	+	-
B11	-	+	-/++	-	-	+	+	-

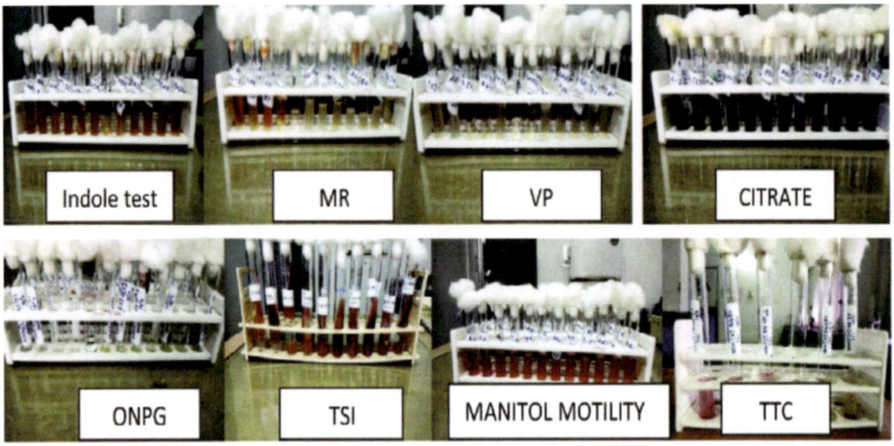

Fig. 12.2: Biochemical tests of the isolates

All of the bacterial isolates exhibited negative results for indole test and voges proskauer test, contrary to the positive results for methyl red and citrate test for the IMViC series of biochemical reaction which were also gas forming and showed positive result for TSI as changed the colour red to yellow, positive results for ONPG & TTC test (**Table 12.3**). Some isolates of Rhizobia are motile so the test for motility was done in the mannitol motility agar medium and among which 9 isolates (B1, B2, B3, B4, B5, B6, B8, B9, B10) were shown to be motile. Our results correspond with the works of Shahzad et al. (2012) who isolated and characterized *Rhizobium meliloti* from *Alfalfa*.

The level of enzymatic activity differentiates one isolates of another in accordance with their physiological characteristics. All the isolates showed positive result for decarboxylase (Arginine) and catalase test (**Table 12.4**; **Figure 12.3**). Whereas mixed results for decarboxylase (Lysine) (positive-B1, B2, B3, B5, B8, B9, B10; negative-B4, B6, B7, B11) and oxidase was observed (positive-B1, B4, B9, B10; negative- B2, B3, B5, B6, B7, B8, B11). Both positive and negative results were also found in enzymatic reactions like gelatinase (positive-B1, B10, B11; negative- B2, B3, B4, B5,B6, B7, B8, B9), urease (positive- B1,B3,B5,B6,B7,B9; negative- B2,B4,B8,B10,B11), and amylase test (positive-B1,B2,B3,B4,B6,B7,B8; negative-B5,B9,B10,B11). All the isolates were observed to be negative for Caseinase test. Sadowsky et al. (1983) observed both the slow and fast growing *Rhizobial* isolates showed positive results for Catalase, Oxidase and Urease and mixed results for Gelatinase.

Among the all sugars, sucrose, maltose, arabinose (with gas formation) and dulcitol was positively utilized by the all the isolates. In case of fructose, dextrose and mannitol, all the isolates showed positive results except B11 (**Table 12.5**; **Figure 12.3**). Both positive and negative results were observed in glucose (Positive-B1,B2,B3,B4,B5,B6,B7,B8,B9; Negative-B10,B11), lactose (positive-

Table 12.4: Sugar utilization test

Orgs	Glucose	lactose	Sucrose	maltose	inositol	dulcitol	arabinose	dextrose	fructose	mannitol
B1	+	++	+	++	++	++	++	++	++	++
B2	++	++	+	++	++	++	++/gas	++	++	++
B3	++	++	++	++	++	++	++	++	++	++
B4	+	-	+	++	++	++	++	++	++	++
B5	++	++	+	++	++	++	++	++	++	++
B6	+	++	++	++	++	++	++	++	++	++
B7	+	+	+	++	-	++	++	++	++	++
B8	+	++	+	++	++	++	++/gas	++	++	++
B9	+	+	+	++	++	++	++	++	++	++
B10	-	-	+	++	++	++	++/gas	++	++	++
B11	-	+	+	++	-	++	++/gas	-	-	-

-:negative, +: mild positive positive, ++:positive

B1,B2,B3,B5,B6,B7,B8,B9,B11; negative-B4,B10) and inositol (positive-B1, B2, B3, B4, B5, B6, B8, B10; negative-B7, B11) as reported by Shahzad et al (2012) during isolation and characterization of *R. meliloti*.

Table 12.5: Plant growth promoting traits exhibited by the bacterial isolates

Orgs	IAA	HCN	PSB	AMMONIA	ANTIBIOSIS	NITRATE
B1	+	-	+	+	+(2.8 cm)	-
B2	-	-	-	+	-	+
B3	-	-	+	+	-	+
B4	+	-	-	+	-	+
B5	-	-	+	+	-	+
B6	-	-	-	+	-	+
B7	-	-	-	+	-	+
B8	-	-	-	+	+(2.9 cm)	+
B9	-	-	-	-	-	+
B10	-	-	-	+	-	+
B11	-	-	-	+	-	+

N.B: diameter of fungi in control – 5 cm

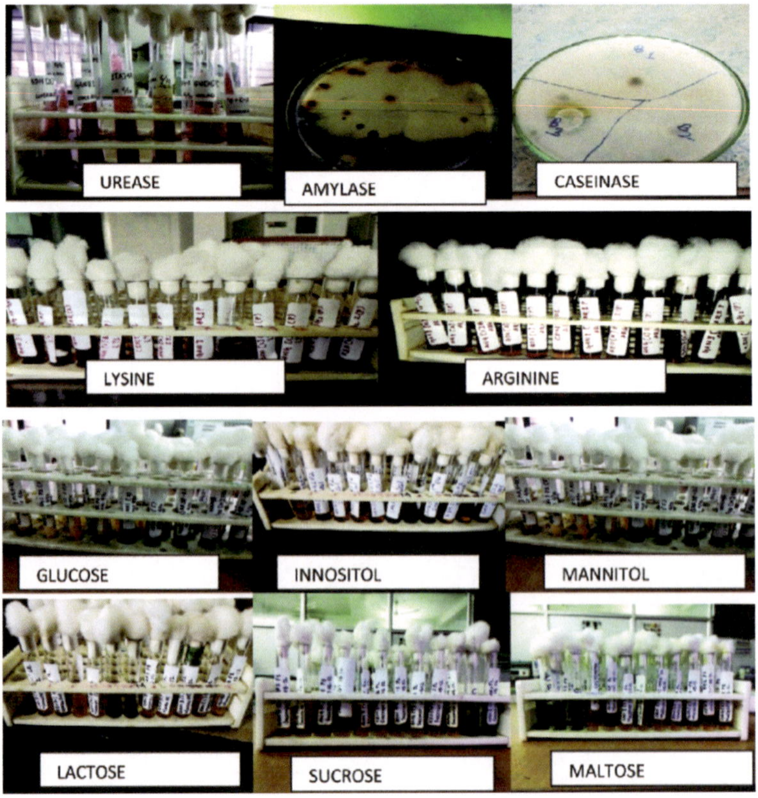

Fig. 12.3: Enzymatic characterization and Sugar Utilization Test of the isolates

In antibiotic sensitivity test, it was found that the organisms were sensitive to Chloramphenicol (except B2), Tetracycline, Rifampicin, Streptomycin and Gentamycin **(Table 12.6)**. All the isolates were Vancomycin resistant. Although some isolates were resistant to Erythromycin (B2, B3, B5, B7, B10, B11) and Ampicillin (B3, B5, B6, B7, B8, B9, B10). Singh et al., (2008) had concluded in their findings that the *Rhizobium* isolates were resistant to erythromycin and ampicillin and susceptible to chloramphenicol and streptomycin.

The bacterial isolates were observed to grow well at room temperature (20-25°C) and a few of the isolates grew well at 15°C. The growth was inhibited at 50°C. It was observed that high soil temperature in tropical areas limit nodulation and dinitrogen fixation by strains of *Rhizobium* (Michiels et al. 1994). When tested with different pH levels, our isolates grew well at pH 6.03, less growth at pH 7.07 and limited growth at around pH 5.02. Fast growing *Rhizobia* are most oftenly considered labile to acidic pH, except *Bradyrhizobia* which is an acid tolerant (Graham et al. 1991).

As regards PGPR characteristics like IAA, PSB (Phosphate Solubilization), Antibiosis, HCN production, Ammonia and Nitrate production of the isolates, B1 and B4 showed positive IAA production. Antibiosis was observed in B1 & B2, against the 5 mm growth of the control fungus (*Fusarium* sp.) our isolates showed 2.8 mm (B1) & 2.9 mm (B8) diameter. PSB positive was shown by B1, B3 & B5 isolates (**Table 12.7; Figure 12.4**). All the isolates were found to be positive for Ammonia (except B9) and nitrate (except B1) production. HCN production was observed to be negative for all the isolates. Inoculation with PGPRs has reported remarkable increase in growth and yield of crops (Kloepper et al. 1993). The isolates that shows PGPR characteristics and can be used as a sustainable development of Agricultural productivity. The stastical analysis was done by using Analysis of Variance (ANOVA) and student t-test the homogeneity of the strains was tested and the value was found to be highly significant at $p < 0.05$ one way ANOVA analysis **(Figure 12.5)**.

DNA of all the 11 isolates was extracted and amplified followed by nif gene amplification. Out of all the 11 isolates, *nif* gene of 6 organisms could be amplified (B2, B4, B5, B6, B7, B8) indicating presence of nif genes. Abo-Elwafa et al., (2018) performed molecular biology studies of nif genes of the isolates, identified based on PCR tool using designed and selected degenerative nif genes primer, which corroborates with these findings.

Table 12.6: Effect of isolates on plant growth Shoot and Root Length

	Seed Soaking							Seed Soaking With Inoculum						
	C	B2	B4	B5	B6	B7	B8	C	B2	B4	B5	B6	B7	B8
Shoot Length	13.6 ± 0.971	18.37 ± 0.691 (+) 35.07%	18.76 ± 1.039 (+) 37.94%	18 ± 0.45 (+) 32.35%	16.8 ± 1.12 (+) 23.52 %	16.9 ± 2.06 (+) 24.46 %	17.03 ± 1.97 (+) 25.22 %	14.02 ± 1.01	17.6 ± 0.85 (+) 25.53 %	16.96 ± 0.49 (+) 20.97 %	13.3 ± 0.84 (-) 5.13 %	14.3 ± 0.98 (+) 1.99 %	19.06 ± 0.717 (+) 35.94 %	19.33 ± 0.088 (+) 37.87 %
Root Length	6 ± 0.28	15.2 ± 0.71 (+) 153.3%	7.93 ± 3.55 (+) 32.16%	5.566 ± 1.05 (-) 7.33%	7.33 ± 0.61 (+) 22.16%	6.23 ± 0.50 (+) 3.83%	6.43 ± 0.71 (+) 16%	6.3 ± 0.40	11.2 ± 2.71 (+) 77.7%	6.83 ± 1.01 (+) 8.41%	13.06 ± 1.15 (+) 107.3%	5.93 ± 0.76 (-) 5.87%	11.2 ± 0.56 (+) 77.77%	6.93 ± 0.23 (+) 10%

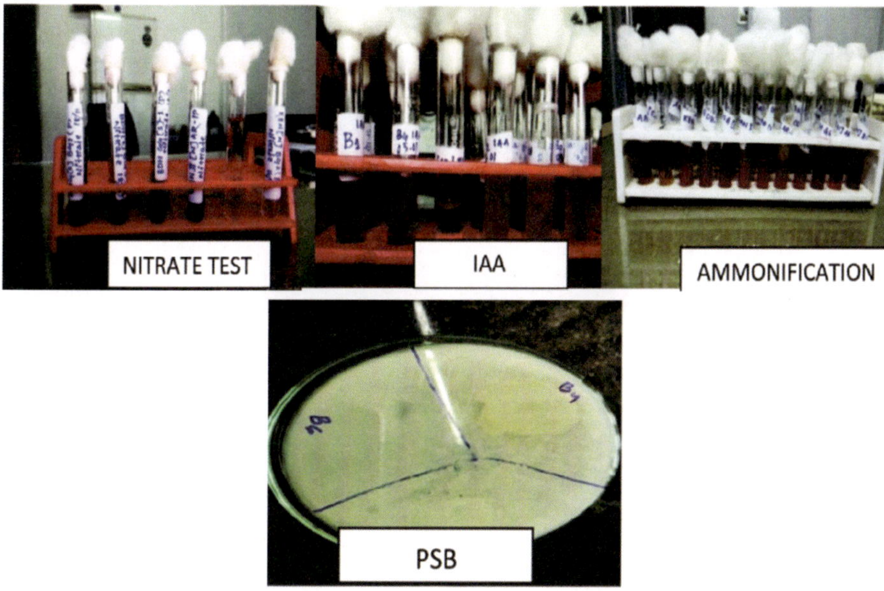

Fig. 12.4: PGPR traits of the isolates.

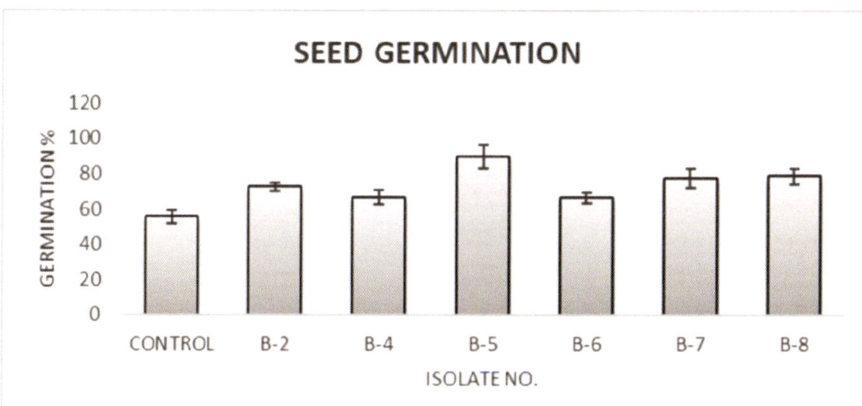

Fig. 12.5: Effect of bacterial isolates on plant growth. Values are mean of three samples & ± represent SEM. Values are highly significant at $p < 0.05$ by one way ANOVA analysis.

Effect of the isolates on Mungbean (*Vigna radiata*) seeds with seed soaking treatment and seed soaking with inoculum were processed on it. The seed germination was found maximum in culture by B8 (83% approx.) **(Figure 12.6)**. In seed soaking treatment the shoot length was found enhanced by B2, B4 & B8 and root length by B2 and B6 **(Table 12.8)**. Whereas in case of seed soaking and inoculum treatment, B2, B7 & B8 showed maximum increase in shoot length and B2, B5 & B7 in root length. Combined effects of *Rhizobium* with certain growth factors like L-trp shows increase in crop yield (Pahari and Mishra, 2017b).

Seed Soaking + Inoculum

Seed Soaking

B2 (Seed Soaking + Inoculum)

B2 (Seed Soaking)

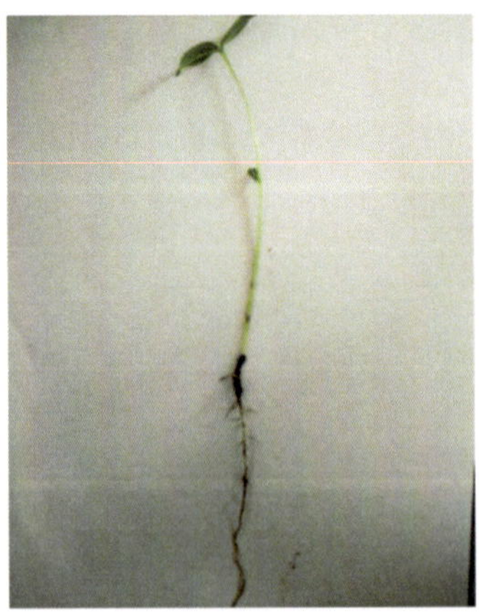

Fig. 12.6: Measurement of root and shoot length after bacterial treatment

12.4 Conclusion

After a series of treatments to the samples (soil and nodule) collected from different regions, physico-chemical analysis of soil, enrichment with YEMB for the isolation of bacteria, followed by gram's reaction, eleven isolates found to be gram negative were further studied for biochemical reactions and PGPR activities. These isolates showed similar characteristics with the family *Rhizobiacae.* Among all the isolates, six isolates were tested to have the nitrogen fixing *nif*-gene in their molecular make up which states that they have the capability to fix atmospheric nitrogen. Besides the presence of *nif*-gene all the isolates showed different plant growth promoting characteristics such as;

production of IAA, production of HCN and phosphate solubilisation. Six isolates which have *nif*-gene and PGPR activity (B2, B4, B5, B6.B7, and B8) were applied to seed and helped in faster germination of seeds and better elongation of root and shoot.

References

Abo-Elwafa AG, El-Refy AM, Hussein HS and El-Feky FA (2018) Characterization and Classification of Rhizobium aegyptiacum based on NifH and NodC Genes. 1st International Scientific Conference "Agriculture and Futuristic Challenges", 1: 532-546

Arun KS (2007) Bio-fertilizers for sustainable agriculture. Mechanism of P-solubilization Sixth Edition, Agrobios publishers, Jodhpur, India,pp 196-197.

Bhardwaj D, Ansari MW, Sahoo RK, and Tuteja N (2014) Biofertilizers function as key player in sustainable agriculture by improving soil fertility, plant tolerance and crop productivity. Microb Cell Fact, 13(1): 66.

Bhatt S, Vyas RV, Shelat HN and Mistry SJ (2013) Isolation and identification of root nodule bacteria of Mung Bean (*Vigna radiate* L.) for Biofertilizer production. Int J Res Pure Appl Microbiol. 3(4): 127-133.

Bromfield ESP, Wheatcroft R and Barran LR (1994) Medium for direct isolation of Rhizobium meliloti from soils. Soil Biol Biochem 26 (4), 423-428.

Deka AK, Azad P (2006) Isolation of Rhizobium strains: Cultural and biochemical characteristics. Legume Res, 29 (3), 209-212.

Donadio S, Monciardini P, Alduina R, Mazza P, Chiocchini C, Cavaletti L, Sosio M, and Puglia AM (2002) Microbial technologies for the discovery of novel bioactive metabolites. J biotechnol, 99(3):187-198.

Glick BR (1995) The enhancement of plant growth by free-living bacteria. Can J Microbiol, 41(2): 109-117.

Graham PH, Sadowsky MJ, Keyser HH, Barnet YM, Bradley RS, Cooper JE, De Ley J, Jarvis BDW, Roslycky EB, Strijdom BW, Young JPW (1991) Proposed minimal standards for the description of new genera and species of root- and stem nodulating bacteria. Int J Syst Bacteriol, 41: 582-587.

Howieson JG, Ewing MA (1986). Acid tolerance in the Rhizobium meliloti- Medicago symbiosis. Aust J Agric Res, 37 (1), 55-64.

Hussain M, Ashraf M, Saleem M, and Hafeez FY (2002) Isolation and characterization of Rhizobial strains from Alfalfa. Pak J AgR Sci, 39:32-34.

Kloepper JW (1993) Plant growth-promoting rhizobacteria as biological control agents. Soil Microb Ecol, 255–274.

Mia MAB, Shamsuddin ZH, Maziah M (2010) Use of plant growth promoting bacteria in banana: a new sustainable banana production. Int J Agric Biol, 12(3), 459-467.

Mia MAB, Shamsuddin ZH, Zakaria W, Marziah M (2005). High yielding and quality banana production through plant growth promoting rhizobacterial inoculation. Fruits 60: 179-185.

Michiels J, Verreth C, Vanderleyden J (1994) Effects of temperature stress on bean nodulating Rhizobium strains. Appl Environ Microbiol, 60:1206-1212.

Pahari A and Mishra BB (2017a) Characterization of siderophore producing rhizobacterial and its effect on growth performance of different vegetables. Int J Curr Microbiol App Sci, 6(5): 1398-1405.

Pahari A and Mishra BB (2017b) Antibiosis of Siderophore Producing Bacterial Isolates against Phythopathogens and Their Effect on Growth of Okra. Int J Curr Microbiol App Sci, 6(8):1925-1929.

Sadowsky MJ, Keyser HH and Bohlool (1983) Biochemical Characterization of Fast- and Slow-Growing Rhizobia Those Nodulate Soybeans. Int J Syst Bacteriol, 33(4):716-722.

Shahzad F, Shafee M, Abbas F, Babar S, Tariq MM, Ahmad Z (2012) Isolation and biochemical characterization of Rhizobium meliloti from the root nodules of Alfalfa (Medico sativa). J Anim Plant Sci, 22(2), 522-4.

Singh B, Kaur R and Singh K (2008) Characterization of Rhizobium strain isolated from the roots of Trigonella foenumgraecum (*Fenugreek*). Afr J Biotechnol, 7(20): 3671-3676.

Tilak KVBR, Ranganayaki N, and Manoharachari C (2006) Synergistic effects of plant-growth promoting rhizobacteria and Rhizobium on nodulation and nitrogen fixation by pigeon pea (*Cajanus cajan*). Eur J Soil Sci, 57:67–71.

Index

β-glucanases, 11, 32
β-1,3-glucanase *see also* β-glucanases
2,4-diacetylphloroglucinol (DAPG), 32, 101

A

Abiotic stresses, 66, 76, 78, 128, 129, 132, 133, 176, 184
ACC (1-aminocyclopropane-1-carboxylate) deaminase, 32, 34, 55, 65, 66, 71-73, 75-78, 83, 94, 103, 130
Acidobacteria, 35
Acidothiobacillus, 17, 156
Actinobacteria or actinomycetes, 27, 32, 34, 35, 37, 140, 141, 215, 216
Agglomeration, 206
Agriculture, 2, 15, 19, 20, 41, 42, 43, 44, 45, 46, 48, 49, 52, 54, 58, 59, 110, 119, 120, 132, 138, 139, 141, 148, 154, 175, 176, 177, 184, 201, 202, 203, 204, 205, 206, 221, 222
Agrobacterium sp., 4, 10, 18, 32, 35, 53, 100, 121, 138, 216
Agroecosystems, 57, 92
Alkalinization, 67
Allelochemicals, 99
Allorhizobium, 10, 139, 142
Alphaproteobacteria, 37
Alternaria sp., 8, 52, 128, 205, 216
Ammonia assimilation, 146,
Ampelomyces, 5, 7, 8
Anabaena azollae 4, 52, 140
Anabaena sp., 3, 4, 13, 52, 123, 140
Anaerobic microbiome, 35
Anammox, 28
Antibiosis, 7, 8, 49, 91, 98, 101, 223, 230, 231
Antibiotics, 7, 19, 20, 22, 27, 49, 52, 75, 92, 95, 99, 101, 107, 130, 155, 164, 215
Antifungal factors, 91
Antifungal metabolites, 130
Antioxidant activity, 75
Antioxidant enzymes, 69, 81
Antioxidant, 69, 72, 75, 81, 93
Antioxidative enzymes 66, 81, 83
Aquaporins, 72
Arbuscular mycorrhizal fungi (AMF), 54, 95, 128, 148, 170, 203, 215
Arbuscules, 5, 128, 216
Artificial fertilizers, 44
Artificial seed, 51
Associative, 36, 51, 124, 137, 140, 141
Asymbiotic, 16, 75, 124, 137, 140
Aspergillus fumigatus, 95, 160
Aspergillus sp., 10, 27, 48, 127, 160, 168, 215, 216, 218
Autochromes microorganism, 37
Autochthonous microorganisms, 27, 30
Autoinducers, 100
Auxin *see also* Indole acetic acid (IAA) 17, 52, 72, 80, 107, 125, 143, 147
Azoarcus sp., 11, 16, 30, 35, 36, 72
Azolla pinnata, 53
Azolla, 12, 46, 48, 52, 53, 123,140, 147, 166
Azospirillum brasilense, 50, 51, 71, 125, 147, 216
Azospirillum lipoferum, 50, 51, 71, 125
Azospirillum sp., 3, 10, 16, 27, 30, 32, 33, 35, 36, 46, 47, 48, 50, 51, 55, 70, 72, 124, 125, 129, 138, 141, 172, 185, 189
Azotobacter chroococcum, 56, 71, 95
Azotobacter sp. 3, 10, 11, 13, 17, 27, 30-36, 46, 50, 52, 55, 56, 70, 71, 91, 95, 100, 107, 108, 126, 127, 129, 138, 140, 145, 150, 164, 171, 172, 181, 185, 189, 214
Azotobactin, 100
Azygospores, 54

B

Bacillus megaterium, 34, 71, 95, 128, 159, 162, 163, 218
Bacillus mucilaginous, 95

Bacillus sp., 27, 33, 47, 56, 73, 78, 80, 91, 159, 176, 181, 182, 183, 216
Bacillus subtilis, 47, 48, 56, 71, 73, 77-80, 82, 181, 187, 216
Bacterial auxins, 80
Bacterial exudates, 179
Bacterial inoculants, 70, 165
Bacteriocins, 91, 99, 108
Bacterization,106
Bacteroid, 142, 144, 146
Bacteroidetes, 33, 37
Beneficial microorganisms, 3, 43, 66, 95
Bioactive metabolites, 44
Bioavailability, 78, 79, 104, 155
Bioavailable, 99, 103
Biocompost, 13
Biocontrol agents, 9, 15, 47, 98, 99, 110, 128
Biocontrol, 9, 10, 15, 34, 55, 66, 75, 91, 94, 98, 99, 101, 110, 128
Biofertilizer 1, 6, 9, 12,15, 18, 20, 41, 48, 53, 56, 58, 94, 95, 119, 120, 121, 124, 127, 129, 130, 132, 133, 137, 146, 147, 148, 154, 155, 161, 162, 166, 170, 171, 172, 175, 176, 177, 179, 180, 184, 185, 186, 188, 190, 192, 201, 202, 203, 204, 205, 206, 222
Bacterial, 2, 3, 8, 13, 15, 20, 31, 33, 34, 47, 49, 53, 55, 70, 76, 78, 80, 82, 83, 91, 97, 98, 99, 100, 102, 104, 106, 108, 109, 122, 126, 130, 138, 142, 144, 147, 159, 160, 162, 163, 164, 165, 167, 168, 170, 179, 181, 184, 185, 186, 189, 190, 218, 222, 223, 224, 225, 228, 230, 231, 233, 234
Cyanobacterial, 123, 124
Fungal, 1, 2, 3, 4, 5, 6, 7, 8, 9, 10, 13, 15, 20, 32, 54, 56, 57, 75, 100, 130, 160, 161, 163, 168, 170, 204, 205, 215, 216, 218
Liquid, 1, 3, 13, 14, 15, 20, 46, 58, 147, 158, 164, 166, 179, 183, 184, 202, 217, 218
Nano, 201-206
Phosphate mobilizing, 6, 14, 55, 202
Phosphorus, 6, 13, 14, 16, 30, 31, 46, 48, 50, 54, 55, 56, 65, 70, 71, 77, 82, 92, 94, 100, 104, 120, 127, 128, 153, 156, 171, 184, 209, 210, 211, 212, 213, 214, 215, 217, 218, 222
Potassium, 1, 2, 6, 11, 13, 16, 17, 45, 46, 48, 54, 55, 125, 153, 154, 155, 156, 157, 158, 159, 160, 161, 162, 163, 164, 167, 168, 170, 175, 176, 177, 178, 179, 181, 183, 184, 185, 187, 189, 192, 210
Biofilm, 35, 80, 179, 184
Biofilters, 103
Bioformulation, 47
Bioinoculant, 106, 185
Bioinputs, 42
Biological health hazards, 28
Biological nitrogen fixation (BNF), 13, 16, 137-141, 144, 222
Biological oxygen demand (BOD), 157
Biological stress, 42
Biomanures, 53
Biopesticide, 43
Bioaugmentation, 103
Bioremediation, 12, 45, 103
Biostimulation, 103
Biosurfactant, 96, 110
Biosynthesis, 18, 75, 77, 99-101, 202, 203
Biotic stresses, 54
Bioventing, 103
Blue-Green Algae (BGA), 13, 95
Bradyrhizobium japonicum, 71, 100, 132
Bradyrhizobium sp., 10, 18, 27, 33, 34, 35, 37, 47, 53, 70, 72, 80, 102, 129, 138, 156, 159, 177, 181, 188
Burkholderia sp., 11, 17, 27, 30, 31, 34-37, 47, 53, 70, 72, 80, 102, 129, 138, 156, 159, 177, 181, 188

C

Calcisol, 49, 50, 129, 130
Capsular polysaccharides (CPS), 55, 161, 179, 183
Carboxylate, 32, 66, 76
Carrier, 3, 14, 31, 58, 95, 120, 146, 147, 165, 166, 167, 184, 192, 205
Carrier-based inoculants, 58, 95
Catalase, 75, 81, 160, 179, 223, 228, 226, 227
Catecholate, 32
Catechols, 100
Chaetomium sp., 5, 8, 216
Chemical fertilizer, 28, 42, 119, 122, 129, 133, 148, 154, 175, 185, 188, 201-204
Chitinase, 11, 33, 82, 94
Choline accumulation, 73
Chromobacterium sp., 121
Cladosporium cucumerinum, 161
Cladosporium sp., 8, 161
Clostridium sp., 4, 56, 140, 161
Coelomycete, 9

Co-inoculation, 48, 130, 132, 148, 162, 185, 189
Colletotrichum sp., 7, 8, 101, 102
Compost, 3, 8, 13, 103, 185, 222
Conventional agriculture, 42, 43, 119, 120
CRISPR-Cas, 175, 192
Cross inoculation groups, 121, 122, 142
Crystal phase, 206
Cyanobacteria, 1, 13, 16, 43, 45, 46, 52, 95, 123, 124, 137, 139-141, 215, 216
Cytokinin, 11, 80, 92, 107, 125, 126, 139, 203
Cytokinin-like substances, 128

D

Dehydroascorbatereductase (DHAR), 81
Denitrification, 6, 139, 204
Desferrioxamine, 32
Diazotrophs, 3, 33, 104, 137, 138, 147
Dinitrogenase reductase, 144, 145
Disease resistance, 2, 32, 156, 210
Dual inoculation, 215

E

Ecofriendly, 1, 172, 206
Economic development of sustainable agriculture (EDSA), 41
Endophytes, 33, 36, 97, 139
Endophytic *Pseudomonas*, 97, 98
Endospore, 160, 179
Enterobacter sp., 4, 16, 17, 18, 25, 31-34, 36, 37, 70, 72, 76, 80, 81, 100, 116, 127, 129, 134, 159, 163, 175, 177, 182, 186, 189, 214, 216
Enterobactin, 32, 100
Environmental pollution, 2, 139, 146, 175, 201, 204
Environmental stresses, 65, 93, 97, 128, 147
Epiphytes, 33, 98
Erwinia sp., 36, 53, 129, 138, 216
Ethylene, 17, 18, 34, 49, 70, 73, 75-77, 92, 93, 130
Exopolysaccharides (EPS), 75, 79, 94, 179, 183
Extracellular enzyme, 10, 16, 161, 217

F

Farm yard Manure (FYM), 169
Farming, 2, 10, 13, 17, 20, 43, 58, 120, 138, 140, 147, 203, 206, 222
Fertilizer, 2, 12, 28, 42, 57, 77, 92, 96, 119, 120, 129, 133, 138, 139, 146, 148, 154, 156, 162, 176, 162, 176, 177, 185, 188, 192, 201, 202, 203, 204, 206, 212, 214
Firmicutes, 33, 34
Flavobacterium sp. 4, 10, 17, 35, 36, 53, 129, 216
Fluorescent pseudomonads (FLPs) 95, 107, 108
Foliar spray, 204
FolinCiocalteu Phenol reagent, 161
Food and Agriculture Organization, 28
Formulation, 2, 10, 14, 27, 58, 125, 146, 147, 154, 166, 172, 177, 184-186, 190, 192, 202, 206
Frankia sp., 3, 16, 130, 139, 141
F-type pyocin, 109
Fumaric acid, 53, 183
Fungicides, 7, 10, 15
Fusarium oxysporum, 8, 35, 216
Fusarium sp., 6, 7, 8, 9, 35, 52, 101, 128, 231

G

Gammaproteobacteria, 37
Genome editing, 175, 176, 192
Germination, 10, 12, 17, 18, 50, 70, 75, 76, 106, 126, 129, 139, 186, 204, 206, 221, 224, 233, 235
Gibberellin, 13, 80, 92, 99, 128
Gliocladium sp., 5, 9
Gliotoxin, 9
Glomalin related soil protein (GRSP), 57
Glomalin, 5, 57
Glomus intraradices, 130
Glomus mosseae, 95, 130
Gluconacetobacter sp., 30, 33
Glutamine oxoglutarate aminotransferase (GOGAT), 30
Glutamine synthetase (GS), 30, 31, 146
Glutathione reductase (GR), 81
Glycine betaine, 73, 75, 81, 82
Granular formulations, 58
Green manure, 53
Growth hormones, 20, 80, 123, 126, 155, 203
Growth promoting substances, 19, 48, 52, 70, 91, 120, 123, 125, 126, 128
Guaicol peroxidase, 81

H

Haber–Bosch process, 30
Herbaspirillum sp., 11, 30, 32, 35, 36
Herbicides, 44, 138, 204
Heteroauxins, 126
High performance liquid chromatographic (HPLC), 183
Hydrogen cyanide (HCN), 55, 92, 99, 101, 130, 223
Hydrophilicity, 206
Hydroxamate, 32, 183
Hyphae, 5, 6, 7,8, 9, 10, 54, 57, 58, 128, 216

I

Immobilization, 209, 210, 214, 217
in planta PGPR, 28
Indigenous rhizobia, 122
Indole acetic acid (IAA) or indole-3-acetic acid, 17, 52, 65, 72, 75, 78, 80, 107, 110, 125, 128, 143, 147

Indole-3-pyruvic acid, 18
Induced systemic resistance (ISR), 75, 91, 99, 101, 110
Insecticides, 10, 15, 43, 44
Integrated Nutrient Management (INM), 42
Integrated Plant Nutrition System (IPNS), 42
Ion homeostasis, 75

J

Jasmonate, 32

K

Klebsiella sp., 4, 33, 34, 36, 140, 182

L

Leghaemoglobin, 144
Legumes, 11, 148, 51, 95, 96, 121, 122, 137, 141, 144, 210, 211, 222
Leguminous plant, 141, 142
Lipoxygenase, 81
Liquid inoculants, 58

M

Macro-aggregates, 57
Macronutrient, 69, 155, 175, 176, 205
Malaysian Rubber Board (MRB), 95
Manure, 3, 53, 154, 165, 169, 185, 212
Mesorhizobium sp., 139, 142
Methanobacterium sp., 35
Methanobrevibacter sp., 35
Methanocella sp., 35
Methanoregula sp., 35
Methanosaeta sp., 35
Methanosarcina sp., 35
Methanosphaerula sp., 35
Microaerophilic, 35, 124, 140
Micro-aggregates, 57
Microbial fertilizer, 96
Microbial inoculants, 2, 3, 95, 147, 163, 164, 166, 168, 202
Microbial siderophores, 31
Microbiome, 28, 35, 37
Micrococcus sp., 4, 10, 53, 73, 76, 216

Micrococcus sp., 4, 10, 53, 73, 76, 216

Micronutrients, 3, 41, 46, 55, 95, 203, 204
Microsymbionts, 51
Mineralization, 19, 52, 72, 95, 210, 211, 214, 217
Mixed culture approach (MCA), 185
Mycofungicide, 8, 9
Mycoparasitism, 7, 8
Mycorrhiza, 1, 4-6, 31, 54, 55, 95, 128, 171, 172, 203, 215
Arbutoid, 5
Ectendo, 5
Ericoid, 4, 5
Monotropoid, 5
Orchid, 4
Phosphate mobilizing, 6, 14, 55, 202
Vesicular arbuscular (VAM), 95, 216
Mycorrhizosphere, 56, 178, 181

N

N-acetylglucosaminyl transferase, 143
N-acyl homoserine lactone, 100
Nanjing feldspar, 55
Nano-coatings, 204
Nano-fungicides, 204
Nano-herbicides, 204
Nano-nutrients, 203, 204, 206
Nano-particles, 203, 204, 206
Nano-sensors, 204
Nano-technology, 202, 206

Necrotic lesion, 101
nif operon, 30
Nitragin, 95
Nitrate reductase, 69
Nitrate reduction, 160, 223
Nitrogen (N_2) fixation, 2, 3, 6, 10 ,11, 12, 14, 28, 30, 33, 48, 51, 55, 71-73, 78, 91, 92, 94, 95, 99, 102, 103, 120-123, 125, 137, 140, 142, 144, 146, 186, 202, 231, 234
Industrial, 30, 52, 102
Non-symbiotic see also Asymbiotic nitrogen (N_2) fixation
Symbiotic, 16, 83, 137, 140, 141
Nitrogen (N_2) fixing bacteria, 121, 145, 146, 148, 163
Nitrogen (N_2) fixing microorganism (NFM), 14, 121, 124
Nitrogen assimilation, 30, 65, 145, 146
Nitrogen cycling bacteria, 97
Nitrogen fertilizer, 12, 28, 137, 139, 141, 144, 146
Nitrogen fixers, 48, 51, 55, 95, 102, 110, 124, 140, 146, 222
Nitrogenase (*nif*) gene, 16, 102, 137, 145, 146, 221, 231, 234, 235
Nitrogenase, 3, 16, 30, 102, 132, 137-139, 141, 142, 144, 145
nod genes, 137, 143
Nodulation, 51, 68, 75, 99, 130, 132, 137, 143, 147, 221, 222, 231
Nodule, 12, 51, 52, 75, 121, 138, 141, 142, 144, 223, 234
Non-enzymatic antioxidants, 69
Non-leguminous plant, 16, 126, 141, 222
Nostoc sp., 3, 13, 123, 124, 140, 216
NPK fertilizer, 187, 188
Nuclease, 109
Nutrient cycling, 43, 128, 133

O

Oceanobacillus sp., 78
Ochratoxin A (OTA), 10
Ochrobactrum sp., 32-34,73
Oomycin A, 101
Opportunistic pathogen, 108
Organic acid, 3, 35, 53, 179, 183, 217
Organic farming, 2, 58, 138, 221, 222
Organic fertilizer, 154
Organic manure, 53
Organo-minerals, 186
Oryza sativa, 28, 47, 148
Osmolyte accumulation, 66, 75
Osmolytes proline, 73
Osmoprotectants, 72
Osmotic stress, 65, 66, 69, 75, 79

P

Paddy organisms, 52
Paenibacillus glucanolyticus, 159, 162, 168, 181
Paenibacillus sp., 17, 27, 80, 130, 156, 159, 162, 168, 175, 177, 181
Penicillium sp., 9, 10, 28, 127, 161, 215, 216
Peribacteroid membrane (PBM), 144, 146
Pesticides, 2, 3, 10, 12, 28, 43, 44, 91, 120, 138, 185, 204
Phenazine, 101
Phenolics, 81
Phenylalanine ammonia-lyase, 81
Phosphatase enzyme, 4
Phosphate mobilization, 3, 4, 6, 49
Phosphate Mobilizing Microbes (PMMs), 14
Phosphate Solubilization, 3, 11, 31, 71, 72, 78, 103, 107, 139, 231
Phosphate Solubilizing Bacteria (PSB), 53, 55, 130, 148, 183, 185, 214
Phosphate solubilizing microorganisms (PSM), 127, 128, 214, 216
Phosphatic fertilizers, 48, 77, 214
Phosphatic fertilizers, 48, 77, 214
Phosphorus mineralization, 209
Phosphotika, 13
Photoautotrophs, 124
Photosynthesis, 3, 13, 16, 51, 65, 69, 81, 176, 189, 210, 211
Phyllobacterium sp., 11, 132
Phylloplane, 34, 37
Phyllosphere, 28, 34, 36, 37
Phytase, 31, 127
Phytic acid, 213
Phytohormones, 16, 17, 32, 34, 49, 66, 70, 72, 73, 75, 77, 80, 91, 92, 99, 103, 106, 107, 110, 126, 128, 129, 143, 147, 181, 186, 215
Phytopathogenic fungi, 7
Phytopathogenic microorganisms, 70, 75
Phytopathogens, 49, 66, 79, 104, 130
Phytoremediation, 92, 103-106, 110
Plant growth promoting microbes (PGPMs), 27, 32, 65

Plant growth promoting *Pseudomonas* (PGPP), 110
Plant Growth Promoting Rhizobacteria (PGPR), 4, 43, 45, 46, 48, 49, 65, 66, 70, 71, 74, 75, 79, 91, 92, 121, 129, 138, 148, 154, 205, 221-223, 231-233
Intracellular PGPR (iPGPR), 10, 138
Plant growth promotion, 27, 28, 45, 49, 50, 70, 92-94, 98, 99, 107, 184,
Plant-soil ecosystem, 2
Poaceae, 28, 141
Polyphenol oxidase, 81
Polyvalent consortia, 37
Potash Mobilizing Microbes, 14
Potassic fertilizer, 176
Potassium solubilizing microorganisms (KSM), 46, 153, 154, 156, 162, 167, 170
Primary Macronutrients, 41, 46
Proline, 35, 71, 73, 75, 81, 82
Promicromonospora sp., 72, 80
Protease, 109
Pseudobactins, 93
Pseudomonas fluorescens, 47, 71, 76, 96, 97, 100, 106, 132
Pseudomonas putida, 50, 71, 72, 97, 100, 129, 181
Pseudomonas sp., 4, 10, 11, 16, 17, 18, 27, 31-36, 47, 48, 49, 50, 53, 55, 70, 71, 72, 73, 75, 76, 83, 91, 92, 94, 96, 97, 98, 98, 99, 100, 102, 103, 104, 106-108, 110, 127, 129, 130, 132, 138, 141, 156, 177, 181, 182, 214-216, 218
Pycnidia, 8
Pyochelin, 32
Pyocin, 108, 109
Pyoluteorin, 101
Pyoverdines, 93
Pyrrolnitrin, 32, 101

Q

Quorum-sensing, 100

R

Reactive oxygen species (ROS), 66, 69, 81
Rhizobacteria, 4, 10, 11, 20, 43, 45, 46, 48, 49, 66, 70, 71, 74, 75, 76, 79, 83, 91, 92, 93, 121, 129, 132, 138, 154, 157, 176, 203, 221, 222, 223
Rhizobiaceae, 51, 121, 139
Rhizobitoxine, 94
Rhizobium sp., 4, 10, 11, 16, 17, 18, 27, 31-35, 46, 48, 50, 51, 53, 55, 73, 79, 91, 95, 104, 121, 122, 127, 129, 130, 137, 138, 139, 142-144, 147, 148, 164, 172, 185, 187, 189, 215, 216, 221, 222, 224, 228, 231, 233
Rhizoplane, 35, 36, 49, 138
Rhizosheath, 71, 79
Rhizosphere bacteria, 91, 100, 132, 156, 184
Rhizosphere microflora, 44
Rhizosphere, 7, 10, 12, 20, 28, 31, 33, 35, 36, 42, 44, 45, 49, 51, 52, 54, 55, 56, 65, 66, 68, 69, 70, 78, 79, 82, 91, 92, 95-97, 99, 101, 103, 104, 120, 121, 126, 130, 138, 141, 142, 154, 156, 159, 162, 163, 166, 169, 177, 179-181, 184, 185, 187, 188, 219, 221, 222
Rhizospheric PGPR, 10, 12, 32, 36, 43
Rhodococcus sp., 17, 214
Ribulosebisphosphate carboxylase, 82
Rock phosphate, 53, 94, 127, 128, 162, 189, 216, 218
Root exudates, 35, 92, 97, 143, 166
Root nodule, 137, 224
R-type pyocin, 109

S

Salicylic acid, 32
Saline soils, 67, 68, 77, 123, 130
Salt stress, 65, 66, 68, 69, 70, 71, 75-83, 130
Secondary Macronutrients, 41, 46
Secondary salinization, 67
Seed treatment, 185, 204
Serratia plymuthica, 100
Serratia sp.,10, 16, 17, 36, 70, 72, 73, 79, 100, 138, 216
Siderophore, 11,16,17, 32, 49, 55, 77, 72, 83, 92, 93, 99, 100, 106, 139, 179, 222
Siderophore-producing bacteria 79, 100
Signal transduction, 30, 107
Silicate bacteria, 186
Silicate solubilizing bacteria, 46, 159, 183
Single-culture approach (SCA), 185
Site-specific recombinases, 100
  Aggregates, 37, 57, 58, 129
  Amendment, 162, 204
  Borne diseases *see also* Soil borne pathogen 13, 47, 203

Borne pathogen, 47
Ecosystem, 2, 10, 19, 20, 58, 18, 119, 140, 154, 159
Erosion, 19, 176, 202
Microbial communities *see also* Soil microflora, 2, 43
Microflora, 2, 19, 35, 44, 49, 58, 97, 148, 202
Organic matter, 5, 8, 19, 52, 95, 97, 120, 123, 155, 161, 177, 186, 201, 222
Pollution, 2, 46, 120, 124, 139, 146, 147, 161, 175, 201, 202, 204
Salinity, 13, 27, 37, 65, 66, 67, 68, 69, 76, 77, 78, 79, 80, 81, 82, 93, 123, 129, 131
Sodicity, 66
Stomatal system, 204
Stress resistant, 204
Strigolactones, 32
S-type pyocin, 109
Sulfate-reducing bacteria, 33
Superoxide dismutase (SOD), 75, 81
Superphosphate, 57, 128
Sustainable agriculture, 2, 27, 41-46, 47, 52, 49, 52, 119, 120, 124, 139, 175-177, 184, 192, 201-203, 206
Suzhou illite, 55
Symbiosis, 6, 51, 54, 123, 128, 222
Symbiosome, 144
Symbiotic, 3-5, 16, 32, 36, 51, 52, 54, 75, 121, 124, 128, 137, 140, 141, 142, 143, 144, 146, 215, 221, 222
Synthetic fertilizer, 139, 201
Systemic acquired resistance (SAR), 98, 101

T

Tensin, 101
Thiobacillus sp., 56, 95, 161, 183, 216
Total soluble sugar (TSS), 82
Trichoderma sp., 5-9, 28, 127, 215, 216
Trichodesmium sp., 123
Tropolone, 101

V

Vascular plants, 128
Vermicompost, 3, 13
Volatile compounds, 32, 49, 129

W

Weedicides, 10, 138

Y

Yeast mannitol broth (YEMB), 224

Z

Zinc solubilizers, 4, 48, 55